Quaternary Alloys Based on III-V Semiconductors

Quaternary Alloys Based on III-V Semiconductors

Vasyl Tomashyk

V. Ye. Lashkaryov Institute of Semiconductor Physics
of NAS of Ukraine, Kyiv

CRC Press
Taylor & Francis Group
Boca Raton London New York

CRC Press is an imprint of the
Taylor & Francis Group, an **informa** business

CRC Press
Taylor & Francis Group
6000 Broken Sound Parkway NW, Suite 300
Boca Raton, FL 33487-2742

First issued in paperback 2020

ISBN-13: 978-0-367-57177-1 (pbk)
ISBN-13: 978-1-4987-7836-7 (hbk)

Library of Congress Cataloging-in-Publication Data
Names: Tomashyk, Vasyl, author.
Title: Quaternary alloys based on III-V semiconductors / Vasyl Tomashyk.
Description: Boca Raton, FL : CRC Press, Taylor & Francis Group, [2018] | Includes bibliographical references and index.
Identifiers: LCCN 2018001535 | ISBN 9781498778367 (hardback : alk. paper) | ISBN 1498778364 (hardback : alk. paper) | ISBN 9780429508080 (master ebook)
Subjects: LCSH: Compound semiconductors--Materials. | Chromium-cobalt-nickel-molybdenum alloys.
Classification: LCC QC611.8.C64 T66 2019 | DDC 621.3815/20284--dc23
LC record available at https://lccn.loc.gov/2018001535

Visit the Taylor & Francis Web site at
http://www.taylorandfrancis.com

and the CRC Press Web site at
http://www.crcpress.com

This book is dedicated to the memory of my sister Oleksandra Bukatyuk (neé Tomashyk, 1952–2008).

Contents

Chapter 6 ■ Systems Based on AlAs   113

Chapter 10 ▪ Systems Based on GaAs 169

List of Symbols and Acronyms

Bu^n	n-butyl, C_4H_9
c-	cubic
DMF	dimethylformamide
DSC	differential scanning calorimetry
DTA	differential thermal analysis
EDX	energy-dispersive X-ray spectroscopy
EPMA	electron probe microanalysis
Et	ethyl, C_2H_5
G	gas
h-	hexagonal
HRTEM	high-resolution transmission electron microscopy
L	liquid
LPE	liquid phase epitaxy
M	mol
MBE	molecular beam epitaxy
Me	methyl, CH_3
mM	mmol/L
MOCVD	metal–organic chemical vapor deposition
MOVPE	metal–organic vapor-phase epitaxy
ppm	parts per million
Pr^i	i-propyl, C_3H_7
Pr^n	n-propyl, C_3H_7
S	solid
SEM	scanning electron microscopy
T	temperature (in kelvins)
TEM	transmission electron microscopy
TG	thermogravimetry
THF	tetrahydrofurane, C_4H_8O
(X)	solid solution based on X
XRD	X-ray diffraction
XPS	X-ray photoelectron spectroscopy

Preface

A SIGNIFICANT VOLUME OF SEMICONDUCTOR DEVICES AND CIRCUITS employ III–V semiconductors, the most commonly used crystal material for integrated circuits. For electronic applications, these semiconductors offer the basic advantage of higher electron mobility, which translates into higher operating speeds. In addition, devices made with III–V semiconductor compounds provide lower-voltage operation for specific functions, radiation hardness (especially important for satellites and space vehicles), and semi-insulating substrates (avoiding the presence of parasitic capacitance in switching devices). Among these semiconductors, nitrides, which have recently been intensively studied, occupy a special place. AlN and related materials are useful in the visible and ultraviolet spectral regions. Since BN has a high thermal conductivity, it is more suitable for the applications in electronic devices. In contrast to II–VI semiconductors, B(Al,Ga,In)N alloys represent a very special class of materials with unique properties due to several remarkable properties that makes them attractive and suitable for reliable applications. The intrinsic group of superhard materials includes cubic boron nitride and some quaternary compounds, which possess an innate hardness. Silicon aluminium oxynitride (SiAlON) and AlON compounds are technical important materials for high-temperature engineering applications. Sialons are well known to exhibit useful physical properties, such as high strength, good wear resistance, high decomposition temperature, good oxidation resistance, excellent thermal shock properties, and resistance to corrosive environments. They have traditionally been investigated with an aim toward applications such as cutting tools, burners, welding nozzles, heat exchanger, and engine applications. AlN is also a ceramic material with mechanical and chemical properties suitable for use in structural applications at elevated temperatures. The alloys formed in the In–E′–E″–Sb quaternary systems are now used as Pb-free solders.

Although ternary phase equilibria in the systems based on boron, aluminium, gallium, and indium pnictides were collected and published in the handbook *Ternary Alloys Based on III–V Semiconductors* by V. Tomashyk (Taylor & Francis, 2017), data pertaining to the phase equilibria in the quaternary systems based on these semiconductor compounds are preferentially dispersed in the scientific literature. This reference book is intended to illustrate the up-to-date experimental and theoretical information about phase relations in based on III–V semiconductor systems with four components. This book critically evaluates many industrially significant systems presented in the form of two-dimensional sections for the condensed phases.

This book includes 495 critically compiled quaternary systems based on III–V semiconductors, including literature data from 728 papers; these data are illustrated in 136 figures and three tables. The information is divided into 15 chapters according to the number of possible combinations of B, Al, Ga, or In with N, P, As, or Sb (BSb compound does not exist). The chapters are structured so that first, group III element numbers in the periodic system are presented in increasing order, that is, from B to In compounds, and then the pnictogen numbers are given in increasing order, that is, from nitrides to antimonides. The same principle is used for further description of the systems in every chapter, that is, in increasing order initially from the second and then from the third components number in the periodic system (the first component is B, Al, Ga, or In and the forth component is N, P, As, or Sb).

Each quaternary system database description contains brief information in the following order: the diagram type, the possible phase transformation and physical–chemical interaction of the components, forming chemical compounds, methods of the equilibrium investigation, thermodynamic characteristics, and the method of the sample preparations. Solid and liquid phase equilibria with vapor are also illustrated in some cases because of their importance for crystal growth from the vapor, from the melt, and by the vapor–liquid–solid technique.

The homogeneity range is of a great importance for governing the crystal defect structure. Therefore, this reference book collects all such data accessible to date. In addition, this book presents data on the baric and temperature dependences of the solubility of the impurities in the lattice of the semiconductors and the liquid phase, as well as the pressure–composition relationship.

Most of the figures are presented in their original form, although some have been slightly modified. If the published data essentially vary, several versions are presented in comparison. The content of system components is presented in mole percent (mol%) (this is not indicated in the figures). If the original phase diagram is given with mass percent (mass %), this is indicated in the figures.

This book will be helpful for researchers in industrial and national laboratories and universities and graduate students majoring in materials science and engineering and solid-state chemistry. It will be also suitable for phase relations researchers, inorganic chemists, and semiconductor and solid-state physicists.

Author

Vasyl Tomashyk is the executive director and head of the department of V. Ye. Lashkaryov Institute for Semiconductor Physics of the National Academy of the Sciences of Ukraine. He graduated from Chernivtsi State University in Ukraine in 1972 (master of chemistry). He is a doctor of chemical sciences (1992), professor (1999), and author of about 630 publications in scientific journals and conference proceedings and nine books (four of them were published by CRC Press), which are devoted to physical-chemical analysis, the chemistry of semiconductors, and chemical treatment of semiconductor surfaces.

Tomashyk is a specialist at the high international level in the field of solid-state and semiconductor chemistry, including physical–chemical analysis and the technology of semiconductor materials. He was head of research topics within the International program "Copernicus." He is a member of Materials Science International Team (Stuttgart, Germany, since 1999), which prepares a series of prestigious reference books under the title *Ternary Alloys* and *Binary Alloys*, and has published 35 chapters in the Landolt–Börnstein New Series. Tomashyk is actively working with young researchers and graduate students, and under his supervision, 20 PhD theses were prepared. For many years, he is also a professor at Ivan Franko Zhytomyr State University in Ukraine.

CHAPTER 1

Systems Based on BN

1.1 BORON–HYDROGEN–LITHIUM–NITROGEN

The **LiNH$_2$BH$_3$** quaternary compound, which crystallizes in the orthorhombic structure with the lattice parameters $a = 711.274 \pm 0.006$, $b = 1394.877 \pm 0.014$, and $c = 515.018 \pm 0.006$ pm, is formed in the B–H–Li–N system (Xiong et al. 2008b). It decomposes directly to hydrogen on heating, with vigorous H$_2$ release at around 92°C.

To obtain the title compound, LiH + NH$_3$BH$_3$ mixture (molar ratio 1:1) and pure NH$_3$BH$_3$ were ball-milled on a planetary mill. Graphite was added to the mixture to enhance ball-milling efficiency. After milling the sample, gases released were identified. All samples were handled in a glove box filled with purified gas.

1.2 BORON–HYDROGEN–SODIUM–NITROGEN

Three quaternary compounds are formed in the B–H–Na–N system.

NaNH$_2$BH$_3$ crystallizes in the orthorhombic structure with the lattice parameters $a = 746.931 \pm 0.007$, $b = 1465.483 \pm 0.016$, and $c = 565.280 \pm 0.008$ pm (Xiong et al. 2008b) [$a = 747.35$, $b = 1464.52$, and $c = 567.39$ pm (Xiong et al. 2008a)]. It decomposes directly to hydrogen on heating at 89°C.

To obtain the title compound, NaH + NH$_3$BH$_3$ mixture (molar ratio 1:1) and pure NH$_3$BH$_3$ were ball-milled on a planetary mill (Xiong et al. 2008b). Graphite was added to the mixture to enhance ball-milling efficiency. After milling the sample, gases released were identified. All samples were handled in a glove box filled with purified gas.

It could be also prepared when the chemical reaction of NH$_3$BH$_3$ with NaH or NaNH$_2$ in tetrahydrofuran (THF) was conducted in a stirring tank reactor (Xiong et al. 2008a). NaH or NaNH$_2$ (8 mM) was added to 30 mL of THF and stirred at constant temperature. As gas was evolved, the pressure variation with time inside the reactor was reduced automatically. The reaction solution was filtered and evaporated under reduced pressure at ambient temperature for crystallization.

Na[(BH$_3$)$_2$NH$_2$] (sodium aminodiboranate) was prepared using two methods (Daly et al. 2010). According to the first method, a solution of NH$_3$·BH$_3$ (97.1 mM) in THF (75 mL) was slowly added to sodium cubes (0.9 M) in THF (75 mL). Gas slowly evolved, and the mixture was stirred at room temperature for 18 h. The mixture was then heated to reflux for 21 h, causing a flocculent white solid to precipitate. The solution was filtered, and the filtrate was evaporated to dryness under vacuum to yield a white solid. The solid was washed with benzene (2 × 40 mL) and pentane (3 × 30 mL) and then was evaporated to dryness under vacuum to yield a free-flowing white powder. The title compound contained some quantity of THF.

In a second method, a solution of $NH_3 \cdot BH_3$ (102 mM) in THF (75 mL) was slowly added to a suspension of $NaNH_2$ (50.0 mM) in THF (75 mL). Gas slowly evolved, and the mixture was stirred at room temperature for 18 h. The mixture was then heated to reflux for 39 h, causing a flocculent white solid to precipitate. The solution was filtered, and the filtrate was evaporated to dryness under vacuum to yield a white solid. The solid was washed with benzene (40 mL), diethyl ether (2 × 25 mL), and pentane (2 × 40 mL) and then was dried under vacuum for 12 h to yield a free-flowing white powder.

All operations were carried out in vacuum or under argon using standard Schlenk techniques. All glassware was dried in an oven at 150°C, assembled hot, and allowed to cool under vacuum before use.

The $B_2H_7N \cdot NaNH_2$ quaternary compound could be obtained at the interaction of $B_2H_7N \cdot NH_3$ with Na in liquid NH_3 (Schlesinger et al. 1938).

1.3 BORON–HYDROGEN–CALCIUM–NITROGEN

The $Ca_2[BN_2]H$ quaternary compound, which crystallizes in the orthorhombic structure with the lattice parameters $a = 920.15 \pm 0.08$, $b = 366.76 \pm 0.02$, and $c = 998.74 \pm 0.12$ pm and a calculated density of 2.365 $g \cdot cm^{-3}$, is formed in this system (Somer et al. 2004c).

The title compound was synthesized from a stoichiometric mixture of CaH_2, BN, and Ca_3N_2. The well-ground educts were sealed into a stainless steel ampoule (placed in a protecting silica tube), heated up to 1000°C within 6 h, annealed at this temperature for 72 h, and cooled to room temperature within 24 h. The starting calcium components and $Ca_2[BN_2]H$ are sensitive to air and moisture. Therefore, all handling was done under inert atmosphere in a glove box.

1.4 BORON–HYDROGEN–CARBON–NITROGEN

Some quaternary compounds are formed in the B–H–C–N system.

Graphite-like materials of composition $BC_{0.9-1.3}N_{0.8-0.9}H_{0.4-0.7}$ and $BC_{3.0-3.2}N_{0.8-1.0}H_{0.2-2}$, which are described as **BCN(H)** and **BC₃N(H)**, were prepared by Kawaguchi et al. (1996). A quartz tube of 40 mm inner diameter and 1000 mm in length was used as a reactor. Acetonitrile vapor for BCN(H) or acrylonitrile vapor for $BC_3N(H)$ (50 $cm^3 \cdot min^{-1}$) carried by N_2 [430 $cm^3 \cdot min^{-1}$ for BCN(H) or 310 $cm^3 \cdot min^{-1}$ for $BC_3N(H)$], BCl_3 (50 $cm^3 \cdot min^{-1}$) gas and H_2 (250 $cm^3 \cdot min^{-1}$) for BCN(H), or only BCl_3 (50 $cm^3 \cdot min^{-1}$) gas for $BC_3N(H)$ was introduced into the hot zone (1000°C) of the reactor. This gas-phase reaction was carried out for 9 h for BCN(H) or 8 h for $BC_3N(H)$. Black plates were deposited in the hot zone, while dark brown powders for BCN(H) or black powders for $BC_3N(H)$ were obtained behind the hot zone. The plates and powders were then heated at 1000°C in a N_2 atmosphere for 1 h.

$(BH_2CN)_n$ was obtained by Györi et al. (1983). The experiments were carried out in dry, oxygen-free solvents under dry oxygen-free nitrogen using the Schlenk technique.

One hundred milliliters of 1.625 M HCl in Me_2S was dropwise added at 0°C to a suspension of $NaBH_3CN$ (167.4 mM) in Me_2S (50 mL). When H_2 evolution had ceased (160.0 mM), the solution was filtered and the NaCl remaining on the filter was washed with Me_2S (3 × 30 mL). The filtrate was evaporated at room temperature, and the syrupy residue was kept under reduced pressure (0.1 Pa) for 1 h. Crystallization of the product was initiated by scratching and was completed within a few hours.

$B_3(CN)_3N_3H_3$ is relatively thermally stable, decomposing in vacuum only at temperatures exceeding 200°C (Brennan et al. 1960).

$[NH_4][B(CN)_4]$ decomposes at 300°C (Bernhardt et al. 2003) and crystallizes in the tetragonal structure with the lattice parameters $a = 713.2 \pm 0.1$, $c = 1474.5 \pm 0.4$ pm, and a calculated density of 1.177 $g \cdot cm^{-3}$ (Küppers et al. 2005). To prepare it, $Ag[B(CN)_4]$ (0.06 M) and NH_4Br (0.08 M) were dissolved in H_2O (200 mL) (Bernhardt et al. 2003). After stirring for 2 h, AgBr was separated by filtration and the water removed under vacuum. The white crude product was recrystallized from acetonitrile.

The title compound could be also obtained, if $K[B(CN)_4]$ (0.08 M) in H_2O (200 mL) was added to the solution of $[NH_4][SiF_6]$ (0.11 M) in H_2O (250 mL) (Bernhardt et al. 2003). The water was removed in vacuum from the turbid reaction mixture. The white residue was extracted with THF (200 and 250 mL), and the combined organic phases were filtered through a glass frit. White residue was obtained after removing all volatiles in vacuo.

Colorless single crystals of $[NH_4][B(CN)_4]$ have been grown from aqueous solutions at room temperature by slow evaporation of H_2O (Küppers et al. 2005).

1.5 BORON–HYDROGEN–OXYGEN–NITROGEN

The $\{[(NH_4)_2\{B_4O_5(OH)_4 \cdot 1.41H_2O\}$ quaternary compound, which crystallizes in the orthorhombic structure with the lattice parameters $a = 719.97 \pm 0.08$, $b = 1069.39 \pm 0.10$, and $c = 726.04 \pm 0.08$ pm at 260 K and an experimental density of 1.531 $g \cdot cm^{-3}$, is formed in this system (Ramachandran and Gagare 2007). It was obtained at the transition metal-catalyzed hydrolysis of NH_3BH_3.

1.6 BORON–HYDROGEN–CHROMIUM–NITROGEN

The $[Cr(NH_3)_6](BH_4)_3$ quaternary compound is formed in the B–H–Cr–N system. To prepare it, stoichiometric amounts of dry $[Cr(NH_3)_6]F_3$ and $NaBH_4$ were placed in the reactor tube, which was attached to the vacuum system (Parry et al. 1958). NH_3 was condensed into the reactor, and metathesis was carried out at 208–228 K. The mixture was stirred, and after 3 to 5 h the content of the tube was frozen with liquid NH_3. $[Cr(NH_3)_6](BH_4)_3 \cdot NH_3$ was isolated from liquid ammonia as long yellow needles. It decomposes at 25°C under high vacuum.

1.7 BORON–HYDROGEN–FLUORINE–NITROGEN

The $BF_3 \cdot NH_3$ quaternary compound, which melts at $163 \pm 1°C$ (Laubengayer and Condike 1948) and crystallizes in the orthorhombic structure with the lattice parameters $a = 822 \pm 3$,

b = 811 ± 3, and c = 931 ± 3 pm (Hoard et al. 1951) and an experimental density of 1.864 g·cm^{-3} (Laubengayer and Condike 1948), is formed in this system. The title compound was prepared by mixing anhydrous BF_3 and NH_3 in a 1 L three-necked flask (Laubengayer and Condike 1948). An outlet tube was provided for excess gases, and the flask was cooled with ice water. The vessel was first swept out with NH_3, and then a slow flow of this gas was maintained. BF_3 was admitted at such a rate that approximately equimolar quantities of the reactants were provided. $BF_3 \cdot NH_3$ accumulated in the reaction flask as a white powder. The heat of formation of solid $BF_3 \cdot NH_3$ from the gases at 0°C is equal to 172.7 kJ·M^{-1}.

1.8 BORON–HYDROGEN–CHLORINE–NITROGEN

Some quaternary compounds are formed in the B–H–Cl–N system.

BH_2ClNH_3 crystallizes in the orthorhombic structure with the lattice parameters a = 670.3 ± 0.1, b = 592.4 ± 0.1, and c = 829.2 ± 0.2 pm at 150 ± 2 K and a calculated density of 0.962 g·cm^{-3} (Lingam et al. 2012). To prepare it, BH_3NH_3 (65 mM) was placed in a two-necked 250 mL flask with a magnetic stir bar (Lingam et al. 2012; Ketchum et al. 1999). The flask was connected to a vacuum line and evacuated, and 150 mL of Et_2O was condensed into it at −78°C. The mixture was allowed to warm to −40°C, and dry HCl was bubbled into the solution for 1 h with constant stirring. When the starting material was completely consumed, the flow of HCl was discontinued. A very small amount of white precipitate was filtered out from the solution. Et_2O was removed from the filtrate under a dynamic vacuum to leave a white solid BH_2ClNH_3.

Decomposition of the title compound was initiated at approximately 30°C under a static vacuum and at 50°C under an Ar atmosphere (Lingam et al. 2012). According to the data of Ketchum et al. (1999), decomposition of the solid BH_2ClNH_3 under NH_3 at 600°C produced amorphous BN and its pyrolysis under vacuum at 1100°C led to the formation of turbostratic BN.

All manipulations were carried out on a high-vacuum line or in a glove box filled with high-purity nitrogen.

BCl_3NH_3 crystallizes in the monoclinic structure with the lattice parameters a = 520.4 ± 0.1, b = 982.5 ± 0.2, and c = 951.6 ± 0.1 pm and β = 92.09° ± 0.01° at 173 K and a calculated density of 1.83 g·cm^{-3} (Avent et al. 1995). In a typical run for preparing the title compound, rigorously dried NH_4Cl (2.88 M) and toluene (1.6 L) were heated under reflux (using a condenser cooled to −78°C) while gaseous BCl_3 (3.24 M) was introduced during 6 h into the gaseous phase. The mixture was cooled to circa 20°C and set aside for 3 days. White crystals of BCl_3NH_3 were filtered off and dried in vacuo. Suitable crystals of this compound were obtained by recrystallization from CH_2Cl_2 after storage at −30°C for 5 months. BCl_3NH_3 is stable at ambient temperature in an inert atmosphere.

$B_3Cl_3N_3H_3$ melts at 83–86.5°C and is extremely sensitive to moisture and sublimes readily (Brown and Laubengayer 1955; Brennan et al. 1960; Avent et al. 1995). It undergoes slow irreversible thermal decomposition above 100°C (Brown and Laubengayer 1955). The heats of sublimation and vaporization are 71.2 and 51.5 kJ·M^{-1}, respectively. The crystals belong to the orthorhombic system and have an experimental density of 1.58 g·cm^{-3}.

The title compound was prepared by the reaction of BCl_3 with NH_4Cl in toluene as the principle product at the obtaining of BCl_3NH_3 (Brown and Laubengayer 1955; Brennan et al. 1960; Avent et al. 1995). It could be also obtained by heating BCl_3NH_3 in toluene under reflux for 2 h (Avent et al. 1995).

1.9 BORON–HYDROGEN–COBALT–NITROGEN

The $[Co(NH_3)_6](BH_4)_3$ quaternary compound is formed in the B–H–Cr–N system. It was prepared by the same way as $[Cr(NH_3)_6](BH_4)_3$ was obtained using $[Co(NH_3)_6]F_3$ instead of $[Cr(NH_3)_6]F_3$ (Parry et al. 1958). $[Co(NH_3)_6](BH_4)_3 \cdot 0.5NH_3$ was isolated from liquid ammonia as long yellow needles. The title compound is stable up to 60°C.

1.10 BORON–LITHIUM–BERYLLIUM–NITROGEN

$LiBe[BN_2]$ quaternary compound was prepared by the same way as $LiCa[BN_2]$ was obtained using Be_3N_2 instead of Ca_3N_2 and Be instead of Ca (Iizuki 1985).

1.11 BORON–LITHIUM–MAGNESIM–NITROGEN

The $LiMg[BN_2]$ and $Li_3Mg_3B_2N_5$ quaternary compounds are formed in the B–Li–Mg–N system. $LiMg[BN_2]$ crystallizes in the tetragonal structure with the lattice parameters $a = 379.8 \pm 0.2$, $c = 891.6 \pm 0.5$ pm, and a calculated density of 1.810 g·cm^{-3} (Herterich et al. 1994; Somer et al. 1997). It was synthesized from a stoichiometric mixture of Li_3N, Mg_3N_2, and BN in a sealed Nb ampoule at 1300–1480°C.

$Li_3Mg_3B_2N_5$ also crystallizes in the tetragonal structure with the lattice parameters $a = 849.87 \pm 0.25$, $c = 889.64 \pm 0.32$ pm, and the calculated and experimental densities of 1.92 and 1.90 ± 0.02 g·cm^{-3}, respectively (Kulinich et al. 2000). It was prepared by the sintering of the mixtures of Li_3N, Mg_3N_2, and α-BN (molar ratio 1:1:2) at 950–1000°C for 2–6 h (Kulinich et al. 1997). It could be also obtained by the sintering of LiMgN and α-BN, $Li_3[BN_2]$ and Mg_3BN_3, $Li_3[BN_2]$, Mg_3N_2 and α-BN, $Li_3[BN_2]$, Mg, and α-BN in N_2 atmosphere or Li_3N, Mg_3BN_3, and α-BN.

1.12 BORON–LITHIUM–CALCIUM–NITROGEN

The $LiCa[BN_2]$ and $LiCa_4[BN_2]_3$ quaternary compounds are formed in the B–Li–Ca–N system. $LiCa[BN_2]$ was prepared by heating a mixture of (1) finely divided Li_3N or metallic Li, (2) finely divided Ca_3N_2 and/or Ca, and (3) h-BN at 800–1300°C in an inert gas or NH_3 atmosphere thereby to react ingredients (1), (2), and (3) with each other in a molten state and then cooling the reaction product to be solidified (Iizuki 1985). According to the data of Kulinich et al. (1996), $LiCa[BN_2]$ does not exist in this system.

$LiCa_4[BN_2]_3$ crystallizes in the cubic structure with the lattice parameter $a = 711.5 \pm 0.1$ pm and a calculated density of 2.615 g·cm^{-3} (Somer et al. 1994, 2000c) [$a = 710.5 \pm 0.3$ pm and a calculated density of 2.626 g·cm^{-3} (Kulinich et al. 1996)]. Pale yellow crystals of this compound were prepared from the binary nitrides Li_3N, Ca_3N_2, and BN (molar ratio 1.5:4:9) in sealed Nb ampoules at 1100°C (Somer et al. 1994, 2000c; Kulinich et al. 1996).

1.13 BORON–LITHIUM–STRONTIUM–NITROGEN

The **LiSr[BN$_2$]** and **LiSr$_4$[BN$_2$]$_3$** quaternary compounds are formed in the B–Li–Sr–N system. The LiSr[BN$_2$] quaternary compound was prepared by the same way as LiCa[BN$_2$] was obtained using Mg$_3$N$_2$ instead of Ca$_3$N$_2$ and Mg instead of Ca (Iizuki 1985).

LiSr$_4$[BN$_2$]$_3$ crystallizes in the cubic structure with the lattice parameter $a = 745.6 \pm 0.1$ pm and a calculated density of 3.797 g·cm^{-3} (Somer et al. 1996d, 2000c). Pale yellow crystals of this compound were prepared from the binary nitrides Li$_3$N, Sr$_3$N$_2$, and BN (molar ratio 1.5:4:9) in sealed Nb ampoules at 1100°C.

1.14 BORON–LITHIUM–BARIUM–NITROGEN

The **LiBa[BN$_2$]** and **LiBa$_4$[BN$_2$]$_3$** quaternary compounds are formed in the B–Li–Ba–N system. LiBa[BN$_2$] quaternary compound was prepared by the same way as LiCa[BN$_2$] was obtained using Ba$_3$N$_2$ instead of Ca$_3$N$_2$ and Ba instead of Ca (Iizuki 1985).

LiBa$_4$[BN$_2$]$_3$ crystallizes in the cubic structure with the lattice parameter $a = 788.0 \pm 0.1$ pm and a calculated density of 4.564 g·cm^{-3} (Somer et al. 2000c). This compound was prepared from the binary nitrides Li$_3$N, Ba$_3$N$_2$, and BN (molar ratio 1.5:4:9) in sealed Nb ampoules at 1100°C.

1.15 BORON–LITHIUM–EUROPIUM–NITROGEN

The **Li$_{0.44}$Eu[BN$_2$]$_3$** and **LiEu$_4$[BN$_2$]$_3$** quaternary compounds are formed in the B–Li–Eu–N system. Li$_{0.44}$Eu[BN$_2$]$_3$ crystallizes in the trigonal structure with the lattice parameters $a = 1202.25 \pm 0.02$, $c = 685.56 \pm 0.02$ pm, and a calculated density of 6.68 g·cm^{-3} (Kokal et al. 2014). This compound was synthesized by using metathesis reaction between Li$_3$[BN$_2$] and EuCl$_3$. Stoichiometric mixtures of Li$_3$[BN$_2$] and EuCl$_3$ were finely ground and mixed homogeneously in an Ar-filled glove box. The mixture was then transferred into a Nb ampoule, which was arc-welded and sealed in a silica tube under vacuum. The reaction ampoule was placed in a muffle furnace and heated up to 850°C in 6 h. After annealing for 12 h, the reaction ampoule was cooled down to room temperature during 4 h. The final product consists of black hexagonally-shaped column-like crystals of the title compound, together with LiEu$_4$[BN$_2$]$_3$ and LiCl as by-products. Li$_{0.44}$Eu[BN$_2$]$_3$ is stable up to 1050°C.

LiEu$_4$[BN$_2$]$_3$ crystallizes in the cubic structure with the lattice parameter $a = 742.54 \pm 0.07$ pm [$a = 742.5 \pm 0.1$ pm (Curda et al. 1994a)] and a calculated density of 5.930 g·cm^{-3} (Somer et al. 2000c). This compound was prepared from the binary nitrides Li$_3$N, Ba$_3$N$_2$, and BN (molar ratio 1.5:4:9) in sealed Nb ampoules at 1100°C (Somer et al. 2000c) [molar ratio 1:4:3 and at 1000°C (Somer et al. 2004a); stoichiometric mixture and at 1680°C (Curda et al. 1994a)].

1.16 BORON–LITHIUM–CARBON–NITROGEN

The **Li[B(CN)$_4$]** and **Li$_2$[B(CN)$_3$]** quaternary compounds are formed in the B–Li–C–N system. Li[B(CN)$_4$] crystallizes in the cubic structure with the lattice parameter $a = 548.15 \pm 0.01$ pm [$a = 547.6 \pm 0.3$ pm (Williams et al. 2000)] and a calculated density of 1.228 g·cm^{-3} (Küppers et al. 2005). To obtain the title compound, in a separating funnel K[B(CN)$_4$] (80.0 mM) was dissolved in H$_2$O (100 mL) (Küppers et al. 2005). The solution was treated with 18.2 mL (80 mM) of Pr^n_3N and 4.0 mL (130 mM) of concentrated HCl and shaken. The resulting white precipitate was extracted three times with CH_2Cl_2 (100, 50, and 50 mL, respectively). The collected organic phases containing $[Pr^n_3NH][B(CN)_4]$ were combined, dried over $MgSO_4$, and filtered through a glass frit. The colorless solution was transferred into a 500 mL round-bottom flask together with 130 mM of $LiOH \cdot H_2O$ in 20 mL of water. The mixture was stirred vigorously for 1 h and finally dried under reduced pressure. The solid residue was extracted with 50 mL of MeCN in a Soxhlet extractor. The resulting solution was taken to dryness in a vacuum and washed with dichloromethane to yield $Li[B(CN)_4]$.

This compound could be also prepared, if Me_3SiCN (0.048 M) was added dropwise to a stirred mixture of $LiBF_4$ (0.011 M) in 50 mL of Bu^n_2O at −78°C (Williams et al. 2000). The mixture was stirred at 22°C for 18 h and then refluxed for 4 h. After filtration, the resulting white solid residue was heated at 250°C for 18 h in vacuo and then extracted with warm MeCN. The solution was evaporated to dryness to yield a colorless solid, which is stable in air. Reactions were performed under prepurified nitrogen using standard Schlenk and dry-box techniques.

Colorless single crystals of $Li[B(CN)_4]$ were prepared from aqueous solutions containing a small amount of Pr^n_3N through slow evaporation of the solvent at room temperature (Küppers et al. 2005).

Preliminary high-pressure experiments on $Li[B(CN)_4]$ in the multianvil have shown that a transparent glassy material, which is extremely hard but amorphous and has a composition close to that of the starting material, is produced at 10 GPa and 400°C (Williams et al. 2000). Further reactions at 900°C and 10 GPa resulted in a crystalline black solid with platelet-like morphology.

To obtain $Li_2[B(CN)_3]$, $Li[B(CN)_4]$ (2.32 mM) was suspended in THF (10 mL) and Bu^nLi (5 mL, 2 M in hexane), and the mixture was stirred for 16 h (Bernhardt et al. 2011a, 2011b). The obtained yellow precipitate was filtered and washed several times with THF. The residue was dried in vacuum.

1.17 BORON–SODIUM–STRONTIUM–NITROGEN

The **NaSr_4[BN_2]_3** quaternary compound, which crystallizes in the cubic structure with the lattice parameter $a = 756.8 \pm 0.1$ pm [$a = 761.1 \pm 0.1$, 754.2 ± 0.1, and 754.1 ± 0.1 pm for the Na/Sr molar ratio in the $NaN_3 + Sr + BN$ reaction mixtures 1–4, 4–4, and 8–4, respectively (Womelsdorf and Meyer 1994b)] and a calculated density of 3.754 g·cm^{-3} (Somer et al. 2000a, 2000c), is formed in the B–Na–Sr–N system. The title compound was synthesized from a mixture of Sr_3N_2, BN, and NaN_3 in sealed Nb ampoules at 900°C (Somer et al. 2000a, 2000c). Single crystals were obtained from the reaction of $Sr_3[BN_2]_2$ with NaBr surplus (molar ratio 1:4) in sealed steel ampoule at 1000°C for 24 h. $SrBr_2$ as well as the excess of NaBr were removed by extraction with liquid NH_3.

1.18 BORON–SODIUM–BARIUM–NITROGEN

The **NaBa_4[BN_2]_3** quaternary compound, which crystallizes in the cubic structure with the lattice parameter $a = 791.68 \pm 0.06$ pm and a calculated density of 4.610 g·cm^{-3} is formed in the B–Na–Ba–N system (Somer et al. 1995k, 2000c). The title compound was synthesized from a mixture of Ba_3N_2, BN, NaN_3, and Na (molar ratio 1:2.5:1:8) in a sealed steel ampoule at 900–1050°C followed by slow cooling (25°C·h^{-1}). The excess Na was removed in high vacuum at 300°C after the reaction.

1.19 BORON–SODIUM–CARBON–NITROGEN

The **Na[B(CN)$_4$]** and **Na$_2$[B(CN)$_3$]** quaternary compounds are formed in the B–Na–C–N system. Na[B(CN)$_4$] melts at 520°C and decomposes at 540°C (Bernhardt et al. 2003) and crystallizes in the cubic structure with the lattice parameter $a = 1168.0 \pm 0.1$ pm and a calculated density of 1.150 g·cm^{-3} (Küppers et al. 2005). To obtain the title compound, NaCN (2.62 mM), LiCl (2.74 mM), and Na[BF$_4$] (0.30 mM) were coarsely ground and mixed together in a dry box (Bernhardt et al. 2003). The reaction mixture was placed in Ni crucible. The crucible was covered with iron lid, transferred to a muffle furnace and heated to 300°C for 1.5 h. After the reaction was ended, the crucible was taken out from the furnace and cooled to room temperature. The cold gray–black porous mixture was transferred into a mortar and coarsely ground. Subsequently, the crushed solid was placed in a 3 L beaker and H_2O_2 (350 mL, 30% aqueous solution, 3 M) was added in a period of 0.5 h by the 30 mL portions under constant stirring. The exothermic reaction with vigorous gas evolution was started, and it was controlled by the addition of ice. Subsequently, it was checked whether any cyanide residues were present in the mixture. The reaction mixture (2.3 L) was divided into two 3 L bearers, and concentrated HCl (circa 300 mL, approximately 3.6 M) was acidified (pH = 5–7) until no more gas evolution was observed. It was then filtered, and the yellow solution with 28 mL of concentrated HCl was stirring. This was followed by the addition of Pr$^n_3$N (63 mL, 0.33 M). After stirring for 15 min, the reaction mixture was diluted with CH_2Cl_2 (250, 150, and 50 mL). The combined organic phases were washed with H_2O (200 mL), and the wash water was extracted with CH_2Cl_2. The combined CH_2Cl_2 phases were dried over $MgSO_4$ and filtered through a glass frit. NaOH (0.63 mM) was dissolved in a little water and added to the organic solution with vigorous stirring. It falls instantly from a beige oily substance and clumps by further stirring (30 min) at the bottom of the vessel. The CH_2Cl_2–Pr$^n_3$N mixture was decanted, and the product was extract with THF (200, 100, and 50 mL) to obtain the residue. The collected THF phases were dried with Na_2CO_3 and all volatile constituents finally removed on a rotary evaporator. Mixture of Na [B(CN)$_4$] and Na[B(CN)$_4$]·THF was obtained after drying at room temperature. NaB(CN)$_4$ was prepared after heating in vacuum at 60°C.

Colorless single crystals of Na[B(CN)$_4$] have been grown from aqueous solutions at room temperature by slow evaporation of H_2O (Küppers et al. 2005).

To prepare Na$_2$[B(CN)$_3$], Na[B(CN)$_4$] (30 mM) and freshly cut Na (60 mM) were loaded in a special reactor with a magnetic stirring bar (Bernhardt et al. 2011a, 2011b). After the transfer of dry, liquid NH_3 (40 mL) in vacuum, the reaction mixture was kept at −40°C, stirred, and shaken. In an exothermic reaction, the blue color of dissolved Na appeared and then vanished, while a yellow solid was formed. The reaction was finished when the mixture consisted of a red solution and a yellow precipitate. Liquid NH_3 was removed under reduced pressure, and the solid residue was dried in vacuum at 50°C. The title compound is accessible also by the reaction of $Na[BF(CN)_3]$ with Na in liquid NH_3 or THF–naphthalene mixture (Landmann et al. 2015).

1.20 BORON–POTASSIUM–CARBON–NITROGEN

The $K[B(CN)_4]$ and $K_2[B(CN)_3]$ quaternary compounds are formed in the B–K–C–N system. $K[B(CN)_4]$ melts at 430°C ± 5°C, decomposes at 510°C (Bernhardt et al. 2003), and crystallizes in the tetragonal structure with the lattice parameters $a = 697.6 \pm 0.1$, $c = 1421.0 \pm 0.3$ pm, and a calculated density of 1.479 g·cm^{-3} (Bernhardt et al. 2000). To obtain the title compound, KCN (2.61 mM), LiCl (2.74 mM), and $K[BF_4]$ (0.29 mM) were coarsely ground and mixed together in a dry box (Bernhardt et al. 2003). The further procedure corresponds to the made in the synthesis of $Na[B(CN)_4]$ (reaction temperature 300°C, reaction time 1 h), but KOH (0.63 mM) was used instead of NaOH and K_2CO_3 was used as drying agent instead of Na_2CO_3. The white product was washed with CH_2Cl_2 and dried in vacuo at room temperature.

K$[B(CN)_4]$ was also prepared as follows. To the solution of $[Bu^n_4N][B(CN)_4]$ (16.8 mM) in MeOH (10 mL), the solution of $AgNO_3$ (17.7 mM) in H_2O (10 mL) was added (Bernhardt et al. 2000). The precipitated $Ag[B(CN)_4]$ was filtered and washed with H_2O and then with warm MeCN (circa 300–500 mL). After that, this solution was added to the solution of KBr (16.8 mM) in H_2O (20 mL). The resulting suspension was evaporated to dryness on a rotary evaporator (60–100°C), H_2O (150 mL) was added, AgBr was filtered, and the title compound was fractionally crystallized.

$K_2[B(CN)_3]$ crystallizes in the orthorhombic structure with the lattice parameters $a = 1345.07 \pm 0.08$, $b = 1053.00 \pm 0.06$, and $c = 895.76 \pm 0.05$ pm at 150 K and a calculated density of 1.749 g·cm^{-3} (Bernhardt et al. 2011a, 2011b). This compound was obtained by the same way as $Na_2[B(CN)_3]$ was prepared using $K[B(CN)_4]$ and K instead of $Na[B(CN)_4]$ and Na, respectively (Bernhardt et al. 2011a, 2011b). KCN was removed from $K_2[B(CN)_3]$ by repeated washing with liquid NH_3 at −70°C. Liquid NH_3 was removed under reduced pressure, and the solid residue was dried in vacuum at 50°C. The title compound is accessible also by the reaction of $K[BF(CN)_3]$ with K in liquid NH_3 or THF–naphthalene mixture (Landmann et al. 2015).

1.21 BORON–RUBIDIUM–CARBON–NITROGEN

The $Rb[B(CN)_4]$ quaternary compound, which melts at 430°C, decomposes at 510°C and crystallizes in the tetragonal structure with the lattice parameters $a = 713.54 \pm 0.02$, $c = 1481.97 \pm 0.06$ pm, and a calculated density of 1.764 g·cm^{-3}, is formed in the B–Rb–C–N system (Küppers et al. 2005). The title compound was prepared by the next procedure. In a 100 mL glass beaker, 10.0 mM of $Na[B(CN)_4]$ was dissolved in H_2O (50 mL). To the vigorously stirred solution, concentrated HCl (2.0 mL, ≈24 mM) and Et_3N (2.8 mL, 20 mM) were added. After a few minutes, the emulsion was transferred into a separating funnel and extracted three times with CH_2Cl_2 (50, 25, and 25 mL, respectively). The organic phases were collected, dried over $MgSO_4$, and filtered through a glass frit. The obtained clear solution was treated under stirring with 50% RbOH (2.5 mL, 21 mM). A colorless suspension appeared, and after 1 h CH_2Cl_2 phase was separated. The solid residue was washed with of CH_2Cl_2 (50 mL), and MeCN (50 mL) was added. The suspension was stirred for 30 min, and the organic phase containing $Rb[B(CN)_4]$ was separated. The extraction was repeated twice with MeCN (50 mL). The combined organic phases were dried over Rb_2CO_3 and filtered through a glass frit, and the solvent was removed under reduced pressure. The obtained solid residue was washed with CH_2Cl_2 and dried in a vacuum to yield $Rb[B(CN)_4]$. Colorless single crystals of this compound have been grown from aqueous solutions at room temperature by slow evaporation of H_2O (Küppers et al. 2005).

1.22 BORON–CESIUM–CARBON–NITROGEN

The $Cs[B(CN)_4]$ quaternary compound, which melts at 420°C and decomposes at 510°C (Bernhardt et al. 2003) and crystallizes in the tetragonal structure with the lattice parameters $a = 730.0 \pm 0.2$, $c = 1534.0 \pm 0.5$ pm, and a calculated density of 2.013 g·cm^{-3}, is formed in the B–Cs–C–N system (Küppers et al. 2005). To obtain this compound, $KB[(CN)_4]$ (0.04 M) was dissolved in a small quantity of H_2O, and concentrated HCl (2 mL, 0.06 M) and Pr^n_3N (12 mL, 0.06 M) were added (Bernhardt et al. 2003). After stirring for 15 min, the mixture was diluted with CH_2Cl_2 (50, 50, and 25 mL). The combined CH_2Cl_2 phases were dried over $MgSO_4$ and filtered through a glass frit. CsOH (0.12 M) was dissolved in a little water and added to the organic solution with vigorous stirring. It falls instantly from a beige oily substance and clumps by further stirring (30 min) at the bottom of the vessel. The CH_2Cl_2–Pr^n_3N mixture was decanted, and the product was extracted with MeCN (200, 100, and 50 mL) to obtain the residue. The collected MeCN phases were dried with Cs_2CO_3 and all volatile constituents finally removed on a rotary evaporator.

1.23 BORON–COPPER–CARBON–NITROGEN

$Cu[B(CN)_4]$ quaternary compound, which decomposes at 470°C (Bernhardt et al. 2003) and crystallizes in the cubic structure with the lattice parameter $a = 543.14 \pm 0.07$ pm and a calculated density of 1.849 g·cm^{-3} (Küppers et al. 2005) [$a = 1070$ pm and the calculated and experimental densities of 1.932 and 1.94 ± 0.01 g·cm^{-3}, respectively (Bessler 1977)], is formed in the B–Cu–C–N system.

The title compound was prepared by the next three methods:

1. $CuSO_4 \cdot 5H_2O$ (10.0 mM), Cu powder (30.5 mM), and $Na[B(CN)_4]$ (19.7 mM) were transferred into a 500 mL round-bottom flask equipped with a valve and a magnetic stirring bar (Küppers et al. 2005). Under an Ar atmosphere, the flask was charged with 25% NH_4OH (150 mL) and the suspension was stirred vigorously for 2 h. Approximately 25 mL of the colorless solution was transferred under an Ar atmosphere into a 50 mL round-bottom flask, and all volatiles were removed under reduced pressure. The obtained colorless solid residue was washed with 10% HCl and dried in a vacuum.

2. $K[B(CN)_4]$ (2.3 mM) was dissolved in H_2O (40 mL) with CuCl (2.0 mM) and heated at 50–80°C for 2 h (Bernhardt et al. 2003). The white suspension was filtered through a glass frit, and the colorless solid was washed with H_2O (20 mL), 6 M HCl (20 mL), THF (20 mL), and diethyl ether (20 mL) and dried in vacuo.

3. The reaction vessel was a 20 cm long vertical glass tube of 2.5 cm internal diameter fused at the bottom of a glass frit of medium porosity (Bessler 1977). The pulverized CuCN was layered on the glass frit. The device was heated in an oil bath. BCl_3 (6.0 g) at 200–220°C was passed at a slow rate in the course of 2 h through solid CuCN (about 15.3 g). After pumping out at 120°C, 20.3 g of strong sintered product was obtained. The crude product was treated for about 30 min with half-concentrated HNO_3. The residue was filtered, washed with distilled H_2O and dried at 120°C. A fine-powder, sand-colored product (6.6 g) was obtained.

Colorless single crystals of $Cu[B(CN)_4]$ were prepared in a small glass U-tube, separated at the bottom by a small glass frit (Küppers et al. 2005). This tube was charged under Ar atmosphere on one side with a saturated aqueous $Li[B(CN)_4]$ solution and on the other side with a saturated CuCl solution in 25% NH_4OH. Crystals formed on the surface of the glass frit while the U-tube was kept at 2°C for 6 weeks.

1.24 BORON–SILVER–CARBON–NITROGEN

The $Ag[B(CN)_4]$ quaternary compound, which crystallizes in the cubic structure with the lattice parameter $a = 573.2 \pm 0.1$ pm and a calculated density of 1.964 g·cm^{-3} (Bernhardt et al. 2000) [$a = 1130$ pm and the calculated and experimental densities of 2.050 and 2.12 $\pm$ 0.01 g·cm^{-3}, respectively (Bessler 1977)], is formed in the B–Cu–C–N system.

The title compound was prepared in the same apparatus, in which $Cu[B(CN)_4]$ was synthesized (Bessler 1977). BCl_3 (4.3 g) at 150–160°C and atmospheric pressure was passed in the course of 2 h through solid AgCN (13.4 g). After pumping out at 120°C, 16 g of a dark-colored, powdery product was obtained. The crude product was extracted five times with 50 mL of concentrated NH_4OH. The residue was then treated to dissolve metallic Ag for about 1 h, with half-concentrated HNO_3. The residue was filtered washed with distilled H_2O and EtOH and dried at 120°C. A fine-powder, sand-colored product (6.6 g) was obtained. There was 2.7 g of a finely powdered ochre product obtained, which still contains some AgCl.

$Ag[B(CN)_4]$ could be also obtained by the interaction of $K[B(CN)_4]$ with $AgNO_3$ (Bernhardt et al. 2000).

1.25 BORON–GOLD–IRON–NITROGEN

The influence of deformation-induced defects on the isothermal precipitation of Au was studied in high-purity Fe–Au–B–N alloys containing 2.83 mass% Au, 0.05 mass% B, and 0.0156 mass% N (Zhang et al. 2014). The samples were aged at 550°C for 64 h in an ultra-high vacuum chamber. Gold precipitation was exclusively observed at the damage region and along grain boundaries. For the cavities and cracks induced by prior room-temperature overloading, preferential Au precipitation was observed at microcracks and cavities at grain boundary triple points in a wide region near the original fracture surface. A strong tendency of Au and h-BN precipitation onto the free (external) surface was demonstrated by X-ray photoelectron spectroscopy.

1.26 BORON–MAGNESIUM–CALCIUM–NITROGEN

The $Mg_2Ca_7[BN_2]_6$ quaternary compound, which crystallizes in the cubic structure with the lattice parameter $a = 715.12$ pm, is formed in the B–Mg–Ca–N system (Häberlen et al. 2002).

1.27 BORON–MAGNESIUM–ALUMINUM–NITROGEN

BN–Mg$_3$N$_2$–AlN. The isothermal sections of this quasiternary system at atmospheric and high pressures were constructed through X-ray diffraction (XRD) and are given in Figure 1.1 (Zhukov et al. 1996). **Mg$_3$AlBN$_4$** [**Mg$_{5.11}$Al$_2$(BN$_2$)$_2$N$_4$**, according to the data of Ströbele et al. (2015)] quaternary compound, which crystallizes in the rhombohedral structure with lattice parameters $a = 343.18 \pm 0.01$ and $c = 3143.6 \pm 0.1$ pm [$a = 343.9 \pm 0.1$ and $c = 3143 \pm 1$ pm (Zhukov et al. 1996)] (in hexagonal setting), and the calculated density of 2.423 g·cm^{-3} [the calculated and experimental densities of 2.58 and 2.53 g·cm^{-3}, respectively (Zhukov et al. 1996)] is formed in this system (Ströbele et al. 2015).

To obtain the title compound, a mixture of $MgCl_2$, AlN, $Li_3(BN_2)$, and Li_3N (molar ratio 3:1:1:1) was homogenized and ground in an agate mortar. Samples were heated at a rate of 2°C·min^{-1} to 1100°C, maintained at this temperature for 48 h before they were cooled to room temperature with the same rate of 2°C·min^{-1}. The tantalum ampoules were opened and products were washed, first with water and then with acetone before they were dried in air at 80°C (Ströbele et al. 2015).

This compound decomposes at the heating to the temperature higher than 1200°C under high pressure (>3 GPa) forming $Mg_3Al_3N_5$, $Mg_3B_2N_4$, and Mg_3BN_3 (Zhukov et al. 1996).

1.28 BORON–MAGNESIUM–GALLIUM–NITROGEN

The $Mg_{3.3}Ga_{0.4}(BN_2)N_2$ quaternary compound, which crystallizes in the rhombohedral structure with lattice parameters $a = 345.60 \pm 0$ and $c = 3179.24 \pm 0.02$ pm (in hexagonal setting) and a calculated density of 2.69 g·cm^{-3}, is formed in the B–Mg–Ga–N system (Dutczak et al. 2016).

To obtain the title compound, a mixture of $MgCl_2$, GaN, Li_3N, and Li_3BN_2 (molar ratio 3:1:1:1) with a total mass of ca. 250 mg was ground in an agate mortar. The mixture was loaded into clean niobium ampoules, sealed by arc-welding, and then fused into evacuated silica ampoules. Ampoules were heated to 1100°C with a heating rate of 2°C·min^{-1}, kept at this temperature for 48 h, and then cooled to room temperature with the same rate. The niobium ampoules were opened and products were washed, first with water and then with acetone before they were dried in air at 80°C. All manipulations of the starting materials were performed in a glove box under dry argon atmosphere.

1.29 BORON–MAGNESIUM–SILICON–NITROGEN

BN–Mg$_3$N$_2$–Si$_3$N$_4$. The isothermal sections of this quasiternary system at atmospheric pressure and under 5.0 GPa pressure were constructed through XRD and are presented in Figure 1.2 (Zhukov et al. 1994). Two quaternary compounds, **Mg$_7$SiB$_2$N$_8$** and **Mg$_{10}$SiBN$_9$**, were determined. $Mg_7SiB_2N_8$ exists in two polymorphic modifications: low-temperature modification transforms into high-temperature modification at temperatures higher than 1130–1230°C under 3.0–7.0 GPa. One of the polymorphic modifications of $Mg_7SiB_2N_8$

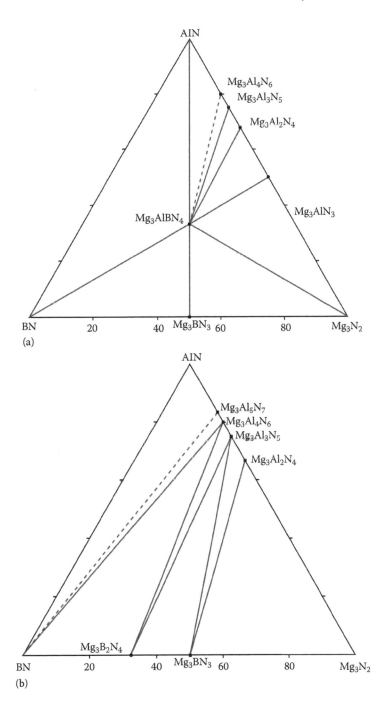

FIGURE 1.1 Isothermal sections of the BN–Mg$_3$N$_2$–AlN quasiternary system at (a) atmospheric and (b) high pressure. (From Zhukov, A.N. et al. *Zhurn. Obshch. Khim.*, 66(7), 1070–1072, 1996. With permission.)

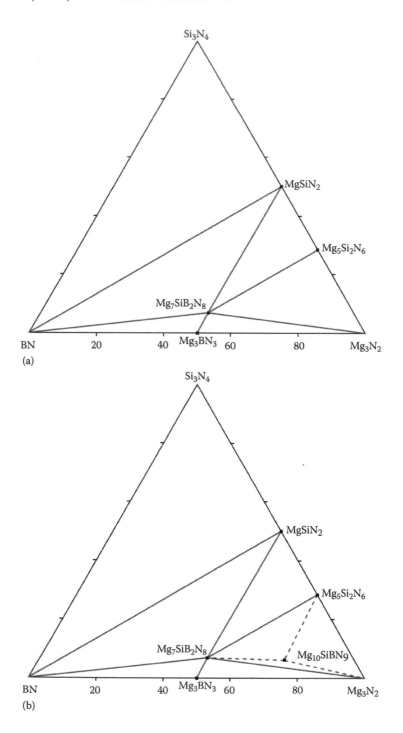

FIGURE 1.2 Isothermal sections of the $BN–Mg_3N_2–Si_3N_4$ quasiternary system at (a) 1200°C and atmospheric pressure and (b) at 1230°C under 5.0 GPa. (From Zhukov, A.N. et al. *Zhurn. Obshch. Khim.*, 64(8), 1242–1245, 1994. With permission.)

crystallizes in the rhombohedral structure with the lattice parameters $a = 343.83 \pm 0.01$ and $c = 3110.7 \pm 0.1$ pm (in hexagonal setting) and calculated density of 2.442 g·cm^{-3} (Ströbele et al. 2015). The second polymorph of this compound crystallizes in the monoclinic structure with the lattice parameters $a = 1055.98 \pm 0.02$, $b = 689.94 \pm 0.01$, $c = 590.82 \pm 0.01$ pm, and $\beta = 100.983 \pm 0.001°$ and a calculated density of 2.43872 g·cm^{-3}.

According to the data of Zhukov et al. (1994), the experimental density of the low pressure modification of $Mg_7SiB_2N_8$ is 2.65 g·cm^{-3}. Its high-pressure modification crystallizes in the orthorhombic structure with the lattice parameters $a = 538.0 \pm 0.5$, $b = 315.5 \pm 0.1$, and $c = 1125.2 \pm 0.3$ pm and the calculated and experimental densities of 2.89 and 2.84 g·cm^{-3}, respectively.

The rhombohedral modification of $Mg_7SiB_2N_8$ was obtained as follows (Ströbele et al. 2015). A mixture of Mg_3N_2, Si_3N_4, BN (molar ratio 7:1:6), and an additional 5 mass% of BN as a flux was homogenized and ground in an agate mortar (total mass of about 200 mg). Samples were heated at a rate of 5°C·min^{-1} to 1150°C and maintained at this temperature for 12 h before they were cooled to room temperature with the same rate. The tantalum ampoules were opened and products were washed, first with water and then with acetone, before they were dried in air. To prepare the monoclinic polymorph of this compound, a mixture of Mg_3N_2, Si_3N_4, and BN (molar ratio 7:1:6) with a total mass of about 200 mg was homogenized and ground in an agate mortar (Ströbele et al. 2015). Samples were heated with a rate of 5°C·min^{-1} to 1150°C, maintained at this temperature for 36 h before they were cooled to room temperature with the same rate. The tantalum ampoules were opened and products were washed, first with water and then with acetone, before they were dried in air.

$Mg_{10}SiBN_9$ exists only at high pressures and crystallizes in the hexagonal structure with the lattice parameters $a = 321.8 \pm 0.2$ and $c = 2652 \pm 1$ pm and the calculated and experimental densities of 2.84 and 2.80 g·cm^{-3}, respectively.

1.30 BORON–MAGNESIUM–CHLORINE–NITROGEN

The **$Mg_2[BN_2]Cl$** quaternary compound crystallizes in the orthorhombic structure with the lattice parameters $a = 661.39 \pm 0.08$, $b = 976.6 \pm 0.1$, and $c = 1060.0 \pm 0.1$ pm and a calculated density of 2.385 g·cm^{-3} (Somer et al. 2004b).

This compound was synthesized from the stoichiometric mixture of $MgCl_2$, BN, and Mg_3N_2. The well-ground and well-mixed educts were sealed into an arc-welded stainless steel ampoule (placed in a protecting silica tube) and heated during 6 h to 1200°C, annealed during 24 h, and cooled down to room temperature within 24 h. The reaction product was a uniform white powder. $Mg_2[BN_2]Cl$ is air and moisture sensitive; therefore, all handling with educts and products was performed under inert conditions.

1.31 BORON–MAGNESIUM–BROMINE–NITROGEN

The **$Mg_2[BN_2]Br$** quaternary compound crystallizes in the monoclinic structure with the lattice parameters $a = 915.09 \pm 0.03$, $b = 662.09 \pm 0.03$, $c = 647.56 \pm 0.02$ pm, and $\beta = 109.353° \pm 0.003°$ and a calculated density of 3.004 ± 0.001 g·cm^{-3} (Kokal et al. 2011b). This compound was synthesized from stoichiometric mixtures of the binaries Mg_3N_2, BN, and $MgBr_2$ in arc-welded Nb ampoules, placed in evacuated silica tubes. The samples were heated during 6 h to 1200°C, annealed for 24 h, and cooled down to room temperature within the next 24 h. The reaction products were single-phase colorless microcrystalline powders.

$Mg_2[BN_2]Br$ and the starting materials Mg_3N_2 and $MgBr_2$ are air and moisture sensitive. Therefore, all handling with starting materials and products had to be performed under inert conditions.

1.32 BORON–MAGNESIUM–IODINE–NITROGEN

The $Mg_2[BN_2]$ **I** quaternary compound crystallizes in the orthorhombic structure with the lattice parameters $a = 1353.5 \pm 0.3$, $b = 935.0 \pm 0.2$, and $c = 1119.4 \pm 0.2$ pm and a calculated density of 2.417 g·cm^{-3} (Somer et al. 2004b). This compound could be obtained from the reaction of the binary phases Mg_3N_2, MgI_2, and BN (molar ratio 5:1:10) at 1200°C. The thermal treatment of the sample was the same as described for $Mg_2[BN_2]Cl$. $Mg_2[BN_2]I$ is air and moisture sensitive; therefore, all handling with educts and products was performed under inert conditions.

1.33 BORON–CALCIUM–CARBON–NITROGEN

The $Ca_{9+0.5x}[BN_2]_{6-x}(CBN)_x$ quaternary compound crystallizes in the cubic structure with the lattice parameters $a = 733.80 \pm 0.04$ pm for $x = 0.36$, $a = 737.09 \pm 0.03$ pm for $x = 1.43$, $a = 739.82 \pm 0.08$ pm for $x = 2$, and a calculated density of 2.5110 g·cm^{-3} for $x = 0.36$, 2.5692 g·cm^{-3} for $x = 1.43$ and 2.5827 g·cm^{-3} for $x = 2$ (Wörle et al. 1998). This compound was obtained in sealed Nb ampoules from Ca_3N_2, BN, graphite, and Ca. The ampoules were filled with the finely ground and well-mixed starting components in a glove box with a purified Ar atmosphere. After sealing under Ar atmosphere, the ampoules were placed into an Ar-filled Al_2O_3 tube and heated to 1200°C for 1 h to yield pure products. Large single crystals were prepared within 3 h at about 1630°C. The title compound forms dark single crystals with metallic luster, which show dark-green to yellow-green color after grinding, depending on the carbon content.

1.34 BORON–CALCIUM–FLUORINE–NITROGEN

The $Ca_2[BN_2]F$ quaternary compound, which crystallizes in the orthorhombic structure with the lattice parameters $a = 918.2 \pm 0.2$, $b = 364.9 \pm 0.1$, and $c = 996.6 \pm 0.2$ pm and a calculated density of 2.745 g·cm^{-3}, is formed in the B–Ca–F–N system (Rohrer and Nesper 1998). This compound was synthesized from stoichiometric amounts of Ca_3N_2, CaF_2, and BN. The basic educts are well mixed and heated in stainless steel ampoules to 1000°C. The temperature was kept for 30 h and then lowered to room temperature by 100°C·h^{-1}. The product is white fine powder, which is stable on exposure to air and moisture. Synthesis at 1150°C yields transparent colorless single crystals of rod-like shape.

1.35 BORON–CALCIUM–CHLORINE–NITROGEN

The $Ca_2[BN_2]Cl$ quaternary compound, which crystallizes in the orthorhombic structure with the lattice parameters $a = 1165.7 \pm 0.1$, $b = 389.1 \pm 0.1$, and $c = 896.5 \pm 0.1$ pm and a calculated density of 2.523 g·cm^{-3} (Rohrer and Nesper 1998) [$a = 1166.7 \pm 0.2$, $b = 390.26 \pm 0.04$, and $c = 899.8 \pm 0.1$ pm and a calculated density of 2.50 g·cm^{-3} (Reckeweg and Meyer

1997)], is formed in the B–Ca–Cl–N system. This compound was synthesized from stoi-chiometric amounts of Ca_3N_2, $CaCl_2$, and BN (Rohrer and Nesper 1998). The basic educts are well mixed and heated in stainless steel ampoules to 1000°C. The temperature was kept for 30 h and then lowered to room temperature by 100°C·h^{-1}. The product is white fine powder, which is moisture sensitive. Synthesis at 850°C yields transparent colorless single crystals of rod-like shape.

The title compound could be also prepared from Ca, $CaCl_2$, and h-BN in sealed Ta ampoules at 1200°C (Reckeweg and Meyer 1997). The crystals obtained were transparent yellow in color.

1.36 BORON–CALCIUM–NICKEL–NITROGEN

The **CaNi(BN)** quaternary compound, which crystallizes in the tetragonal structure with the lattice parameters $a = 353.59 \pm 0.01$ and $c = 763.94 \pm 0.05$ pm (Imamura et al. 2012) [$a = 353.24 \pm 0.03$ and $c = 763.59 \pm 0.09$ pm and a calculated density of 4.309 g·cm^{-3} (Blaschkowski and Meyer 2002); $a = 353.38 \pm 0.04$ and $c = 763.76 \pm 0.01$ pm (Neukirch et al. 2006)], is formed in the B–Ca–Ni–N system. To obtain CaNi(BN), the reaction was performed in a Ta container starting from mixtures of Ca, Ni, and h-BN (molar ratio 1:1:1) (Blaschkowski and Meyer 2002). Ni powder was thoroughly mixed with h-BN and then transferred into a Ta container together with chunks of Ca metal. The tantalum container was sealed under Ar and then placed into a vacuum-sealed silica ampoule. The sample was heated to 1000°C within 8 h and remained at this temperature for 60 h. After cooling the furnace by its natural cooling rate, a brass-colored crystalline material of the title com-pound was obtained as reaction product. CaNi(BN) could be also obtained using Ca_3N_2, Ni, and B or Ca_3N_2, Ni, B, and BN powders as the starting materials (Imamura et al. 2012). They were mixed well and loaded in a Ta capsule in an Ar-filled glove box. The Ta con-tainer was then welded inside the glove box and heat-treated at 5 GPa and 1400°C for 1 h. This compound is unstable in air, but its dense sintered pellets prepared under high pressure were stable enough under an ambient atmosphere.

According to the calculated band structure, an extremely narrow band gap is present in CaNi(BN) (Blaschkowski and Meyer 2002).

1.37 BORON–CALCIUM–PALLADIUM–NITROGEN

The **CaPd(BN)** quaternary compound, which crystallizes in the tetragonal structure with the lattice parameters $a = 377.38 \pm 0.01$ and $c = 760.95 \pm 0.01$ pm, is formed in the B–Ca–Pd–N system (Blaschkowski and Meyer 2002). It was obtained by the same way as CaNi(BN) was prepared using Pd instead of Ni. After cooling the furnace by its natural cooling rate, a black microcrystalline powder of the title compound was obtained as reaction product. This compound is unstable in air. All manipulations for the synthesis of CaPd(BN) were performed in an Ar-filled glove box.

1.38 BORON–STRONTIUM–BARIUM–NITROGEN

The **$SrBa_8[BN_2]_6$** quaternary compound, which crystallizes in the cubic structure with the lattice parameter $a = 791.3 \pm 0.1$ pm, is formed in the B–Sr–Ba–N system (Öztürk et al.

2005). It was synthesized from the binaries Sr_2N, Ba_2N, and h-BN [molar ratio (2–3):8:6] in arc-welded Nb ampoules at 1100°C. This compound is pale yellow crystalline powder, which is sensitive to air and moisture.

1.39 BORON–STRONTIUM–FLUORINE–NITROGEN

The **$Sr_2[BN_2]F$** quaternary compound, which crystallizes in the orthorhombic structure with the lattice parameters $a = 989.1 \pm 0.2$, $b = 390.4 \pm 0.1$, and $c = 1019.3 \pm 0.2$ pm and a calculated density of 3.933 g·cm^{-3}, is formed in the B–Sr–F–N system (Rohrer and Nesper 1998). This compound was synthesized from stoichiometric amounts of Sr_3N_2, SrF_2, and BN. The basic educts are well mixed and heated in stainless steel ampoules to 1000°C. The temperature was kept for 30 h and then lowered to room temperature by 100°C·h^{-1}. The product is white fine powder, which is moisture sensitive. Synthesis at 850°C yields transparent colorless single crystals of rod-like shape.

1.40 BORON–STRONTIUM–CHLORINE–NITROGEN

The **$Sr_2[BN_2]Cl$** quaternary compound, which crystallizes in the orthorhombic structure with the lattice parameters $a = 1240.8 \pm 0.1$, $b = 416.1 \pm 0.1$, and $c = 917.0 \pm 0.1$ pm and a calculated density of 3.501 g·cm^{-3} (Rohrer and Nesper 1998) [$a = 1242.8 \pm 0.1$, $b = 416.75 \pm 0.04$, and $c = 920.8 \pm 0.1$ pm and a calculated density of 3.48 g·cm^{-3} (Reckeweg and Meyer 1997)], is formed in the B–Sr–Cl–N system. This compound was synthesized from stoichiometric amounts of Sr_3N_2, $SrCl_2$, and BN (Rohrer and Nesper 1998). The basic educts are well mixed and heated in stainless steel ampoules to 1000°C. The temperature was kept for 30 h and then lowered to room temperature by 100°C·h^{-1}. The product is white fine powder, which is moisture sensitive. Synthesis at 850°C yields transparent colorless single crystals of rod-like shape.

The title compound could be also prepared from Sr, $SrCl_2$, and h-BN in sealed Ta ampoules at 1200°C (Reckeweg and Meyer 1997). The crystals obtained were blue in color.

1.41 BORON–STRONTIUM–BROMINE–NITROGEN

The **$Sr_2[BN_2]Br$** quaternary compound, which crystallizes in the trigonal structure with the lattice parameters $a = 411.692 \pm 0.002$ and $c = 2646.11 \pm 0.02$ pm and a calculated density of 3.77 g·cm^{-3}, is formed in the B–Sr–Br–N system (Kokal et al. 2011a). The title compound was obtained as white crystalline powder from the reaction of the mixture of $Sr_3[BN_2]_2$ and $SrBr_2$ (molar ratio 1:1). The educts were transferred into Nb ampoule, which was sealed and in turn jacketed in Ar-filled quartz ampoule. The sample was heated to 950°C within 6 h, annealed for 24 h, and cooled down to room temperature within 8 h. This compound is sensitive to air and moisture.

1.42 BORON–STRONTIUM–IODINE–NITROGEN

The **$Sr_2[BN_2]I$** quaternary compound, which crystallizes in the monoclinic structure with the lattice parameters $a = 1028.4 \pm 0.7$, $b = 422.4 \pm 0.3$, $c = 1324.6 \pm 0.9$ pm, and $\beta = 90.87° \pm 0.06°$ and a calculated density of 3.936 g·cm^{-3}, is formed in the B–Sr–I–N system (Rohrer and Nesper 1999b). This compound was synthesized from stoichiometric amounts of Sr_3N_2, SrI_2, and BN. The precursors were well mixed and heated to 530°C in a stainless steel ampoule for 25 h. The temperature was then increased by 50°C·h^{-1} to 880°C, lowered to 750°C by 1°C·h^{-1}, and kept there for 50 h. The product is a light yellow powder. Single crystals are thin, transparent platelets.

1.43 BORON–BARIUM–EUROPIUM–NITROGEN

The $EuBa_8[BN_2]_6$ quaternary compound, which crystallizes in the cubic structure with the lattice parameter $a = 783.9 \pm 0.1$ pm, is formed in the B–Ba–Eu–N system (Öztürk et al. 2005). It was synthesized from the binaries EuN, Ba_2N, and h-BN [molar ratio (2–3):8:6] in arc-welded Nb ampoules at 1100°C. This compound is black crystalline powder, which is sensitive to air and moisture.

1.44 BORON–BARIUM–SILICON–NITROGEN

The $Ba_{26}B_{12}Si_5N_{27}$ quaternary compound, which crystallizes in the orthorhombic structure with the lattice parameters $a = 1769.42 \pm 0.04$, $b = 3414.37 \pm 0.06$, and $c = 1004.10 \pm 0.02$ pm and a calculated density of 4.620 g·cm^{-3}, is formed in the B–Ba–Si–N system (Takayuki et al. 2015). The title compound was prepared as follows. In an Ar-filled glove box, Ba chunks (1.0 mM), Si powder (0.5 mM), La flakes (0.25 mM), and Na chips (3.6 mM) were weighed and put into a crucible of sintered BN. The crucible was sealed in a stainless steel tube and heated at 900°C for 2 h and then cooled to 300°C at a rate of 4°C·h^{-1}. After cooling to room temperature by shutting off the power to the furnace, the tube was cut open in the glove box and the crucible was washed with liquid NH_3 to dissolve away Na. Single crystals of the title compound were grown on the wall of the crucible.

1.45 BORON–BARIUM–OXYGEN–NITROGEN

The $Ba_4[BN_2]_2O$ quaternary compound, which crystallizes in the orthorhombic structure with the lattice parameters $a = 1575.3 \pm 0.1$, $b = 729.1 \pm 0.4$, and $c = 731.9 \pm 0.1$ pm and a calculated density of 5.081 g·cm^{-3}, is formed in the B–Ba–O–N system (Curda et al. 1994b; Somer et al. 1997). This compound was originally formed due to the presence of oxygen impurity in commercially available Ba_3N_2. It could be also prepared from a stoichiometric mixture of BaO, Ba_3N_2, and BN in a sealed Nb ampoule at 1080°C.

1.46 BORON–BARIUM–FLUORINE–NITROGEN

The $Ba_8[BN_2]_5F$ quaternary compound, which crystallizes in the triclinic structure with the lattice parameters $a = 420.4 \pm 0.3$, $b = 2092 \pm 2$, and $c = 2095 \pm 2$ pm and $\alpha = 91.74° \pm 0.06°$, $\beta = 90.03° \pm 0.06°$, and $\gamma = 93.12° \pm 0.07°$ and a calculated density of 4.739 g·cm^{-3}, is formed in the B–Ba–F–N system (Rohrer and Nesper 1999a). This compound was synthesized from stoichiometric amounts of Ba_3N_2, BaF_2, and BN. The basic educts are well mixed and heated in stainless steel ampoules to 1000°C. The temperature was kept for 10 h and then lowered to room temperature by 50°C·h^{-1}. The product was a mixture of yellow needles, which were intergrown and light yellow crystals of rod-like shape.

1.47 BORON–BARIUM–CHLORINE–NITROGEN

The $Ba_2[BN_2]Cl$ quaternary compound, which crystallizes in the cubic structure with the lattice parameter a = 1462.88 ± 0.01 pm at 173 ± 2 K and a calculated density of 4.44 g·cm^{-3}, is formed in the B–Ba–Cl–N system (Reckeweg et al. 2008). The title compound was prepared as follows. Ba (3.02 mM), BaCl$_2$ (1.01 mM), h-BN (2.01 mM), and NaN$_3$ (0.69 mM) were intimately ground and arc-welded into a clean Ta container. The metal container was sealed into an evacuated silica tube. The tube was placed upright in a box furnace and heated to 930°C within 12 h. After a 3-day reaction time, the furnace was switched off and allowed to cool to room temperature. The product contained as the main product transparent, pale-blue chunky crystals of Ba$_2$[BN$_2$]Cl and some cube-shaped NaCl crystals. Crystals of Ba$_2$[BN$_2$]Cl are moderately air and moisture sensitive; therefore, all manipulations were performed in a glove box under purified argon.

1.48 BORON–ALUMINUM–GALLIUM–NITROGEN

BN–AlN–GaN. Thermodynamic calculations based on a strictly regular solution model and experimental results have predicted an unstable region of mixing to occur in this system (Wei and Edgar 2000; Takayama et al. 2001). The spinodal isotherms at 600–1000°C are plotted in Figure 1.3 (Wei and Edgar 2000). It is shown that the unstable composition region is reduced as the growth temperature increases.

The calculated structural and electronic properties of the $B_xAl_yGa_{1-x-y}N$ solid solutions for different B and Al compositions show a linear behavior of the lattice constant and a nonlinear behavior of the direct and indirect band gaps (Djoudi et al. 2012). The incorporation of B decreases the lattice parameter by 15.72% and increases the energy gap by 81.53%, while the incorporation of Al reduces the lattice parameter by 2.45% and increases the energy gap by 24.82%. The calculated band-gap energy of $B_xAl_yGa_{1-x-y}N$ is in the range of 3.121–5.122 eV.

1.49 BORON–ALUMINUM–INDIUM–NITROGEN

BN–AlN–InN. The unstable mixing region was calculated from the free energy of mixing using a strictly regular solution model (Takayama et al. 2001). The calculated spinodal isotherms for this quasiternary system between 1000°C and 3000°C are shown in Figure 1.4.

1.50 BORON–ALUMINUM–CARBON–NITROGEN

The $B_6Al_{0.185}CN_{0.256}$ quaternary compound, which crystallizes in the orthorhombic structure with the lattice parameters a = 568.5 ± 0.2, b = 890.3 ± 0.3, and c = 912.2 ± 0.3 pm and a calculated density of 2.459 g·cm^{-3}, is formed in the B–Al–C–N system (Rizzol et al. 2002).

1.51 BORON–ALUMINUM–TITANIUM–NITROGEN

Conductive TiB_2–**AlN** ceramics were prepared by self-propagating high-temperature synthesis in a filtration combustion regime at high nitrogen pressures (Bunin et al. 2002). The effect of N$_2$ pressure on the combustion rate and the Al metal content of the reaction product were studied. The combustion rate was also shown to depend on the composition

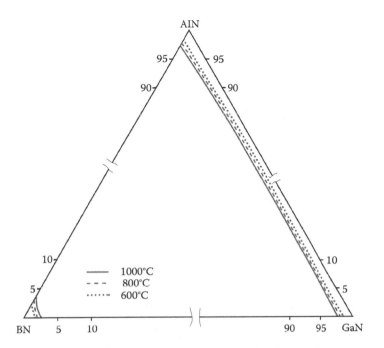

FIGURE 1.3 Calculated spinodal isotherms for the BN–AlN–GaN quasiternary system at the temperature range of 600–1000°C. (Reprinted from *J. Cryst. Growth*, 208(1–4), Wei, C.H., and Edgar, J.H., Unstable composition region in the wurtzite $B_{1-x-y}Ga_xAl_yN$ system, 179–182, Copyright (2000), with permission from Elsevier.)

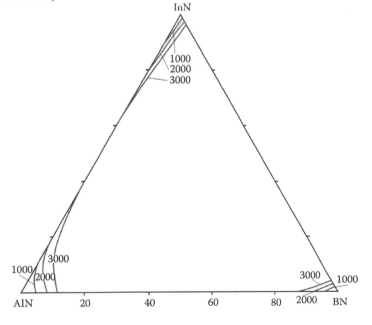

FIGURE 1.4 Calculated spinodal isotherms for the BN–AlN–InN quasiternary system between 1000°C and 3000°C. (Reprinted from *J. Cryst. Growth*, 222(1–2), Takayama, T. et al., Analysis of phase-separation region in wurtzite group III nitride quaternary material system using modifed valence force field model, 29–37, Copyright (2001), with permission from Elsevier.)

of the starting mixture. The phase composition of the ceramics was determined by XRD, chemical analysis, and microstructural examination.

1.52 BORON–GALLIUM–INDIUM–NITROGEN

BN–GaN–InN. The unstable mixing region was calculated from the free energy of mixing using a strictly regular solution model (Takayama et al. 2001). The calculated spinodal isotherms for this quasiternary system between 1000°C and 3000°C are shown in Figure 1.5.

1.53 BORON–GALLIUM–GERMANIUM–NITROGEN

The **B₂GaGeN** quaternary compound, which crystallizes in the monoclinic structure with the lattice parameters $a = 724.9 \pm 0.1$, $b = 421.0 \pm 0.1$, $c = 931.4 \pm 0.1$ pm, and $\beta = 108.87° \pm 0.01°$ and a calculated density of 5.322 ± 0.001 g·cm^{-3}, is formed in the B–Ga–Ge–N system (Clarke and DiSalvo 1997). This compound was prepared from NaN₃, Na, Ba, Ge, and Ga in the Na/Ba/Ge/Ga/N molar ratio about 18:4:1:1:18 in Nb tubes sealed inside quartz tubes under vacuum. The samples were placed in a muffle furnace, and the temperature was raised to 760°C over 15 h, maintained at 760°C for 24 h, and then lowered linearly to 100°C over 200 h, whence the furnace was turned off. B₂GaGeN decomposes very readily in air; therefore, all materials were handled in a dry box, in which the Ar atmosphere was constantly circulated through molecular sieves.

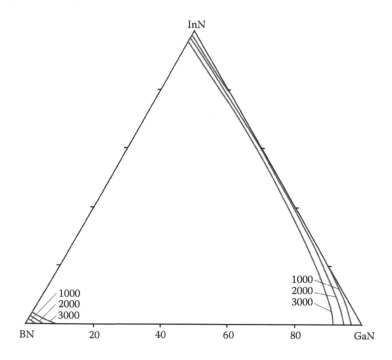

FIGURE 1.5 Calculated spinodal isotherms for the BN–GaN–InN quasiternary system between 1000°C and 3000°C. (Reprinted from *J. Cryst. Growth*, 222(1–2), Takayama, T. et al., Analysis of phase-separation region in wurtzite group III nitride quaternary material system using modifed valence force field model, 29–37, Copyright (2001), with permission from Elsevier.)

1.54 BORON–THALLIUM–CARBON–NITROGEN

$Tl[B(CN)_4]$ quaternary compound, which melts at 370°C and decomposes at 470°C and crystallizes in the tetragonal structure with the lattice parameters $a = 706.55 \pm 0.02$, $c = 1467.91 \pm 0.04$ pm, and a calculated density of 2.895 g·cm^{-3}, is formed in the B–Tl–C–N system (Küppers et al. 2005). To obtain this compound, $Na[B(CN)_4]$ (10.0 mM) was dissolved in H_2O (50 mL) and treated with Et_3N (2.80 mL, 20 mM) as well as concentrated HCl (2.0 mL, ≈24 mM). The clear solution was transferred into a separating funnel and extracted four times with CH_2Cl_2 (75, 50, 50, and 25 mL, respectively). Afterward, the combined organic phases were dried over $MgSO_4$, filtered, and evaporated under reduced pressure. The obtained $[NEt_3H][B(CN)_4]$ was suspended in a solution Tl_2CO_3 (10.0 mM) in H_2O (100 mL). Subsequently, all volatiles were removed at 50°C under reduced pressure. H_2O (100 mL) was added, and the suspension was evaporated again. This procedure was repeated three times, and by washing of the colorless solid residue with CH_2Cl_2, the last traces of Et_3N were removed. The solid was suspended in MeCN (250 mL) and filtered through a glass frit to separate excess Tl_2CO_3, and the solvent was removed under reduced pressure. Colorless $Tl[B(CN)_4]$ was obtained. Single crystals of this compound have been grown from aqueous solutions at room temperature by slow evaporation of H_2O.

1.55 BORON–SCANDIUM–CARBON–NITROGEN

The $Sc_{1-x}B_{15.5}CN$ quaternary compound, which crystallizes in the trigonal structure with the lattice parameters $a = 556.8 \pm 0.2$ and $c = 1075.6 \pm 0.2$ pm and a calculated density of 2.842 g·cm^{-3}, is formed in the B–Sc–C–N system (Leithe-Jasper et al. 2004). In a typical experiment for obtaining the title compound, about 1 g of a "ScB$_{17}$C" master alloy was powdered and mixed with Sn powder followed by cold isostatic pressing into a cylindrical bar. This rod was placed in a crucible made from sintered BN, which was inserted into a graphite susceptor with a cap. Under a flow of Ar, this setup was quickly (within 1 h) heated up to 1600°C, kept there for 6–8 h, and cooled down to room temperature (100°C·h^{-1}). The tin matrix was dissolved by leaching in warm concentrated HCl. Silver gray agglomerates of intergrown hexagonal plates with a metallic luster were isolated. Crystals can be also grown in a Si flux under similar experimental conditions.

1.56 BORON–YTTRIUM–CARBON–NITROGEN

The $YB_{15.5}CN$ and $YB_{22}C_2N$ quaternary compounds are formed in the B–Y–C–N system. $YB_{15.5}CN$ crystallizes in the trigonal structure with the lattice parameters $a = 559.19 \pm 0.06$ and $c = 1087.3 \pm 0.5$ pm (Zhang et al. 2001b; Leithe-Jasper et al. 2004). Starting materials for the title compound production were powders of YB_{12}, graphite, and h-BN (Leithe-Jasper et al. 2004). The mixture was fired at 1600°C in a graphite susceptor under a flow of Ar. Single crystals of $YB_{15.5}CN$ were prepared by a high-temperature solution growth method using Cu or Sn as the flux (Zhang et al. 2001b).

$YB_{22}C_2N$ also crystallizes in the trigonal structure with the lattice parameters $a = 562.3$ and $c = 4478.5 \pm 0.3$ pm and the calculated and experimental densities of 2.97 and 2.60–2.85 g·cm^{-3}, respectively (Zhang et al. 2001a) [$a = 562.5$ and $c = 4478.6$ pm (Mori and Nishimura 2006), $a = 559.82 \pm 0.02$, and $c = 4482.3 \pm 0.2$ pm for $Y_{0.74}B_{22.25}C_{1.75}N$

composition (Sologub and Mori 2013)]. This compound was prepared using the mixture of YB_{12}, B, graphite, and h-BN (Mori and Nishimura 2006). The mixture was heated to around 1600°C.

It could be also obtained as follows. YB_{12}, amorphous B, graphite, and BCN in various molar ratios were well mixed in acetone and pressed into pellets with hydrostatic pressure of 250 MPa or so before heating (Zhang et al. 2001a). The pellets reacted in a BN or graphite crucible, which was inserted into a heated graphite susceptor. The reaction process was performed in vacuum and maintained at a temperature of about 1700°C for 8–12 h.

For obtaining $Y_{0.74}B_{22.25}C_{1.75}N$ composition, the powdery YB_4, amorphous B, and graphite were used (Sologub and Mori 2013). In order to introduce nitrogen into the initial mixture of elements, BN or BCN was added. The starting powdery components were well mixed in acetone and compacted by cold isostatic pressing into a cylindrical bar applying the pressure of 250 MPa. This rod was placed in a crucible made from sintered BN, which was inserted into a graphite susceptor. The reaction process was performed under a flow of Ar at a predetermined temperature of about 1700°C for 8 h; afterward, the setup was cooled down in 1 h to room temperature. After synthesis, the samples were powderized, compacted, and annealed for 10 h at 1500°C.

1.57 BORON–YTTRIUM–NICKEL–NITROGEN

The **YNi(BN)** quaternary compound, which crystallizes in the tetragonal structure with the lattice parameters $a = 347.81 \pm 0.03$ and $c = 757.29 \pm 0.07$ pm, is formed in the B–Y–Ni–N system (Neukirch et al. 2006). To obtain this compound, the reaction was performed in a Ta container starting from mixtures of Y, Ni, and h-BN (molar ratio 1:1:1), using a total mass of about 400 mg. Alternatively, YN and NiB were used as starting materials Reaction was performed as follows: Yttrium powder was thoroughly mixed with h-BN and then transferred into a Ta container. The container was sealed under Ar and then placed into a vacuum-sealed silica ampoule. The sample were heated to temperatures between 1000°C and 1200°C within 1 h to 3 days and remained at this temperature for 2–14 days. After cooling the furnace by its natural cooling rate, black crystalline powders of YNi(BN) were obtained as reaction product. All manipulations during the synthesis were performed in an Ar-filled dry box.

1.58 BORON–LANTHANUM–OXYGEN–NITROGEN

The **$La_6(BN_3)O_6$** quaternary compound, which crystallizes in the orthorhombic structure with lattice parameter $a = 366.88 \pm 0.03$, $b = 2509.2 \pm 0.3$, and $c = 1101.1 \pm 0.1$ pm and a calculated density of 6.437 g·cm^{-3}, is formed in the B–La–O–N system (Jing and Meyer 2002). The title compound was prepared by two methods. According to the first method, Li_3BN_2, Li_3N, and LaOCl were used (molar ratio 1: 1: 6) as starting materials. The reaction mixture was heated slowly to 950°C and cooled after 3 days by switching off the furnace. According to the second method, a mixture of $LaBO_3$, LaN, and La_2O_3 (molar ratio 1:3:1) was slowly heated to 1100°C, kept at this temperature for 4 days, and then cooled by switching off the oven. Light brown crystals of $La_6(BN_3)O_6$ were obtained.

1.59 BORON–LANTHANUM–COBALT–NITROGEN

The **La$_3$Co$_2$B$_2$N$_3$** quaternary compound, which crystallizes in the tetragonal structure with the lattice parameters $a = 371.52 \pm 0.08$ and $c = 2039.9 \pm 0.7$ pm (Imamura et al. 2012), is formed in the B–La–Co–N system. The title compound was obtained from the mixture of LaN, Ni, and B. The starting materials were mixed well and loaded in a Ta capsule in an Ar-filled glove box. The Ta container was then welded inside the glove box and heat-treated at 5 GPa and 1400°C for 2 h. The dense sintered pellets prepared under high pressure were stable enough under an ambient atmosphere. La$_3$Co$_2$B$_2$N$_3$ could be also obtained by the same way as La$_3$Ni$_2$B$_2$N$_3$ was prepared using Co powder instead of Ni powder and pyrolysis at 1200°C (Wideman et al. 1996).

1.60 BORON–LANTHANUM–NICKEL–NITROGEN

The **LaNi(BN)** and **La$_3$Ni$_2$B$_2$N$_3$** quaternary compounds are formed in the B–La–Ni–N system. LaNi(BN) crystallizes in the tetragonal structure with the lattice parameters $a = 372.66 \pm 0.01$ and $c = 757.59 \pm 0.04$ pm (Imamura et al. 2012) [$a = 371.96 \pm 0.03$ and $c = 758.23 \pm 0.09$ pm (Neukirch et al. 2006); $a = 372.5 \pm 0.2$ and $c = 759.0 \pm 0.4$ pm (Cava et al. 1994); $a = 373$ and $c = 764$ pm (Zandbergen et al. 1994)]. This compound was synthesized using LaN, Ni, and B powders as the starting materials (Imamura et al. 2012). They were mixed well and loaded in a Ta capsule in an Ar-filled glove box. The Ta container was then welded inside the glove box and heat-treated at 5 GPa and 1200°C for 0.5 h. The obtained pellets prepared under high pressure were stable enough under an ambient atmosphere. LaNi(BN) and La$_3$Ni$_2$B$_2$N$_3$ often coexist when stoichiometric mixtures were used as starting materials. Therefore, the nominal chemical composition was deviated from ideal stoichiometry to reduce these competing phases. In particular, LaNi(BN) was prepared from a mixture of the raw materials with the nominal atomic ratio LnN/Ni/B = 0.9:1:1.

LaNi(BN) could be also obtained by the same way as YNi(BN) was synthesized using La and LaN instead of Y and YN, respectively (Neukirch et al. 2006). LaNi(BN) could be also prepared by arc-melting and annealing (Cava et al. 1994; Zandbergen et al. 1994). The starting materials were La, Ni, and B. LaN powder could also be employed as starting material if care was taken to avoid significant air exposure. Nitrogen was introduced to the samples by arc-melting under N$_2$.

La$_3$Ni$_2$B$_2$N$_3$ also crystallizes in the tetragonal structure with the lattice parameters $a = 371.23 \pm 0.01$ and $c = 2053.8 \pm 0.1$ pm (Imamura et al. 2012) [$a = 372.1 \pm 0.2$ and $c = 2051.6 \pm 0.5$ pm (Cava et al. 1994); $a = 373$ and $c = 2067$ pm (Zandbergen et al. 1994); $a = 372.512 \pm 0.005$ and $c = 2051.72 \pm 0.04$ pm (Huang et al. 1995); $a = 372.07 \pm 0.06$ and $c = 2051.4 \pm 0.6$ pm (Michor et al. 1996); $a = 372.95 \pm 0.02$ and $c = 2056.3 \pm 0.2$ pm and a calculated density of 6.942 g·cm^{-3} (Blaschkowski and Meyer 2003); $a = 372$ and $c = 2052$ pm (Ali et al. 2011b); $a = 370.95 \pm 0.02, 370.97 \pm 0.02, 371.06 \pm 0.02, 371.39 \pm 0.02,$ and 371.88 ± 0.02 and $c = 2051.93 \pm 0.2, 2051.68 \pm 0.2, 2051.07 \pm 0.02, 2051.02 \pm 0.02,$ and 2052.22 ± 0.02 pm for $\delta = 0.1$ at 4, 30, 80, 180, and 300 K, respectively (Ali et al. 2010, 2017)]. Polycrystalline samples of La$_3$Ni$_2$B$_2$N$_{3-\delta}$ were prepared by inductive levitation melting (Ali et al. 2010, 2017). The starting materials were La, Ni, B, and nitrogen gas. In the first preparation step, stoichiometric amounts of Ni and B were melted several times in Ar atmosphere to produce the binary compound NiB. In the next step, La metal was premelted in vacuum and then melted with NiB to prepare a $La_3Ni_2B_2$ precursor alloy. In a third step, these alloys were repeatedly melted in Ar/N_2 atmosphere such that the N stoichiometry is slowly increased to reach projected stoichiometry near $La_3Ni_2B_2N_{2.7}$ for one set of samples and projected stoichiometry near $La_3Ni_2B_2N_{2.9}$ for a second set of samples. As cast samples were initially heat-treated in a high-vacuum furnace at 1100°C for 1 week. For a second and final annealing, samples projected as $La_3Ni_2B_2N_{2.7}$ were sealed in quartz ampoules under 20 kPa argon atmosphere and then annealed at 1130°C for 42 h and, finally, rapidly quenched in water. For samples with compositions projected as $La_3Ni_2B_2N_{2.9}$, the analogous final annealing procedure was applied at 1150°C.

This compound could be synthesized as follows. La powder (39.8 mM), Ni powder (26.4 mM), and borazinyl $[B_3N_3H_{\sim4}]_x$ (~14.5 mM) were charged under an inert atmosphere into a 100 mL one-piece glass reaction flask equipped with a high-vacuum stopcock (Wideman et al. 1996). The flask was sealed, removed to a vacuum line, and evacuated, and ~10 mL of glyme was vacuum-transferred into the flask at 77 K. Upon warming to room temperature, the polymer dissolved and the mixture was sonicated for 30 min (67 kHz, 200 W) at room temperature while the solvent was slowly evaporated. The sample was further dried at 70°C for 12 h in vacuo. The material was then removed from the flask under an inert atmosphere and isolated as a dark gray solid. Pyrolysis of a 1.72 g sample of the polyborazylene/metal dispersion at 1000°C under argon for 48 h produced $La_3Ni_2B_2N_3$ in high ceramic yields.

Large samples of $La_3Ni_2B_2N_3$ were prepared at melting under N_2 the mixture of La, Ni, and h-BN (Michor et al. 1996). For annealing, the samples were wrapped in protective Mo foil, sealed in evacuated silica capsules, and heated for 150 h at 1100°C. This compound could be also prepared by arc-melting and annealing (Cava et al. 1994; Zandbergen et al. 1994; Huang et al. 1995). The starting materials were La, Ni, and B. LaN powder could be also employed as starting material if care was taken to avoid significant air exposure. Nitrogen was introduced to the samples by arc-melting under N_2. The samples were annealed for successive periods of 10–12 h at 1000°C and 1050°C. This treatment was followed by a final annealing of 1 week at 1000°C. Crystalline samples of $La_3Ni_2B_2N_3$ were also synthesized using solid-state metathesis reactions from combinations of La, $LaCl_3$, and $NiCl_2$ together with Li_3BN_2 at 1025°C for 1 week (Jing et al. 2002; Blaschkowski and Meyer 2003). Single-crystal growth was obtained by the aid of a LiCl flux that is formed during the reaction (Jing et al. 2002).

This compound could be also prepared by the same way as LaNi(BN) was synthesized, but the mixture was heat-treated at 1400°C for 2 h (Imamura et al. 2012).

$La_3Ni_2B_2N_3$ revealed a remarkable stability and hardly changed even when stored for several weeks under normal conditions (Michor et al. 1996). This compound was determined, using local-density approximations, to be a three-dimensional metal (Singh and Pickett 1995).

1.61 BORON–LANTHANUM–RUTHENIUM–NITROGEN

The **LaRu(BN)** and **La₃Ru₂B₂N₃** quaternary compounds are formed in the B–La–Ru–N system. LaRu(BN) crystallizes in the tetragonal structure with the lattice parameters $a = 376.46 \pm 0.04$ and $c = 795.14 \pm 0.13$ pm (Imamura et al. 2012). This compound was synthesized using LaN, Ru, and B powders as the starting materials. They were mixed well and loaded in a Ta capsule in an Ar-filled glove box. The Ta container was then welded inside the glove box and heat-treated at 2 GPa and 1200°C for 1 h. The obtained pellets prepared under high pressure were stable enough under an ambient atmosphere. LaRu(BN) and La₃Ru₂B₂N₃ often coexist when stoichiometric mixtures were used as starting materials. Therefore, the nominal chemical composition was deviated from ideal stoichiometry to reduce these competing phases. In particular, LaRu(BN) and La₃Ru₂B₂N₃ were synthesized from the nominal stoichiometry LnN/Ru/B = 0.9:1:1 and 3.3:1.9:2, respectively.

La₃Ru₂B₂N₃ also crystallizes in the tetragonal structure with the lattice parameters $a = 374.54 \pm 0.02$ and $c = 2122.2 \pm 0.1$ pm (Imamura et al. 2012). It was prepared by the same way as LaRu(BN) was synthesized, but the mixture was heat-treated at 1600°C.

1.62 BORON–LANTHANUM–RHODIUM–NITROGEN

The **LaRh(BN)** quaternary compound, which crystallizes in the tetragonal structure with the lattice parameters $a = 386.70 \pm 0.04$ and $c = 748.90 \pm 0.11$ pm (Imamura et al. 2012), is formed in the B–La–Rh–N system. This compound was synthesized using LaN, Rh, and B powders as the starting materials. They were mixed well and loaded in a Ta capsule in an Ar-filled glove box. The Ta container was then welded inside the glove box and heat-treated at 5 GPa and 1400°C for 1 h. The obtained pellets prepared under high pressure were stable enough under an ambient atmosphere.

1.63 BORON–LANTHANUM–IRIDIUM–NITROGEN

The **LaIr(BN)** quaternary compound, which crystallizes in the tetragonal structure with the lattice parameters $a = 384.58 \pm 0.01$ and $c = 763.92 \pm 0.07$ pm (Imamura et al. 2012), is formed in the B–La–Ir–N system. This compound was synthesized by the same way as LaRh(BN) was prepared. The obtained pellets prepared under high pressure were stable enough under an ambient atmosphere.

1.64 BORON–LANTHANUM–PLATINUM–NITROGEN

The **LaPt(BN)** quaternary compound, which crystallizes in the tetragonal structure with the lattice parameters $a = 382.40 \pm 0.02$ and $c = 786.49 \pm 0.06$ pm (Imamura et al. 2012), is formed in the B–La–Pt–N system. This compound was synthesized by the same way as LaRh(BN) was prepared. The obtained pellets prepared under high pressure were stable enough under an ambient atmosphere.

1.65 BORON–CERIUM–NICKEL–NITROGEN

The **CeNi(BN)** and **Ce₃Ni₂B₂N₃** quaternary compounds are formed in the B–Ce–Ni–N system. CeNi(BN) crystallizes in the tetragonal structure with the lattice parameters $a =$

358.49 ± 0.03 and c = 751.41 ± 0.09 pm (Neukirch et al. 2006). This compound could be obtained by the same way as YNi(BN) was synthesized using Ce and CeN instead of Y and YN, respectively. It was also prepared from reactions of equimolar amounts of CeN and NiB, which were carried out in fused Ta tubes near 1200°C for about 3 days (Glaser et al. 2008). All manipulations of the starting materials were performed in an Ar-filled glove box.

$Ce_3Ni_2B_2N_3$ also crystallizes in the tetragonal structure with the lattice parameters a = 358.17 ± 0.01 and c = 2028.3 ± 0.1 pm (Glaser et al. 2008) [a = 357 and c = 2025 pm (Ali et al. 2011a, 2011b)]. It was synthesized through a modified metathesis reaction by heating a carefully ground mixture of 225.5 mg $CeCl_3$, 79.1 mg $NiCl_2$, 21.2 mg Li, 7.6 mg BN, and 18.21 MG Li_3BN_2 (molar ratio 3:2:10:1:1) in a fused Ta ampoule at 1040°C for 1 week (Glaser et al. 2008). When the ampoule was cooled down to room temperature, it was opened in air and the product was washed, first with a mixture of acetone/water and then with acetone. All manipulations of the starting materials were also performed in an Ar-filled glove box.

1.66 BORON–PRASEODYMIUM–NICKEL–NITROGEN

The **PrNi(BN)** and **$Pr_3Ni_2B_2N_3$** quaternary compounds are formed in the B–Pr–Ni–N system. PrNi(BN) crystallizes in the tetragonal structure with the lattice parameters a = 363.05 ± 0.02 and c = 761.31 ± 0.06 pm (Neukirch et al. 2006). This compound could be obtained by the same way as YNi(BN) was synthesized using Pr and PrN instead of Y and YN, respectively.

$Pr_3Ni_2B_2N_3$ also crystallizes in the tetragonal structure with the lattice parameters a = 362 and c = 2051 pm (Ali et al. 2011b).

1.67 BORON–NEODYMIUM–NICKEL–NITROGEN

The **$Nd_3Ni_2B_2N_3$** quaternary compound, which crystallizes in the tetragonal structure with the lattice parameters a = 360 and c = 2049 pm, is formed in the B–Nd–Ni–N system (Ali et al. 2011b).

1.68 BORON–EUROPIUM–BROMINE–NITROGEN

The **$Eu_2[BN_2]Br$** quaternary compound, which crystallizes in the trigonal structure with the lattice parameters a = 407.28 ± 0.03 and c = 2658.9 ± 0.3 pm, is formed in the B–Eu–Br–N system (Höhn et al. 2009; Kokal et al. 2011a). The title compound was obtained as orange red plates from the reaction of the mixture of EuN, h-BN, and $EuBr_2$ (molar ratio 3:2:1.1). The educts were transferred into Nb ampoule, which was sealed and in turn jacketed in Ar-filled quartz ampoule. The sample was heated to 1000°C within 6 h [to 750°C within 8 h (Höhn et al. 2009)], annealed for 24 h, and cooled down to room temperature within 8 h. This compound is sensitive to air and moisture. Therefore, all manipulations were performed under inert conditions in a glove box.

1.69 BORON–EUROPIUM–IODINE–NITROGEN

The **$Eu_2[BN_2]I$** quaternary compound, which crystallizes in the monoclinic structure with the lattice parameters a = 1025.48 ± 0.06, b = 415.87 ± 0.03, c = 1312.34 ± 0.09 pm, and

$\beta = 91.215° \pm 0.004°$ and a calculated density of 5.57 g·cm^{-3}, is formed in the B–Eu–I–N system (Kokal et al. 2011a). The title compound was obtained as orange red plates from the reaction of the mixture of EuN, h-BN, and EuI$_2$ (molar ratio 3:2:1.1). The educts were transferred into Nb ampoule, which was sealed and in turn jacketed in Ar-filled quartz ampoule. The sample was heated to 1000°C within 6 h, annealed for 24 h, and cooled down to room temperature within 8 h. This compound is sensitive to air and moisture.

1.70 BORON–HOLMIUM–CARBON–NITROGEN

The **HoB$_{15.5}$CN**, **HoB$_{17}$CN**, and **HoB$_{22}$C$_2$N** quaternary compounds are formed in the B–Ho–C–N system. HoB$_{15.5}$CN crystallizes in the trigonal structure with the lattice parameters $a = 558.83 \pm 0.07$ and $c = 1087.8 \pm 0.6$ pm (Zhang et al. 2001b; Leithe-Jasper et al. 2004). Starting materials for the title compound production were powders of HoB$_{12}$, graphite, and h-BN (Leithe-Jasper et al. 2004). The mixture was fired at 1600°C in a graphite susceptor under a flow of Ar. Single crystals of HoB$_{15.5}$CN were prepared by a high-temperature solution growth method using Cu or Sn as the flux (Zhang et al. 2001b).

HoB$_{17}$CN also crystallizes in the trigonal structure with the lattice parameters $a = 558.6$ and $c = 1082.7$ pm (Mori and Nishimura 2006). This compound was prepared using the mixture of HoB$_{12}$, B, graphite, and h-BN. The mixture was heated to around 1600°C.

HoB$_{22}$C$_2$N also crystallizes in the trigonal structure with the lattice parameters $a = 561.4 \pm 0.9$ and $c = 4462.5 \pm 0.5$ pm (Zhang et al. 2001a). It could be obtained as follows. HoB$_{12}$, amorphous B, graphite, and BCN in various molar ratios were well mixed in acetone and pressed into pellets with hydrostatic pressure of 250 MPa or so before heating. The pellets reacted in a BN or graphite crucible, which was inserted into a heated graphite susceptor. The reaction process was performed in vacuum and maintained at a temperature of about 1700°C for 8–12h.

1.71 BORON–ERBIUM–CARBON–NITROGEN

The **ErB$_{15.5}$CN** or **ErB$_{17}$C$_{1.3}$N$_{0.6}$** and **ErB$_{22}$C$_2$N** quaternary compounds are formed in the B–Er–C–N system. ErB$_{17}$C$_{1.3}$N$_{0.6}$ [according to the data of Zhang et al. (2001b), the composition of this compound is ErB$_{15.5}$CN] crystallizes in the trigonal structure with the lattice parameters $a = 558.89 \pm 0.05$ and $c = 1088.0 \pm 0.6$ pm (Zhang et al. 2001b; Leithe-Jasper et al. 2004). Starting materials for the title compound production were powders of ErB$_{12}$, graphite, and h-BN (Leithe-Jasper et al. 2004). The mixture was fired at 1600°C in a graphite susceptor under a flow of Ar. Its single crystals were prepared by a high-temperature solution growth method using Cu or Sn as the flux (Zhang et al. 2001b).

ErB$_{22}$C$_2$N also crystallizes in the trigonal structure with the lattice parameters $a = 562.6 \pm 0.4$ and $c = 4468.1 \pm 0.9$ pm (Zhang et al. 2001a) [$a = 563.0$ and $c = 4470.4$ pm (Mori and Nishimura 2006)]. This compound was prepared using the mixture of ErB$_{12}$, B, graphite, and h-BN (Mori and Nishimura 2006). The mixture was heated to around 1600°C.

It could be also obtained as follows. ErB$_{12}$, amorphous B, graphite, and BCN in various molar ratios were well mixed in acetone and pressed into pellets with hydrostatic pressure of 250 MPa or so before heating (Zhang et al. 2001a). The pellets reacted in a BN or graphite crucible, which was inserted into a heated graphite susceptor. The reaction process was performed in vacuum and maintained at a temperature of about 1700°C for 8–12 h.

1.72 BORON–THULIUM–CARBON–NITROGEN

The **TmB$_{15.5}$CN** and **TmB$_{22}$C$_2$N** quaternary compounds are formed in the B–Tm–C–N system. TmB$_{15.5}$CN crystallizes in the trigonal structure with the lattice parameters $a =$ 558.0 ± 0.1 and $c =$ 1085.0 ± 0.6 pm (Zhang et al. 2001b; Leithe-Jasper et al. 2004). Starting materials for the title compound production were powders of TmB$_{12}$, graphite, and h-BN (Leithe-Jasper et al. 2004). The mixture was fired at 1600°C in a graphite susceptor under a flow of Ar. Single crystals of TmB$_{15.5}$CN were prepared by a high-temperature solution growth method using Cu or Sn as the flux (Zhang et al. 2001b).

TmB$_{22}$C$_2$N also crystallizes in the trigonal structure with the lattice parameters $a =$ 563.1 ± 0.5 and $c =$ 4473.7 ± 0.9 pm (Zhang et al. 2001a). It could be obtained as follows. TmB$_{12}$, amorphous B, graphite, and BCN in various molar ratios were well mixed in acetone and pressed into pellets with hydrostatic pressure of 250 MPa or so before heating (Zhang et al. 2001a). The pellets reacted in a BN or graphite crucible, which was inserted into a heated graphite susceptor. The reaction process was performed in vacuum and maintained at a temperature of about 1700°C for 8–12 h.

1.73 BORON–THULIUM–NICKEL–NITROGEN

The **TmNi(BN)** quaternary compound, which crystallizes in the tetragonal structure with the lattice parameters $a =$ 344.06 ± 0.06 and $c =$ 750.7 ± 0.2 pm, is formed in the B–Tm–Ni–N system (Neukirch et al. 2006). This compound could be obtained by the same way as YNi(BN) was synthesized using Tm and TmN instead of Y and YN, respectively.

1.74 BORON–YTTERBIUM–NICKEL–NITROGEN

The **YbNi(BN)** quaternary compound, which crystallizes in the tetragonal structure with the lattice parameters $a =$ 343.10 ± 0.02 and $c =$ 748.71 ± 0.06 pm, is formed in the B–Yb–Ni–N system (Neukirch et al. 2006). This compound could be obtained by the same way as YNi(BN) was synthesized using Yb and YbN instead of Y and YN, respectively.

1.75 BORON–LUTETIUM–CARBON–NITROGEN

The **LuB$_{15.5}$CN** and **LuB$_{22}$C$_2$N** quaternary compounds are formed in the B–Lu–C–N system. LuB$_{15.5}$CN crystallizes in the trigonal structure with the lattice parameters $a =$ 557.71 ± 0.09 and $c =$ 1083.9 ± 0.4 pm (Zhang et al. 2001b; Leithe-Jasper et al. 2004). Starting materials for the title compound production were powders of LuB$_{12}$, graphite, and h-BN (Leithe-Jasper et al. 2004). The mixture was fired at 1600°C in a graphite susceptor under a flow of Ar. Its single crystals were prepared by a high-temperature solution growth method using Cu or Sn as the flux (Zhang et al. 2001b).

LuB$_{22}$C$_2$N also crystallizes in the trigonal structure with the lattice parameters $a =$ 559.5 ± 0.7 and $c =$ 4446.4 ± 0.6 pm (Zhang et al. 2001a) [$a =$ 561.4 and $c =$ 4459.0 pm (Mori and

Nishimura 2006)]. This compound was prepared using the mixture of LuB$_{12}$, B, graphite, and h-BN (Mori and Nishimura 2006). The mixture was heated to around 1600°C.

It could be also obtained as follows. LuB$_{12}$, amorphous B, graphite, and BCN in various molar ratios were well mixed in acetone and pressed into pellets with hydrostatic pressure of 250 MPa or so before heating (Zhang et al. 2001a). The pellets reacted in a BN or graphite crucible, which was inserted into a heated graphite susceptor. The reaction process was performed in vacuum and maintained at a temperature of about 1700°C for 8–12 h.

1.76 BORON–CARBON–SILICON–NITROGEN

The isothermal sections of this quaternary system at 1400°C and 2000°C were calculated by Seifert et al. (2001). No quaternary compounds were found. In both cases, the concentration tetrahedron is divided into different distorted tetrahedra indicating the four-phase equilibria at the specific temperature: nine four-phase equilibria at 1400°C (SiC+BN+ B$_4$C+C, Si$_3$N$_4$+SiC+C+BN, G+Si$_3$N$_4$+C+BN, Si$_3$N$_4$+SiC+Si+BN, SiC+BN+Si+L, SiC+BN+ B$_4$C+L, B$_4$C+L+BN+SiB$_6$, B$_4$C+BN+SiB$_6$+SiB$_n$, and B$_4$C+B+SiB$_n$+BN) and six four-phase equilibria at 2000°C (SiC+BN+B$_4$C+C, G+SiC+C+BN, G+SiC+L+BN, L+BN+SiC+ B$_4$C, L+B$_4$C+SiB$_n$+BN, and B+BN+SiB$_n$+B$_4$C). Concentration sections through the 1400°C and 2000°C tetrahedra at 10 at% B were calculated and are shown in Figure 1.6.

BN–SiC. The phase relations in this system under 0.1 MPa of N$_2$ pressure is shown in Figure 1.7 (Kasper 1996). XRD of the hot-pressed blanks revealed the presence of several SiC polytypes, including 21R, 15R, 6H, 4H, and 3C (Ruh et al. 1984). With this multiphase assemblage, it was not possible to quantify the amount of each phase present, but the SiC patterns were similar regardless of the BN content. Boron nitride was always present as the hexagonal graphite structure, and no evidence of reaction or formation of solid solution between SiC and BN was observed. BN stabilizes β-SiC and low-hexagonality polytypes (Zangvil and Ruh 1985).

1.77 BORON–CARBON–TITANIUM–NITROGEN

BN–TiC. According to the data of XRD, c-BN does not interact with TiC at the temperatures 800–1400°C under pressure up to 8 GPa (Bondarenko et al. 1978; Novikov et al. 1997).

1.78 BORON–CARBON–TANTALUM–NITROGEN

BN–TaC. From thermodynamic calculations, it follows that TaC reacts with c-BN forming tantalum boride (Benko et al. 2004). XRD determined the presence of TaB in this system.

1.79 BORON–CARBON–OXYGEN–NITROGEN

BN$_{0.5}$O$_{0.4}$C$_{0.1}$ quaternary phase is formed in the B–C–O–N system (Garvie et al. 1999). To obtain this phase, amorphous B, B$_2$O$_3$, and C were mixed in proportions according to 16B + xC + yB$_2$O$_3$ which were x = 2 or 4 and y = 1 or 3. The mixtures were pressed into a pellet and enclosed in a h-BN capsule. The encapsulated material was then placed in a furnace, pressurized at 7.5 GPa, and heated at 1700°C for 30 min to 5 h. The title phase is a minor

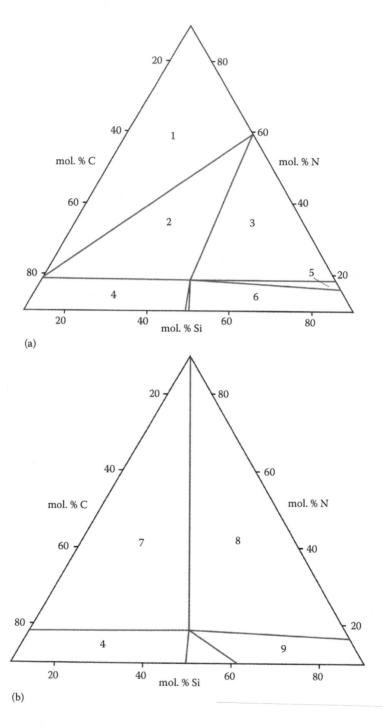

FIGURE 1.6 Concentration sections in the B–C–Si–N system at constant boron content (10 at%) at (a) 1400°C and (b) 2000°C: 1, G + Si_3N_4+ C + BN; 2, Si_3N_4+ SiC + C + BN; 3, Si_3N_4+ SiC + Si + BN; 4, SiC + BN + B_4C + C; 5, SiC + Si + BN + L; 6, SiC + BN + B_4C + L; 7, G + SiC + C + BN; 8, G + SiC + L + BN; 9, L + SiC + BN. Seifert, H.J. et al.: Phase equilibria of precursor-derived Si–(B–)C–N ceramics.

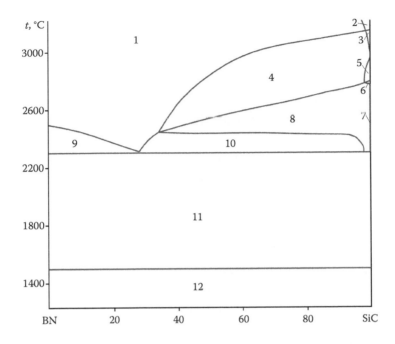

FIGURE 1.7   Phase relations in the BN–SiC system under 0.1 MPa of N_2 pressure: 1, L + G; 2, G; 3, G + graphite; 4, L + G + graphite; 5, L + graphite; 6, L + graphite + SiC; 7, G + graphite + SiC; 8, G + L + graphite + SiC; 9, G + L + BN, 10, G + L + SiC; 11, G + graphite + BN + SiC; 12, G + BN + SiC + Si_3N_4. (From Kasper, B., *Phasengleichgewichte im System B–C–N–Si*, Thesis, Max-Planck-Institut, Stuttgart, 1, 1996. With permission.)

constituent of the product and occurs as small, <30 nm diameter, rounded pseudo-hexagonal particles adhering to materials with the α-rhombohedral B structure.

1.80 BORON–CARBON–SULFUR–NITROGEN

The **B(SCN)₃** quaternary compound (boron thiocyanate) is formed in the B–C–S–N system (Cocksedge 1908). It was obtained as follows. BBr_3 reacts with AgSCN with the formation of $B(SCN)_3$, which was extracted from the products by means of cold benzene. Traces of hydrolysis occur unless extreme care is taken. The title compound separates from benzene in short, rhombic crystals or sometimes in the form of radiating needles.

1.81 BORON–CARBON–CHROMIUM–NITROGEN

BN–Cr₃C₂. According to the data of XRD, *c*-BN does not interact with Cr_3C_2 at the temperatures 800–1400°C (Bondarenko et al. 1978).

1.82 BORON–CARBON–PLATINUM–NITROGEN

At 1200°C under 3–5 GPa, Pt reacts readily with the pyrolyzed B/C/N precursors to form **Pt(BₓCᵧNᵤ)**, which crystallizes in the cubic structure with the lattice parameter *a* = 388 pm (Nicolich et al. 2001).

1.83 BORON–PHOSPHORUS–CHLORINE–NITROGEN

The $[Cl_2BNPCl_2NPCl_3]_2$ quaternary compound crystallizes in the monoclinic structure with the lattice parameters $a = 961.5 \pm 0.1$, $b = 830.8 \pm 0.1$, $c = 1517.0 \pm 0.2$ pm, and $\beta = 107.32° \pm 0.01°$ and a calculated density of 2.003 g·cm^{-3} (Jäschke and Jansen 2002a). It was prepared by the following procedure. To a cooled to −78°C solution of BCl_3 (0.15 M) in CH_2Cl_2 (400 mL) was slowly added a solution of $Me_3SiNPCl_3$ (0.15 M) in CH_2Cl_2 (200 mL). The heating of the reaction mixture was gradual in such a way that the reaction solution was held for 2 h at −50°C and further stirred at −35°C for 2 h before leaving finally warm to room temperature. After removal of the solvent, a white solid was obtained. For purification, it was recrystallized from CH_2Cl_2. At 5°C, large, colorless, tabular-like crystals of the title compound were separated out, which were isolated by filtration.

1.84 BORON–VANADIUM–NIOBIUM–NITROGEN

Substitution of Nb in Nb_2BN by V in powder compacts heat-treated under ≈0.1 MPa of Ar for 100 h at 1400°C resulted in the formation of rather limited solid solutions with only small changes in the unit-cell dimensions (Rogl et al. 1988). With respect to the unit-cell dimension of Nb_2BN, the lattice parameters of $\mathbf{Nb_{2-x}V_xBN}$ at the solubility limit of $x \approx 0.1$ were $a = 317.16 \pm 0.03$, $b = 1784.1 \pm 0.6$, and $c = 311.35 \pm 0.04$ pm.

1.85 BORON–NIOBIUM–TANTALUM–NITROGEN

Substitution of Nb in Nb_2BN by Ta in powder compacts heat-treated under ≈0.1 MPa of Ar for 100 h at 1400°C resulted in the formation of rather limited solid solutions with only small changes in the unit-cell dimensions (Rogl et al. 1988). With respect to the unit-cell dimension of Nb_2BN, the lattice parameters of $\mathbf{Nb_{2-x}Ta_xBN}$ at the solubility limit of $x \approx 0.2$ were $a = 317.14 \pm 0.03$, $b = 1784.6 \pm 0.3$, and $c = 311.37 \pm 0.03$ pm.

CHAPTER 2

Systems Based on BP

2.1 BORON–LITHIUM–OXYGEN–PHOSPHORUS

The **Li$_2$B$_3$PO$_8$** and **Li$_3$BP$_2$O$_8$** quaternary compounds are formed in the B–Li–O–P system.

Li$_2$B$_3$PO$_8$ crystallizes in the triclinic structure with the lattice parameters a = 869.65 ± 0.03, b = 1073.54 ± 0.05, c = 1282.82 ± 0.05 pm, α = 90.8057 ± 0.0014°, β = 90.6311 ± 0.0012°, and γ = 90.0940 ± 0.0013° and a calculated density of 2.277 g cm^{-3} (Hasegawa and Yamane 2014a). The sample obtained by heating a mixture of Li$_4$P$_2$O$_7$ and H$_3$BO$_3$ at 550°C for 12 h was polycrystalline Li$_2$B$_3$PO$_8$. The polycrystalline sample was melted by heating at 670°C for 1 h. When the melt was cooled to 600°C at a rate of 10°C h^{-1}, and then cooled to room temperature in the furnace, the obtained sample was a glass phase, which was confirmed by powder X-ray diffraction (XRD). A polycrystalline sample with coarse grains of the title compound was prepared by cooling the melt from 670°C to 600°C at a rate of 2°C h^{-1}. Single crystals of Li$_2$B$_3$PO$_8$ were obtained by slow cooling of the melt. This compound is unstable at the temperatures higher than 710°C (Tien and Hummel 1961).

Li$_3$BP$_2$O$_8$ [Li$_{22}$B$_{11}$P$_{13}$O$_{60}$ (Tien and Hummel 1961)] also crystallizes in the triclinic structure with the lattice parameters a = 518.88 ± 0.05, b = 741.18 ± 0.07, c = 767.35 ± 0.07 pm, α = 101.18 ± 0.01°, β = 105.07 ± 0.01°, and γ = 90.34 ± 0.01° and a calculated density of 2.637 g cm^{-3} (Hasegawa and Yamane 2014b). To prepare this compound, H$_3$BO$_3$, Li$_2$CO$_3$, and NH$_4$H$_2$PO$_4$ were used as starting powders, which were weighed with a molar ratio of Li/B/P = 22:11:13, mixed, and pressed into a pellet. The pellet was placed on a Pt plate and heated at 200°C for 9 h in air. After cooling, the product was powdered, pelletized, and heated at 550°C for 12 h. For the preparation of single crystals, the pellet sample obtained by heating at 550°C was heated again at 660°C for 1 h, cooled to 600°C for 6 h, and then cooled to room temperature in the furnace. This compound is also unstable at temperatures higher than 710°C (Tien and Hummel 1961).

2.2 BORON–LITHIUM–CHLORINE–PHOSPHORUS

BP–LiCl. According to the data of metallography and XRD, the solubility of BP in LiCl melt was not detected (Goryunova et al. 1964).

2.3 BORON–LITHIUM–BROMINE–PHOSPHORUS

BP–LiBr. According to the data of metallography and XRD, the solubility of BP in LiBr melt was not detected (Goryunova et al. 1964).

2.4 BORON–SODIUM–POTASSIUM–PHOSPHORUS

The **NaK_2BP_2** quaternary compound, which crystallizes in the monoclinic structure with the lattice parameters $a = 1256.1 \pm 0.2$, $b = 500.31 \pm 0.09$, $c = 1209.2 \pm 0.2$ pm, and $\beta = 120.94 \pm 0.01°$, is formed in the B–Na–K–P system (Somer et al. 1991d, 2000b). It could be synthesized from a stoichiometric mixture of the elements or from BP, NaP, and K in an evacuated and sealed Nb or steel ampoule at 730°C.

2.5 BORON–SODIUM–OXYGEN–PHOSPHORUS

The **$Na_3BP_2O_8$**, **$Na_3B_6PO_{13}$**, and **$Na_5B_2P_3O_{13}$** quaternary compounds are formed in the B–Na–O–P system. $Na_3BP_2O_8$ is stable up to 850°C and crystallizes in the monoclinic structure with the lattice parameters $a = 1256.7 \pm 0.4$, $b = 1029.0 \pm 0.3$, $c = 1021.0 \pm 0.3$ pm, and $\beta = 92.492 \pm 0.005°$ and a calculated density of 2.72 g cm^{-3} (Xiong et al. 2007). The title compound was prepared via low-temperature molten salt technique in Teflon-lined stainless-steel autoclave (50 mL) using H_3BO_3 or NaH_2PO_4 as flux with a filling degree of 20–30% by the initial powder reactants. The reactants were homogenized by grinding before they were placed into the autoclave. After a reaction time of typically 3 days at 260°C, the autoclave was removed from the furnace and allowed to cool to room temperature. The final product containing large single crystals were washed with hot water (50°C) until the residual flux and other products in the form of powder were completely removed and then dried in air at 60°C. $Na_3BP_2O_8$ was also synthesized from a mixtures of $La[B_8O_{11}(OH)_5]$ (0.650 g) and NaH_2PO_4 (3.570 g) in a molar ratio of 1:23. The reaction of H_3BO_3, NaH_2PO_4, NaF, and Na_2CO_3 (molar ratio 4:12:2:1) also resulted to this compound but with much lower yield.

Na$_3$B$_6$PO$_{13}$ is stable up to 700°C, transformed into amorphous glassy state at 800°C and crystallizes in the orthorhombic structure with the lattice parameters $a = 937.27 \pm 0.04$, $b = 1623.07 \pm 0.07$, and $c = 673.23 \pm 0.03$ pm and a calculated density of 2.42 g cm^{-3} (Xiong et al. 2007). This compound was also prepared via low-temperature molten salt technique by the same procedure that was used for $Na_3BP_2O_8$ obtaining. It was synthesized from mixtures of H_3BO_3 (5.600 g), NaH_2PO_4 (1.190 g), NaF (0.420 g), and Na_2CO_3 (0.530 g) in a molar ratio of 18:2:2:1.

$Na_5B_2P_3O_{13}$ congruently melts at 747°C [787°C (Li et al. 2003)], solidifies during cooling by the formation of a homogeneous glasses, and crystallizes in the monoclinic structure with the lattice parameters $a = 671.1 \pm 0.9$, $b = 1161.8 \pm 1.2$, $c = 768.6 \pm 0.9$ pm, and $\beta = 115.17 \pm 0.02°$ (Hauf et al. 1995, 1998). An experimental density of this compound is 2.68 g cm^{-3} (Li et al. 2003). It was obtained by hydrothermal synthesis starting from mixture of $NaBO_2 \cdot 4H_2O$ and $NaH_2PO_4 \cdot 2H_2O$ (molar ratio 2:3) (Hauf et al. 1995, 1998). After the addition of H_2O (15 mL), the system was heated up to 90°C, and the total volume was reduced to 10 mL by evaporation of H_2O. The highly viscous solution was filled into

Teflon autoclave (degree of fill 50%) and treated at 150°C for 2 days. The crystalline reaction product was separated from the solution by filtration and finally washed with H_2O. The reaction can be also carried out by using the peroxoborate $NaBO_3 \cdot 4H_2O$ as a borate source.

For microwave-assisted synthesis of $Na_5B_2P_3O_{13}$, a dry and ground solid mixture $NaBO_2 \cdot 4H_2O$ and $NaH_2PO_4 \cdot 2H_2O$ (molar ratio 2:3) was filled into a Teflon autoclave and was treated in a conventional kitchen microwave system (2.450 GHz; max 1650 W) for the short period of only 2 min (Hauf et al. 1998). The crystallinity of the reaction product was improved by additional annealing at 100°C for 2 h.

$Na_5B_2P_3O_{13}$ could be also prepared by a solid-state reaction using Na_2CO_3, H_3BO_3, and $NH_4H_2PO_4$ (molar ratio 2.5:2:3) (Li et al. 2003; Rahab et al. 2007). The mixed powder was preheated at 300°C for 12 h and at 700°C for 12 h. Then, the temperature was increased to 750°C and kept for 24 h to complete the reaction. Single crystals have been grown by the Czochralski method in air (Li et al. 2003) or by the heat exchanger method (Rahab et al. 2007).

BP–$Na_2B_4O_7$. According to the data of metallography and XRD, the solubility of BP in $Na_2B_4O_7$ melt was not detected (Goryunova et al. 1964).

2.6 BORON–SODIUM–CHLORINE–PHOSPHORUS

BP–NaCl. According to the data of metallography and XRD, the solubility of BP in NaCl melt is less than 0.01 mass% (Goryunova et al. 1964).

2.7 BORON–SODIUM–BROMINE–PHOSPHORUS

BP–NaBr. According to the data of metallography and XRD, the solubility of BP in NaBr melt is less than 0.01 mass% (Goryunova et al. 1964).

2.8 BORON–POTASSIUM–FLUORINE–PHOSPHORUS

BP–KF. According to the data of metallography and XRD, the solubility of BP in KF melt is approximately 1 mass% (Goryunova et al. 1964).

2.9 BORON–CESIUM–ARSENIC–PHOSPHORUS

$Cs_3[BPAs]$ quaternary compound, which crystallizes in the monoclinic structure with the lattice parameters $a = 998.5 \pm 0.1$, $b = 980.9 \pm 0.1$, $c = 1001.6 \pm 0.1$ pm, and $\beta = 110.18 \pm 0.01°$ and a calculated density of 3.782 $g \cdot cm^{-3}$, is formed in the B–Cs–As–P system (Somer et al. 1996b, 2000b). This compound was prepared from a mixture of BP, CsAs, and Cs (molar ratio 1:1:3) in sealed Nb ampoules at 730°C (Somer et al. 1996b). The excess of Cs was removed in high vacuum at 250°C after the reaction. Yellow–red prismatic crystals of the title compound were obtained from BP and Cs_3As (Somer et al. 2000b). $Cs_3[BPAs]$ is very sensitive against oxidation and hydrolysis (Somer et al. 2000b).

2.10 BORON–COPPER–IRON–PHOSPHORUS

The evolution of the structural and soft magnetic properties of the $Fe_{80.8-84.8}Cu_{1.2}B_{8-10}P_{6-8}$ amorphous alloy during annealing has been investigated by Urata et al. (2011) and Cao et al.

(2017). Using differential scanning calorimetry, XRD, Mössbauer spectroscopy, and magnetometry, it was shown that atomic regions in the amorphous structure tend to transform from Fe_3B- to FeB-like chemical short-range order with annealing before crystallization, which results in the increasing number of Fe clusters in the amorphous matrix and contributes to the optimization of magnetic properties during structural relaxation. Moreover, Mössbauer spectroscopy also indirectly manifests the formation of CuP clusters during the preannealing process.

$Fe_{80.8-84.8}Cu_{1.2}B_{8-10}P_{6-8}$ amorphous alloys were prepared by arc-melting using Fe and Cu metals and P–Fe and B–Fe alloys in an Ar atmosphere (Cao et al. 2017). It was remelted four times with electromagnetic mixing. The melt-spun alloys form a heteroamorphous structure (Urata et al. 2011). A single-roller melt-spinning method in air was used to rapidly produce the solidified ribbons.

2.11 BORON–MAGNESIUM–OXYGEN–PHOSPHORUS

The $\mathbf{Mg_3BPO_7}$ quaternary compound, which melts at circa 1230°C and is characterized by two polymorphic modifications, exists in the B–Mg–O–P system (Liebertz and Stähr 1982). The transition of the high-temperature form into the low-temperature form is prompt, and β-Mg_3BPO_7 cannot be quenched to room temperature. α-Mg_3BPO_7 crystallizes in the orthorhombic structure with the lattice parameters $a = 849.7 \pm 0.5$, $b = 488.0 \pm 0.5$, and $c = 1255.8 \pm 0.5$ pm and a calculated density of 2.89 g·cm^{-3} (Liebertz and Stähr 1982) [$a = 849.5 \pm 0.3$, $b = 488.6 \pm 0.1$, $c = 1256.5 \pm 0.4$ pm (Gözel et al. 1998)]. α-Mg_3BPO_7 has been synthesized by the solid-state reactions of $MgHPO_4 \cdot H_2O$ with $MgCO_3$ and H_3BO_3 at 1200°C, $Mg_3B_2O_6$ with $MgCO_3$ and $(NH_4)_2HPO_4$ at 1100°C, and MgO with B_2O_3 and P_2O_5 at 1100°C with the molar ratios of 1:2:1, 1:3:2, and 6:1:1, respectively (Gözel et al. 1998).

2.12 BORON–CALCIUM–OXYGEN–PHOSPHORUS

The $\mathbf{CaBPO_5}$ $\mathbf{(2CaO \cdot P_2O_5 \cdot B_2O_3)}$ and $\mathbf{Ca_{9.64}(P_{5.73}B_{0.27}O_{24})(BO_2)_{0.73}}$ quaternary compounds are formed in the B–Ca–O–P system. $CaBPO_5$ incongruently melts at 1045°C [at 1025°C (Kniep et al. 1994a, 1994b)] decomposing into $Ca_2P_2O_7$ and molten B_2O_3 (Bauer 1965) and crystallizes in the trigonal structure with the lattice parameters $a = 668.4 \pm 0.2$ and $c = 661.6 \pm 0.2$ pm and a calculated density of 3.15 g·cm^{-3} (Baykal et al. 2000a) [$a = 667.99 \pm 0.02$ and $c = 661.21 \pm 0.03$ pm and a calculated density of 2.53 g·cm^{-3} (Kniep et al. 1994a, 1994b) and in the hexagonal structure with the lattice parameters $a = 668.8$ and $c = 1323.4$ pm and an experimental density of 3.106 g·cm^{-3} (Bauer 1965, 1966)]. The synthesis of this compound was attempted with three different sets of the starting materials:

1. $CaHPO_4 \cdot 2H_2O$ and crystalline H_3BO_3 were mixed (molar ratio 1:1.1), and the mixture was heated in a Pt crucible to 600°C (Bauer 1965; Kniep et al. 1994a, 1994b; Baykal et al. 2000a). This temperature was maintained until the complete elimination of H_2O (about 2 h) and the sintered product was heated in a Pt crucible in an O_2 stream at 900°C for 3 days.

2. In the second method, B_2O_3 was used instead of H_3BO_3 (Baykal et al. 2000a).

3. In the third method, $CaBPO_5$ was prepared by heating $CaCO_3$ and BPO_4 in the same conditions (Baykal et al. 2000a).

$Ca_{9.64}(P_{5.73}B_{0.27}O_{24})(BO_2)_{0.73}$ also crystallizes in the trigonal structure with the lattice parameters $a = 945.6 \pm 0.1$ and $c = 690.5 \pm 0.1$ pm and a calculated density of 3.11 $g \cdot cm^{-3}$ (Ito et al. 1988). Single crystals of this compound were grown by standard flux growth with excess B_2O_3 as a flux. A mixture with composition 35 mass% CaO + 5 mass% P_2O_5 + 60 mass% B_2O_3 was heated at 1200°C for 10 h and then cooled at a rate of 8.3°C·h^{-1}.

2.13 BORON–STRONTIUM–OXYGEN–PHOSPHORUS

The $SrBPO_5$ ($2SrO \cdot P_2O_5 \cdot B_2O_3$), $Sr_3P(BO_3)$, $Sr_6BP_5O_{20}$, and $Sr_{10}[(PO_4)_{5.5}(BO_4)_{0.5}](BO_2)$ quaternary compounds are formed in the B–Sr–O–P system. $SrBPO_5$ incongruently melts at 890°C [at 1092°C (Kniep et al. 1994a, 1994b)] decomposing into $Sr_2P_2O_7$ and molten B_2O_3 (Bauer 1965) and crystallizes in the trigonal structure with the lattice parameters $a = 688.50 \pm 0.10$ and $c = 687.00 \pm 0.14$ pm and a calculated density of 3.699 $g \cdot cm^{-3}$ (Pan et al. 2003) [$a = 684.88 \pm 0.01$ and $c = 681.59 \pm 0.02$ pm and a calculated density of 3.19 $g \cdot cm^{-3}$ (Kniep et al. 1994a, 1994b); $a = 684.88 \pm 0.01$ and $c = 681.59 + 0.02$ pm and an experimental density of 3.77 $g \cdot cm^{-3}$ (Baykal et al. 2000b); in the hexagonal structure with the lattice parameters $a = 685.7$ and $c = 1365.7$ pm (Bauer 1965, 1966)]. Polycrystalline samples of this compound were obtained by using solid-state reaction techniques (Bauer 1965; Kniep et al. 1994a, 1994b; Pan et al. 2003). The stoichiometric mixture of $SrCO_3$ and $NH_4H_2PO_4$ [$(NH_4)_2HPO_4$ could be also used (Bauer 1965; Kniep et al. 1994a, 1994b)] and crystalline H_3BO_3 was ground thoroughly and then packed into a Pt crucible. The temperature was raised to 500°C at a rate of 2°C·min^{-1}. After preheating at 500°C for 10 h, the products were cooled to room temperature and ground up again; the mixture was heated at 900°C for 24 h and then cooled to room temperature. The crystal was grown using BPO_4 as a flux by the top-seeded solution growth method (Pan et al. 2003).

$SrBPO_5$ was also prepared by hydrothermal synthesis starting from mixtures of $Sr(OH)_2$, $NaBO_2 \cdot 2H_2O$, and $NaH_2PO_4 \cdot 2H_2O$ in a molar ratio of 4:3:3 (totally 8 g) (Baykal et al. 2000b). HNO_3 (4 M; 17 mL) and distilled H_2O (5 mL) was used to dissolve this solid mixture by heating up to 90°C, and the total volume was reduced to 10 mL by evaporation of H_2O. The highly viscous solution (pH < 1) was filled into the Teflon-coated autoclave (degree of filling 60%) and treated at 160°C for 3 days.

$Sr_3P(BO_3)$ crystallizes in the hexagonal structure with the lattice parameters $a = 525.9 \pm 0.1$ and $c = 1270.6 \pm 0.4$ pm (Somer et al. 1995i). It was synthesized from Sr_3P_2, SrO, and B_2O_3 (molar ratio 1:3:1) in sealed Nb ampoules at 1500°C.

$Sr_6BP_5O_{20}$ crystallizes in the tetragonal structure with the lattice parameters $a = 978.95 \pm 0.07$ and $c = 1903.2 \pm 0.3$ pm and a calculated density of 3.683 $g \cdot cm^{-3}$ (Ehrenberg et al. 2006) [$a = 978.392 \pm 0.002$ and $c = 1901.318 \pm 0.003$ pm and a calculated density of

3.691 ± 0.001 g·cm^{-3} (Shin et al. 2005)]. To obtain it, SrHPO$_4$ (5.030 M), SrCO$_3$ (0.970 M), and H$_3$BO$_3$ (1.050 M) were mixed (Ehrenberg et al. 2006). This mixture was loaded into silica crucible, covered, and fired in a furnace. It was heated to 1200°C over 5 h, held at 1200°C for next 5 h, and then cooled to room temperature over 6 h.

This compound was also produced by a solution-based synthesis method (Shin et al. 2005). The raw powders Sr(NO$_3$)$_2$, H$_3$BO$_3$, and (NH$_4$)$_2$HPO$_4$ were dissolved in deionized H$_2$O, and then a stoichiometric amount of each solution was collected in a quartz container. The solution in the container was stirred and then dried at 100°C for 48 h, followed by further drying at 600°C for 6 h. The dried samples were pulverized and successfully fired at 1200°C under a reducing atmosphere; a white crystalline powder of Sr$_6$BP$_5$O$_{20}$ resulted. A small amount of Sr$_3$(PO$_4$)$_2$ was detected as an impurity.

Sr$_{10}$[(PO$_4$)$_{5.5}$(BO$_4$)$_{0.5}$](BO$_2$) crystallizes in the trigonal structure with the lattice parameters $a = 979.73 \pm 0.08$ and $c = 730.56 \pm 0.08$ pm and a calculated density of 4.043 g·cm^{-3} (Chen et al. 2010). The title compound was first prepared by solid-state reactions from SrCO$_3$ (3 mM), NH$_4$H$_2$PO$_4$ (1 mM), and H$_3$BO$_3$ (2 mM). The mixed raw materials were heated to 500°C in a closed alumina crucible in air for 3 h. Then the preheated mixture was ground again and subsequently heated to 1150°C for 8 h in an alumina crucible. The final reaction product is a white powder. Excess boron results in the formation of a glassy phase. White transparent single crystals were obtained after cooling from 1550°C (corundum crucible, heating and cooling rates 10°C·min^{-1}, Ar atmosphere).

2.14 BORON–BARIUM–OXYGEN–PHOSPHORUS

The **BaBPO$_5$ (2BaO·P$_2$O$_5$·B$_2$O$_3$), Ba$_3$P(BO$_3$), Ba$_3$BPO$_7$ [Ba$_3$(BO$_3$)(PO$_4$)], and Ba$_3$BP$_3$O$_{12}$ [Ba$_3$(PO$_4$)$_2$BPO$_4$]** quaternary compounds are formed in the B–Ba–O–P system. BaBPO$_5$ incongruently melts at 1210°C (Bauer 1966) and crystallizes in the trigonal structure with the lattice parameters $a = 713.29 \pm 0.10$ and $c = 703.68 \pm 0.14$ pm (Pan et al. 2002) [in the hexagonal structure with the lattice parameters $a = 711.1$ and $c = 1397.7$ pm (Bauer 1966); $a = 710.9 \pm 0.1$ and $c = 699.0 \pm 0.2$ pm (Shi et al. 1998)]. This compound was prepared by thermal synthesis from BaCO$_3$, (NH$_4$)$_2$HPO$_4$, and H$_3$BO$_3$ (molar ratio 1:1:1.1) (Bauer 1966; Shi et al. 1998). The mixture was heated at 1000°C for 2 h in a Pt crucible. Single crystals have been grown by the top-seeded solution growth method using Li$_4$P$_2$O$_7$ as the flux (Pan et al. 2002). Thermal decomposition of BaBPO$_5$ leads to the formation of Ba$_3$BP$_3$O$_{12}$ and Ba$_2$P$_2$O$_7$ (Shi et al. 1998).

Ba$_3$P(BO$_3$) crystallizes in the hexagonal structure with the lattice parameters $a = 550.2 \pm 0.1$ and $c = 1350.6 \pm 0.4$ pm (Somer et al. 1995h). It was synthesized from Ba$_3$P$_2$, BaO, and B$_2$O$_3$ (molar ratio 1:3:1) in sealed Nb ampoules at 1300°C.

Ba$_3$BPO$_7$ also crystallizes in the hexagonal structure with the lattice parameters $a = 548.98 \pm 0.01$ and $c = 1475.51 \pm 0.01$ pm and a calculated density of 4.883 g·cm^{-3} (Ma et al. 2004). Polycrystalline compound was prepared by high-temperature solid-state reactions. BaCO$_3$, H$_3$BO$_3$, and NH$_4$H$_2$PO$_4$ were mixed in a molar ratio of 3:1:1 and grounded, heated at 1200°C in a Pt crucible for 48 h, and then cooled to room temperature.

Ba$_3$BP$_3$O$_{12}$ crystallizes in the orthorhombic structure with the lattice parameters $a = 708.59 \pm 0.13$, $b = 1429.03 \pm 0.22$, and $c = 2218.68 \pm 0.32$ pm (Shi et al. 1998) [$a = 2221.1 \pm$

0.8, $b = 1429.6 \pm 0.6$, and $c = 710.0 \pm 0.4$ pm and a calculated density of 4.17 g·cm^{-3} (Kniep et al. 1994a, 1994b); $a = 706.6 \pm 0.1$, $b = 1426.8 \pm 0.1$, and $c = 2215.9 \pm 0.4$ pm and a calculated density of 4.209 g·cm^{-3} (Park and Bluhm 1995)]. The title compound was obtained as follows (Kniep et al. 1994a, 1994b). Mixtures of $BaCO_3$, H_3BO_3, and $(NH_4)_2HPO_4$ (molar ratio 2:1:3) were heated to 1300°C over 4 h in a Pt crucible and were kept at this temperature for 8 h. After a further 24 h at 1000°C, the mixtures were cooled to room temperature over 4 h. A glassy matrix was formed, in which were embedded elongated orthorhombic prisms of $Ba_3BP_3O_{12}$.

2.15 BORON–ZINC–OXYGEN–PHOSPHORUS

The Zn_3BPO_7 [$Zn_3(BO_3)(PO_4)$] quaternary compound, which melts at circa 930°C and is characterized by two polymorphic modifications, exists in the B–Zn–O–P system (Liebertz and Stähr 1982). The transition of the high-temperature form into the low-temperature form should be at around 600°C and is very sluggish. α-Zn_3BPO_7 crystallizes in the orthorhombic structure with the lattice parameters $a = 843.8 \pm 0.5$, $b = 488.4 \pm 0.5$, and $c = 1274.6 \pm 0.5$ pm and a calculated density of 4.42 g·cm^{-3}, and β-Zn_3BPO_7 crystallizes in the hexagonal structure with the lattice parameters $a = 843.9 \pm 0.3$ and $c = 1303.0 \pm 0.3$ pm and the experimental and calculated densities of 4.33 and 4.34 g·cm^{-3}, respectively (Liebertz and Stähr 1982) [$a = 843.5 \pm 0.4$ and $c = 1303.2 \pm 0.6$ pm (Wang et al. 2002); $a = 846.24 \pm 0.03$ and $c = 1306.90 \pm 0.07$ pm and a calculated density of 4.301 g·cm^{-3} (Zhang et al. 2011)]. β-Zn_3BPO_7 was synthesized by a standard solid-state reaction of the starting components, using ZnO, H_3BO_3, and $NH_4H_2PO_4$ in the molar ratio of 3:1:1 (Zhang et al. 2011). A Pt crucible filled with Zn_3BPO_7 was heated to 1000°C, kept at that temperature for 12 h, and was then cooled to the saturation temperature. A seed crystal of β-Zn_3BPO_7 attached to a Pt rod was inserted into the solution, and then the temperature was cooled at a rate of 0.3°C per day until the end of the growth. The obtained crystal was pulled out of the surface of the solution, cooled to 430°C at a rate of 20°C·h^{-1}, and then rapidly cooled to 350°C in 1.5 h. Finally, the crystal was removed out of the furnace and cooled to room temperature.

According to the data of Bluhm and Park (1997), the title compound also crystallizes in the monoclinic structure with the lattice parameters $a = 972.5 \pm 0.2$, $b = 1272.0 \pm 0.3$, $c = 487.4 \pm 0.3$ pm, and $\beta = 119.80 \pm 0.04°$ and a calculated density of 4.442 g·cm^{-3}. To obtain it, $ZnCO_3$, B_2O_3, and P_2O_5 (molar ratio 6:1:1) were mixed and heated in a Pt crucible at 1050°C for 6 h. The product was then cooled at a rate of 20°C·h^{-1} and removed from the furnace. It is colorless, finally crystalline, and nonmoisture sensitive.

2.16 BORON–ZINC–CHLORINE–PHOSPHORUS

BP–ZnCl₂. According to the data of metallography and XRD, the solubility of BP in $ZnCl_2$ melt was not detected (Goryunova et al. 1964).

2.17 BORON–CADMIUM–CHLORINE–PHOSPHORUS

BP–CdCl₂. According to the data of metallography and XRD, the solubility of BP in $CdCl_2$ melt is less than 0.01 mass% (Goryunova et al. 1964).

2.18 BORON–GALLIUM–OXYGEN–PHOSPHORUS

The $(B_{0.29}Ga_{0.71})PO_4$ quaternary compound, which crystallizes in the tetragonal structure with the lattice parameters $a = 473.43 \pm 0.01$ and $c = 708.96 \pm 0.04$ pm and a calculated density of 3.084 g·cm^{-3}, is formed in the B–Ga–O–P system (Huang et al. 2010). The title compound has been synthesized via high-temperature solid-state reaction method. A mixture of H_3BO_3, $NH_4H_2PO_4$, and Ga_2O_3 with molar ratio of B/Ga/P = 12:3:4 was well ground and reacted first at 700°C for 4 h, then cooled down to room temperature and reground again, pressed into pellets and reacted at 1000°C for 8 h, and, at last shut off the furnace, and cooled down to room temperature. The extra B_2O_3 in the products were washed out by hot water. White crystals of $(B_{0.29}Ga_{0.71})PO_4$ were obtained.

2.19 BORON–INDIUM–OXYGEN–PHOSPHORUS

The $In_2B(PO_4)_3$ quaternary compound, which crystallizes in the hexagonal structure with the lattice parameters $a = 816.98 \pm 0.06$ and $c = 773.75 \pm 0.11$ pm and a calculated density of 3.901 g·cm^{-3}, is formed in the B–In–O–P system (Zhang et al. 2010). The title compound was obtained by reaction a mixture of In_2O_3, H_3BO_3, and $NH_4H_2PO_4$ in a molar ratio of 1:1:3 at 880°C.

2.20 BORON–LANTHANUM–OXYGEN–PHOSPHORUS

The $La_7O_6(BO_3)(PO_4)_2$ quaternary compound, which crystallizes in the monoclinic structure with the lattice parameters $a = 701.9$, $b = 1791.5$, $c = 1265.3$ pm, and $\beta = 97.52°$ (Shi et al. 1997) [$a = 701.72$, $b = 1791.25$, $c = 1265.00$ pm, and $\beta = 97.516°$ (Shi et al. 1996)], is formed in the B–La–O–P system. The title compound was synthesized by heating the mixture of La_2O_3, H_3BO_3, and $NH_4H_2PO_4$ according to stoichiometric composition (Palkina et al. 1984; Shi et al. 1997). The starting materials were mixed and pressed at about 0.1 GPa into pellets and heated to 400°C for 4 h in covered Pt crucibles. The specimens thus obtained were reground and pressed into pellets, then calcined at 1350°C ± 10°C for about 72 h, and furnace-cooled to room temperature. $La_7O_6(BO_3)(PO_4)_2$ could be also prepared by heating mixtures of $La_2(CO_3)_3 \cdot 8H_2O$, H_3BO_3 and $NH_4H_2PO_4$ at 1150–1300°C for 48 h (Shi et al. 1996).

2.21 BORON–CERIUM–OXYGEN–PHOSPHORUS

The $Ce_7O_6(BO_3)(PO_4)_2$ quaternary compound is formed in the B–Ce–O–P system (Palkina et al. 1984). The title compound was synthesized by heating mixture of Ce_2O_3, H_3BO_3, and $NH_4H_2PO_4$ according to stoichiometric composition. The temperature of last annealing stage was 1375°C ± 25°C.

2.22 BORON–PRASEODYMIUM–OXYGEN–PHOSPHORUS

The $Pr_7O_6(BO_3)(PO_4)_2$ quaternary compound, which crystallizes in the monoclinic structure with lattice parameters $a = 689.39 \pm 0.04$, $b = 1766.2 \pm 0.1$, $c = 1244.2 \pm 0.1$ pm, and $\beta = 97.24 \pm 0.01°$ (Ewald et al. 2004) [$a = 688.3 \pm 0.5$, $b = 1765.1 \pm 0.9$, $c = 1245.0 \pm 0.6$ pm, and $\beta = 97.25 \pm 0.06°$ and an experimental density of 6.26 g.cm^{-3} (Palkina et al. 1984)], is formed in the B –Pr –O –P system. Green transparent prismatic single crystals of the title compound were obtained by a two step high-temperature synthesis (Pt crucible, air) from mixtures of Pr_2O_3 with H_3BO_3 and $NH_4H_2PO_4$ (molar ratio 7:2:4) (Ewald et al. 2004). The reaction mixture (total mass 0.5 g) was intensively ground and pressed into pellets, which were then slowly heated to 450°C and kept at that temperature for 24 h to remove all volatile components. In the second step, the samples (reground and pressed into pellets again) were heated up to 1400°C and kept at that temperature for 96 h before being slowly cooled down to room temperature with a cooling rate of approximately 0.5°C·min^{-1}.

Single crystals of $Pr_7O_6(BO_3)(PO_4)_2$ were grown from the melt of PbO containing 1 mol% Pr_2O_3 and 4 mol% BPO_4 within the temperature interval from 900°C to 1100°C (Palkina et al. 1984).

2.23 BORON–NEODYMIUM–OXYGEN–PHOSPHORUS

The $Nd_7O_6(BO_3)(PO_4)_2$ quaternary compound, which crystallizes in the monoclinic structure with the lattice parameters $a = 686.2$, $b = 1759.1$, $c = 1237.5$ pm, and $\beta = 97.18°$ (Shi et al. 1997) [$a = 683.8 \pm 0.2$, $b = 1751.3 \pm 0.7$, $c = 1232.7 \pm 0.5$ pm, and $\beta = 97.19 \pm 0.03°$ and a calculated density of 6.14 g·cm^{-3} (Palkina et al. 1984)], is formed in the B–Nd–O–P system. The title compound was synthesized by heating mixture of Nd_2O_3, H_3BO_3, and $NH_4H_2PO_4$ according to stoichiometric composition (Palkina et al. 1984, Shi et al. 1997). The starting materials were mixed and pressed at about 0.1 GPa into pellets and heated to 400°C for 4 h in covered Pt crucibles (Shi et al. 1997). The specimens thus obtained were reground and pressed into pellets, then calcined at 1350°C $\pm$ 10°C for about 72 h, and furnace-cooled to room temperature. Single crystals of $Nd_7O_6(BO_3)(PO_4)_2$ were grown from the melt of PbO containing 1 mol% Nd_2O_3 and 4 mol% BPO_4 within the temperature interval from 900°C to 1100°C (Palkina et al. 1984).

2.24 BORON–PROMETHIUM–OXYGEN–PHOSPHORUS

The $Pm_7O_6(BO_3)(PO_4)_2$ quaternary compound is formed in the B–Pm–O–P system (Palkina et al. 1984). The title compound was synthesized by heating mixture of Pm_2O_3, H_3BO_3, and $NH_4H_2PO_4$ according to stoichiometric composition. The temperature of last annealing stage was 1375°C $\pm$ 25°C.

2.25 BORON–SAMARIUM–OXYGEN–PHOSPHORUS

The $Sm_7O_6(BO_3)(PO_4)_2$ quaternary compound, which crystallizes in the monoclinic structure with the lattice parameters $a = 677.81 \pm 0.07$, $b = 1739.6 \pm 0.1$, $c = 1221.8 \pm 0.1$ pm, and $\beta = 96.96 \pm 0.01°$ (Ewald et al. 2004), is formed in the B –Sm –O –P system. Green transparent prismatic single crystals of the title compounds were obtained by the same way as $Pr_7O_6(BO_3)(PO_4)_2$ was prepared using Sm_2O_3 instead of Pr_2O_3 (Palkina et al. 1984; Ewald et al. 2004).

2.26 BORON–EUROPIUM–OXYGEN–PHOSPHORUS

The $Eu_7O_6(BO_3)(PO_4)_2$ quaternary compound is formed in the B–Eu–O–P system (Palkina et al. 1984). The title compound was synthesized by heating mixture of Eu_2O_3, H_3BO_3, and $NH_4H_2PO_4$ according to stoichiometric composition. The temperature of last annealing stage was 1375°C $\pm$ 25°C.

2.27 BORON–GADOLINIUM–OXYGEN–PHOSPHORUS

The $Gd_7O_6(BO_3)(PO_4)_2$ quaternary compound, which crystallizes in the monoclinic structure with the lattice parameters $a = 670.4$, $b = 1729.9$, $c = 1210.0$ pm, and $\beta = 96.94°$ (Shi et al. 1997), is formed in the B–Gd–O–P system. The title compound was synthesized by heating mixture of Gd_2O_3, H_3BO_3, and $NH_4H_2PO_4$ according to stoichiometric composition (Palkina et al. 1984; Shi et al. 1997). The starting materials were mixed and pressed at about 0.1 GPa into pellets and heated to 400°C for 4 h in covered Pt crucibles. The specimens thus obtained were reground and pressed into pellets, then calcined at 1350°C ± 10°C for about 72 h, and furnace-cooled to room temperature.

2.28 BORON–TERBIUM–OXYGEN–PHOSPHORUS

$Tb_7O_6(BO_3)(PO_4)_2$ quaternary compound is formed in the B–Tb–O–P system (Palkina et al. 1984). The title compound was synthesized by heating mixture of Tb_2O_3, H_3BO_3, and $NH_4H_2PO_4$ according to the stoichiometric composition. The temperature of last annealing stage was 1375°C ± 25°C.

2.29 BORON–DYSPROSIUM–OXYGEN–PHOSPHORUS

The $Dy_7O_6(BO_3)(PO_4)_2$ quaternary compound, which crystallizes in the monoclinic structure with the lattice parameters $a = 662.3$, $b = 1717.2$, $c = 1196.0$ pm, and $\beta = 96.76°$ (Shi et al. 1997), is formed in the B–Dy–O–P system. The title compound was synthesized by heating mixture of Dy_2O_3, H_3BO_3, and $NH_4H_2PO_4$ according to stoichiometric composition (Palkina et al. 1984; Shi et al. 1997). The starting materials were mixed and pressed at about 0.1 GPa into pellets and heated to 400°C for 4 h in covered Pt crucibles. The specimens thus obtained were reground and pressed into pellets, then calcined at 1350°C ± 10°C for about 72 h, and furnace-cooled to room temperature.

2.30 BORON–THORIUM–OXYGEN–PHOSPHORUS

The $Th_2[BO_4][PO_4]$ quaternary compound, which crystallizes in the monoclinic structure with the lattice parameters $a = 846.65 ± 0.02$, $b = 795.52 ± 0.02$, $c = 822.97 ± 0.01$ pm, and $\beta = 103.746 ± 0.001°$ and a calculated density of 7.820 g·cm^{-3}, is formed in the B–Th–O–P system (Lipp and Burns 2011). To obtain colorless and platy crystals of the title compound, $Th[NO_3]_4 \cdot 4H_2O$, Ta_2O_5, and P_2O_5 were combined in a molar ratio of 2:1:1. These were then heated with B_2O_3 added as a fluxing agent in a furnace to 1200°C for 12 h. The temperature was then lowered at a rate of 5.5°C·h^{-1} to 350°C, after which the cooling rate was increased to ~100°C·h^{-1} until room temperature was reached. During heating, the reactants were contained in a Pt crucible with a Pt lid. This resided inside a high-alumina crucible with a high-alumina lid. B_2O_3 was removed from the products by dissolving it in demineralized water.

2.31 BORON–URANIUM–OXYGEN–PHOSPHORUS

The $U_2[BO_4][PO_4]$ quaternary compound, which crystallizes in the monoclinic structure with the lattice parameters $a = 854.6 ± 0.2$, $b = 775.3 ± 0.2$, $c = 816.3 ± 0.2$ pm, and

$\beta = 102.52 \pm 0.03°$ and a calculated density of 8.12 g·cm^{-3}, is formed in the B–U–O–P system (Hinteregger et al. 2013). It was synthesized by a two-stage synthesis. The synthesis of the precursor was achieved under high-pressure and high-temperature conditions of 7.0 GPa and 700°C, respectively, from a nonstoichiometric mixture of UO_3 (46.63 mg), H_3BO_3 (30.13 mg), and partially hydrolyzed P_4O_{10} (23.14 mg). The starting materials were finely ground, inserted into a gold capsule, and placed in a BN crucible. The crucible was placed into an 18/11-assembly and compressed by eight tungsten carbide cubes. To apply the pressure, a 1000 ton multianvil press with a Walker-type module was used. In detail, the 18/11-assembly was compressed to 7.0 GPa in 200 min and heated to 700°C in the following 10 min, kept there for 10 min, and cooled down to 300°C in 30 min at constant pressure. After natural cooling to room temperature by switching off the heating, a decompression period of 10 h was required. The recovered MgO octahedron (pressure transmitting medium) was broken apart, and the sample was carefully separated from the surrounding graphite and BN crucible. The precursor was gained in the form of a yellow tough bulk. The bulk was dried for 3 h in a compartment dryer at 80°C. XRD showed an amorphous phase. This precursor was the starting material for the synthesis of $U_2[BO_4][PO_4]$. For the synthesis, high-pressure and high-temperature conditions of 12.5 GPa and 1000°C, respectively, were required by using an 14/11-assembly. In detail, the 14/11-assembly was compressed to 12.5 GPa in 380 min and heated to 1000°C in the following 10 min, kept there for 5 min, and cooled to 450°C in 25 min at constant pressure. After natural cooling to room temperature by switching off the heating, a decompression period of 12 h was required. $U_2[BO_4][PO_4]$ was gained in form of emerald green air- and water-resistant crystals in an emerald green amorphous matrix.

2.32 BORON–CARBON–IRON–PHOSPHORUS

Glassy alloy with composition $B_4C_5Fe_{80}P_{11}$ was prepared by induction melting a mixture of pure Fe, prealloyed Fe–C and Fe–P ingots, and pure B crystal in an Ar atmosphere (Inoue et al. 1995). The difference between glass transition temperature (T_g) and crystallization temperature (T_x) is $\Delta T_x = 24°C$ for this alloy.

2.33 BORON–SILICON–IRON–PHOSPHORUS

An amorphous phase exhibiting the glass transition phenomenon is formed in the composition range of 7–12 at% P, 3–5 at% B, and 3–7 at% Si in the $Fe_{80}(B,Si,P)_{20}$ alloys (Inoue and Park 1996). The largest ΔT_x value of 36°C was observed for the $B_4Si_4Fe_{80}P_{12}$ amorphous alloys.

Obvious glass transition was observed at the limited $Fe_{76}(Si_xB_yP_z)_{24}$ ($x + y + z = 1$) compositions in the amorphous-forming range (Figure 2.1), and the large ΔT_x of over 40°C was observed in the range of $x = 0.3$–0.5, $y = 0.2$–0.5 and $z = 0.2$–0.04 (Makino et al. 2007, 2008). The largest ΔT_x is 52°C for $B_{9.6}Si_{9.6}Fe_{76}P_{4.8}$ and $B_{10}Si_9Fe_{75}P_5$.

Quaternary B–Si–Fe–P alloys were prepared by induction melting a mixture of pure Fe, prealloyed Fe–P ingots, and pure B and Si crystal in an Ar atmosphere (Inoue and Park 1996; Makino et al. 2007).

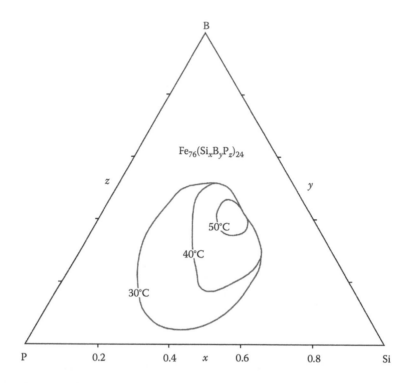

FIGURE 2.1 Glass-forming region in the B–Si–Fe–P quaternary system at 76 at% Fe [$Fe_{76}(Si_xB_yP_z)_{24}$]. (From Makino, A. et al., *Mater. Trans.*, 48(11), 3024–3027, 2007. With permission.)

2.34 BORON–LEAD–OXYGEN–PHOSPHORUS

The **$PbBPO_5$**, **$Pb_3[(PO_4)_2BPO_4]$**, and **$Pb_6(PO_4)[B(PO_4)]_4$** quaternary compounds are formed in the B–Pb–O–P system. $PbBPO_5$ crystallizes in the trigonal structure with the lattice parameters $a = 685.1 \pm 8.0$ and $c = 682.1 \pm 9.5$ pm (Ben Ali et al. 2001). Polycrystalline samples of the title compound have been synthesized by solid-state reaction between high-purity $Pb(NO_3)_2$, H_3BO_3, and $(NH_4)_2HPO_4$ as starting materials. The stoichiometric mixture was first ground and progressively heated in a platinum crucible at 400°C for 12 h. The resulting compound was ground again and heated for a further 72 h at 680°C. An appropriate excess of H_3BO_3 was then added to compensate the B_2O_3 losses.

$Pb_3[(PO_4)_2BPO_4]$ crystallizes in the orthorhombic structure with the lattice parameters $a = 694.6 \pm 0.1$, $b = 1419.9 \pm 0.3$, and $c = 2111.6 \pm 0.2$ pm and a calculated density of 5.851 g·cm^{-3} (Park and Bluhm 1995). Single crystals of this compound were obtained by melting B_2O_3, P_2O_5, and PbO at 960°C.

$Pb_6(PO_4)[B(PO_4)]_4$ crystallizes in the tetragonal structure with the lattice parameters $a = 695.1 \pm 0.2$ and $c = 973.0 \pm 0.4$ pm and a calculated density of 8.114 + 0.008 g·cm^{-3} (Belokoneva et al. 2001). It was obtained by hydrothermal synthesis starting from the mixture of B_2O_3 and $Pb(H_2PO_4)_2$ (molar ratio 1:2) at 250°C and 0.1 GPa. $NH_2H_2PO_4$ was used as a mineralizer.

BP–PbO. BP reacts with PbO melt (Goryunova et al. 1964).

2.35 BORON–LEAD–FLUORINE–PHOSPHORUS

BP–PbF$_2$. BP reacts with PbF$_2$ melt (Goryunova et al. 1964).

2.36 BORON–LEAD–CHLORINE–PHOSPHORUS

BP–PbCl$_2$. According to the data of metallography and XRD, the solubility of BP in PbCl$_2$ melt was not detected (Goryunova et al. 1964).

2.37 BORON–LEAD–BROMINE–PHOSPHORUS

BP–PbBr$_2$. According to the data of metallography and XRD, the solubility of BP in PbBr$_2$ melt is less than 0.01 mass% (Goryunova et al. 1964).

2.38 BORON–BISMUTH–CHLORINE–PHOSPHORUS

BP–BiCl$_3$. According to the data of metallography and XRD, the solubility of BP in BiCl$_3$ melt was not detected (Goryunova et al. 1964).

2.39 BORON–VANADIUM–OXYGEN–PHOSPHORUS

The **V$_2$[B(PO$_4$)$_3$]** quaternary compound, which crystallizes in the hexagonal structure with the lattice parameters $a = 1398.82 \pm 0.06$ and $c = 745.15 \pm 0.06$ pm and a calculated density of 3.137 g·cm^{-3}, is formed in the B–V–O–P system (Meisel et al. 2004). It was obtained by the reaction of BPO$_4$ and VO$_2$ at 1050°C.

2.40 BORON–NIOBIUM–IRON–PHOSPHORUS

The glassy metal **Fe$_{77}$Nb$_3$B$_{13}$P$_7$** and **Fe$_{78}$Nb$_4$B$_{10}$P$_8$** were prepared by arc-melting in an Ar atmosphere (Matsumoto et al. 2009; Urata et al. 2011). A single-roller melt-spinning method in air was used to rapidly produce the solidified ribbons. The glassy metal alloy powders were made by a water atomize method. The glassy metal alloys present a wide supercooled liquid region. These glassy metal alloys have high stability in amorphous structure.

2.41 BORON–OXYGEN–CHROMIUM–PHOSPHORUS

The **Cr$_2$B(PO$_4$)$_3$** quaternary compound, which crystallizes in the trigonal structure with the lattice parameters $a = 795.419 \pm 0.006$ and $c = 736.130 \pm 0.001$ pm, is formed in the B–O–Cr–P system (Mi et al. 2000). It was synthesized by high-temperature reaction (1000°C) of mixtures of Cr$_2$O$_3$ with BPO$_4$ (molar ratio 1:3) and CrPO$_4$·4H$_2$O with BPO$_4$·xH$_2$O (molar ratio 2:1), respectively. The syntheses were carried out under air.

2.42 BORON–OXYGEN–IRON–PHOSPHORUS

The **Fe$_2$B(PO$_4$)$_3$** quaternary compound, which crystallizes in the hexagonal structure with the lattice parameters $a = 803.06 \pm 0.09$ and $c = 740.89 \pm 0.13$ pm and a calculated density of 3.270 g·cm^{-3} (Zhang et al. 2010) [$a = 803.47 \pm 0.08$ and $c = 741.63 \pm 0.13$ pm and a calculated density of 3.263 g·cm^{-3} (Li et al. 2010); in the trigonal structure with the lattice parameters $a = 802.703 \pm 0.006$ and $c = 740.168 \pm 0.009$ pm and a calculated density of

3.2758 ± 0.0001 g·cm^{-3} (Chen et al. 2004)], is formed in the B–O–Fe–P system. The title compound was obtained by reaction a mixture of Fe_2O_3, H_3BO_3, and $NH_4H_2PO_4$ in a molar ratio of 1:1:3 at 880°C (Zhang et al. 2010) [Fe/P/B molar ratio of 2:3:1 at 1000°C for 11 h (Chen et al. 2004)]. Single crystals of $Fe_2B(PO_4)_3$ have been prepared by the high-temperature solution growth method in air (Li et al. 2010). A powder mixture of Fe_2O_3, B_2O_3, and $NaPO_3$ at the molar ratio of Fe/B/Na/P = 1:5:10:10 was ground and transferred to a Pt crucible. The sample was gradually heated in air at 900°C for 24 h. After that, the intermediate product was slowly cooled to 400°C at the rate of 2°C·h^{-1} and then quenched to room temperature. The obtained crystals were light red and of prismatic shape.

2.43 BORON–OXYGEN–COBALT–PHOSPHORUS

The **Co_3BPO_7** and **$Co_5BP_3O_{14}$** quaternary compounds are formed in the B–O–Co–P system. Co_3BPO_7 is stable up to 1100°C and crystallizes in the monoclinic structure with the lattice parameters a = 977.4 ± 0.2, b = 1268.8 ± 0.2, c = 490.57 ± 0.08 pm, and β = 119.749 ± 0.002° (Yilmaz et al. 2001). The title compound was made in a flux of H_3BO_3 (fivefold excess) from a stoichiometric mixture of $CoCO_3 \cdot xH_2O$ and $NH_4H_2PO_4$. The mixture was calcined at 400°C for 4 h and ground well before heating at 1050°C for 3 h. The heated sample was subsequently cooled to 600°C for a period of 6 h and then cooled to room temperature. The reaction product was washed with distilled water several times to remove excess boron. The crystals were platelike and purple colored.

$Co_5BP_3O_{14}$ also crystallizes in the monoclinic structure with the lattice parameters a = 640.5 ± 0.1, b = 823.8 ± 0.1, c = 1895.5 ± 0.4 pm, and β = 96.93 ± 0.03° (Bontchev and Sevov 1996). This compound was prepared in a flux of B_2O_3 (circa 10-fold excess) from a stoichiometric mixture of $CoCO_3 \cdot xH_2O$ and 98% H_3PO_4. Heating the mixture at 950°C for 12 h and then slow cooling it (1°C·min^{-1}) to room temperature provide relatively large crystals with deep blue color and with no particular shape.

2.44 BORON–CHLORINE–MANGANESE–PHOSPHORUS

$BP–MnCl_2$. According to the data of metallography and XRD, the solubility of BP in $MnCl_2$ melt was not detected (Goryunova et al. 1964).

2.45 BORON–CHLORINE–IRON–PHOSPHORUS

$BP–FeCl_2$. BP reacts with $FeCl_2$ melt (Goryunova et al. 1964).

2.46 BORON–IRON–NICKEL–PHOSPHORUS

Flux-melting and water-quenching techniques were used to prepare bulk glassy **$B_6Fe_{40}Ni_{40}P_{14}$** alloys (Shen and Schwartz 2001). To synthesize such alloys, Fe, Fe_2B, and FeP powders and Ni were used. The bulk glassy alloy of this composition has a wide supercooled liquid region (ΔT_x) of about 42°C.

CHAPTER 3

Systems Based on BAs

3.1 BORON–CALCIUM–OXYGEN–ARSENIC

The **CaBAsO$_5$ (2CaO·As$_2$O$_5$·B$_2$O$_3$)** quaternary compound, which incongruently melts at 975°C decomposing into Ca$_2$As$_2$O$_7$ and molten B$_2$O$_3$ and crystallizes in the hexagonal structure with the lattice parameters a = 685.5 and c = 1356.1 pm and an experimental density of 3.70 g·cm^{-3}, is formed in the B–Ca–O–As system (Bauer 1965, 1966). To prepare the title compound, CaHAsO$_4$·2H$_2$O and crystalline H$_3$BO$_3$ were mixed (molar ratio 1:1.1), and the mixture was heated in a Pt crucible to 600°C (Bauer 1965). This temperature was maintained until the complete elimination of H$_2$O (about 2 h), and the sintered product was heated in a Pt crucible in an O$_2$ stream at 900°C for 3 days.

3.2 BORON–STRONTIUM–OXYGEN–ARSENIC

The **SrBAsO$_5$ (2SrO·As$_2$O$_5$·B$_2$O$_3$)** quaternary compound, which incongruently melts at 1100°C decomposing into Sr$_2$As$_2$O$_7$ and molten B$_2$O$_3$ and crystallizes in the hexagonal structure with the lattice parameters a = 705.5 and c = 1380.5 pm, is formed in the B–Sr–O–As system (Bauer 1965, 1966). A simultaneous thermal decomposition of primarily formed distrontium arsenate to tristrontium arsenate occurs. To prepare this compound, SrCO$_3$ was dissolved in the equivalent amount of arsenic acid (Bauer 1965). The obtained solution was evaporated to dryness and heated at 800°C. Boric acid in molar ratio of 2SrO·As$_2$O$_5$/H$_3$BO$_3$ = 1:2.2 was added, and the mixture was heated at 900°C for 3 days.

3.3 BORON–BARIUM–OXYGEN–ARSENIC

The **BaBAsO$_5$ (2BaO·As$_2$O$_5$·B$_2$O$_3$)** quaternary compound, which incongruently melts at 1070°C and crystallizes in the hexagonal structure with the lattice parameters a = 729.1 and c = 1423.5 pm, is formed in the B–Ba–O–As system (Bauer 1966). This compound was prepared by thermal synthesis from 2BaO·As$_2$O$_5$ and H$_3$BO$_3$ (molar ratio 1:2.2). The mixture was slowly heated to 800°C and kept at that temperature for 3 days.

3.4 BORON–ALUMINUM–OXYGEN–ARSENIC

The **$BAl_6As_3O_{15}$** quaternary compound (mineral szklaryite), which crystallizes in the orthorhombic structure with the lattice parameters a = 470.01, b = 1182.8, and c = 2024.3 pm, is formed in the B–Al–O–As system (Pieczka et al. 2013).

3.5 BORON–GALLIUM–INDIUM–ARSENIC

BAs–GaAs–InAs. The spinodal decomposition ranges with and without the coherency stress were calculated for $B_xGa_yIn_{1-x-y}As$ solid solution lattice matched to GaAs ($0 \leq x < 0.1$, $y = 1 - 3.175x$) (Asomoza et al. 2001b). The estimations showed that the spinodal decomposition range increases almost linearly with temperature of up to 1000°C at $x = 0.1$. Solid solutions with $x < 0.01$ were obtained by metal–organic chemical vapor deposition using $GaEt_3$, $InMe_3$, and AsH_3 (Geisz et al. 2001). A common 3 vol% of B_2H_6 in H_2 mixture was used as boron source.

$B_xGa_{1-x-y}In_yAs$ solid solutions have also been grown by low-pressure metalorganic vapor phase epitaxy (Gottschalch et al. 2003). The epitaxial layers were grown on GaAs using the precursors BEt_3, $GaMe_3$, $InMe_3$, and AsH_3. Boron and indium compositions of the alloys were varied from $0 \leq x \leq 0.04$ and $0 \leq y \leq 0.35$, respectively.

3.6 BORON–GALLIUM–ANTIMONY–ARSENIC

The lattice constant of quaternary **$B_xGa_{1-x}As_{1-y}Sb_y$** alloy can be expressed by Vegard's law (Wang et al. 2013). Using first-principles density functional calculations, it was shown that the band-gap $B_xGa_{1-x}As_{1-y}Sb_y$ are direct, and it will slightly reduce with the increment of boron and antimony composition. The calculated optical properties indicate the significant change induced by the incorporation of B and Sb.

3.7 BORON–LEAD–OXYGEN–ARSENIC

The **$Pb(BAsO_5)$** and **$Pb_6(AsO_4)[B(AsO_4)_4]$** quaternary compounds are formed in the B–Pb–O–As system. $Pb(BAsO_5)$ crystallizes in the trigonal structure with the lattice parameters a = 712.2 ± 0.1 and c = 693.5 ± 0.2 pm and a calculated density of 6.098 g·cm^{-3}, and $Pb_6(AsO_4)[B(AsO_4)_4]$ crystallizes in the tetragonal structure with the lattice parameters a = 715.4 ± 0.1 and c = 976.2 ± 0.3 pm and a calculated density of 6.476 g·cm^{-3} (Park and Bluhm 1996). Single crystals of these compounds were obtained by melting B_2O_3, As_2O_5, and PbO at 800°C.

CHAPTER 4

Systems Based on AlN

4.1 ALUMINUM–HYDROGEN–SODIUM–NITROGEN

The **NaAl(NH$_2$)$_4$** quaternary compound, which crystallizes in the monoclinic structure with the lattice parameters $a = 732.4 \pm 0.6$, $b = 605.0 \pm 0.5$, and $c = 1318 \pm 1$ pm and $\beta = 94°00' \pm 0°15'$, is formed in the Al–H–Na–N system (Molinie et al. 1973). This compound was obtained by the action of sodium solution in liquid NH$_3$ on Al metallic in tubes sealed at ambient temperature.

4.2 ALUMINUM–HYDROGEN–POTASSIUM–NITROGEN

The **KAl(NH$_2$)$_4$** quaternary compound, which exists in two polymorphic modifications, is formed in the Al–H–K–N system (Molinie et al. 1973). α-KAl(NH$_2$)$_4$ crystallizes in the orthorhombic structure with the lattice parameters $a = 1000 \pm 1$, $b = 580 \pm 1$, and $c = 1014 \pm 1$ pm and the experimental and calculated densities of 1.45 ± 0.02 and 1.47 g·cm^{-3}, respectively. This compound was obtained by the action of potassium solution in liquid NH$_3$ on Al metallic in tubes sealed at ambient temperature.

4.3 ALUMINUM–HYDROGEN–CESIUM–NITROGEN

The **CsAl(NH$_2$)$_4$** quaternary compound, which crystallizes in the tetragonal structure with the lattice parameters $a = 757.9 \pm 0.7$ and $c = 537.5 \pm 0.5$ pm, is formed in the Al–H–Cs–N system (Molinie et al. 1973). This compound was obtained by the action of cesium solution in liquid NH$_3$ on Al metallic in tubes sealed at ambient temperature.

4.4 ALUMINUM–HYDROGEN–OXYGEN–NITROGEN

The **[Al(H$_2$O)$_6$](NO$_3$)$_3$·3H$_2$O** quaternary compound, which crystallizes in the monoclinic structure with the lattice parameters $a = 1389.2 \pm 0.2$, $b = 960.7 \pm 0.1$, and $c = 1090.7 \pm 0.2$ pm and $\beta = 95.51° \pm 0.02°$ and a calculated density of 1.719 g·cm^{-3} (Lazar et al. 1991) [$a = 1390.1 \pm 1.5$, $b = 963.6 \pm 1.5$, $c = 1090.3 \pm 1.5$ pm and $\beta = 84°48'$ and the experimental and calculated densities of 1.714 and 1.713 g·cm^{-3}, respectively (Kannan and Viswamitra 1965)], is formed in the Al–H–O–N system. Repeated crystallization from saturated aqueous solution and also from alcohol under slightly varying rates of evaporation and temperature gave only monoclinic crystals (Kannan and Viswamitra 1965).

4.5 ALUMINUM–LITHIUM–SILICON–NITROGEN

AlN–LiSi$_2$N$_3$. Solid solutions of the composition **Li$_x$Al$_{12-3x}$Si$_{2x}$N$_{12}$** ($1 \leq x \leq 3$) have been prepared via a precursor route (Ischenko et al. 2002). The polymeric amide precursors have been used. According to X-ray diffraction (XRD) studies, all phases obtained have wurtzite-type substructures with an orthorhombic superstructure related to LiSi$_2$N$_3$. However, observed powder patterns cannot be adequately described in orthorhombic symmetry, and the significant diffusiveness of the superstructure reflections indicates the presence of disorder. High-resolution electron microscopy emphasizes the highly defective nature of the microstructure and the presence of nanoscale domains of apparently varying symmetry. The composition **LiAlSi$_2$N$_4$** (at $x = 3$) crystallizes in the orthorhombic structure with the lattice parameters $a = 926.0$, $b = 533.3$, and $c = 486.9$ pm.

4.6 ALUMINUM–LITHIUM–OXYGEN–NITROGEN

During the reaction of Li$_2$O with AlN, and of Li$_3$N with Al$_2$O$_3$, the formation of **Li$_2$AlNO** quaternary compound has been observed (Łapinski and Podsiadło 1987). It is stable up to 1100°C and then decomposes according to the next scheme: 2Li$_2$AlNO = LiAlO$_2$ + AlN + Li$_3$N.

4.7 ALUMINUM–SODIUM–OXYGEN–NITROGEN

(1.2 + y)Na$_2$O·(11 − x/2)Al$_2$O$_3$·xAlN ($x = 0$–0.7) ceramics were obtained by hot-pressing technique using AlN as the nitrogen source (Harata et al. 1981). Samples were prepared by hot-pressing technique using Na$_2$CO$_3$, α-Al$_2$O$_3$, and AlN as starting materials. At the first stage, Na$_2$CO$_3$ and α-Al$_2$O$_3$ were mixed well in various batch compositions, (1.2 + y)Na$_2$O·(11 − x/2)Al$_2$O$_3$ ($y = x/2$ in most cases), and then calcined at 1400°C for 1 h in closed Pt containers to avoid Na$_2$O loss by vaporization. At the second state, x moles of AlN were added to the calcined powder of batch composition, (1.2 + y)Na$_2$O·(11 − x/2) Al$_2$O$_3$, and mixed again in a highly pure alumina pot mill using Bu^{n}OH as a medium. At the final stage, mixed powders were hot-pressed under 29.4 MPa at a temperature between 1650°C and 1750°C for 0.5 h in a graphite die.

4.8 ALUMINUM–COPPER–TITANIUM–NITROGEN

The alloys of the Al–Cu–Ti–N quaternary system were annealed at 850°C for 120 h and quenched in water. The composition of the phases was determined by energy-dispersive X-ray spectroscopy on a scanning electron microscope or on a microprobe. The structural analysis was done by XRD. A number of three-phase equilibria and 18 four-phase equilibria were determined and listed. A network of three-phase spaces and 29 four-phase spaces at 850°C was proposed by Durlu et al. (1997) and included in the literature evaluation by Raghavan (2006b). The quaternary η-type cubic phase is the dominating phase in the system, occurring in 21 four-phase spaces, 14 of which are experimentally observed. Carim (1989) first reported this phase with the general formula **Ti$_{3-x}$Cu$_{3-x}$Al$_{2x}$N** and lattice parameter of 1130 ± 20 pm. When in equilibrium with various phases of this quaternary system,

Durlu et al. (1997) found the η-phase to have a wide homogeneity range and a lattice parameter of 1127.9–1136.8 pm and obtained a single-phase material at the composition **Ti₃Cu₂AlN₀.₈** with a = 1133.9 pm. This compound may be stable at 900°C (Carim 1989).

4.9 ALUMINUM–BERYLLIUM–SILICON–NITROGEN

AlN–Be₃N₂–Si₃N₄. The isothermal section of this quasiternary system at 1760°C is shown in Figure 4.1 (Schneider et al. 1979, 1980). Complete solid solutions exist between AlN and BeSiN₂. With increasing amount of AlN, the lattice parameters increase linearly. $Al_{2-2x}Be_xSi_xN_2$ solid solution is in thermodynamic equilibrium with β-Si₃N₄ and β-Be₃N₂. The samples were heated to 1760°C in a vacuum-tight N₂-flushed induction furnace. Temperature was kept for 45 min. During thermal treatment, a uniaxial pressure of 28 MPa was applied.

4.10 ALUMINUM–MAGNESIUM–SILICON–NITROGEN

The **MgAlSiN₃** quaternary compound, which crystallizes in the orthorhombic structure with the lattice parameters a = 940.657 ± 0.009, b = 542.569 ± 0.004, and c = 488.985 ± 0.003 pm, is formed in the Al–Mg–Si–N system (Han et al. 2011) [a = 945.95, b = 546.15, and c = 489.57 pm (Mikami et al. 2008)]. To obtain this compound, the mixed raw materials (AlN, Mg₃N₂, and α-Si₃N₄) were fired at 1800°C for 2 h under a pressurized N₂ environment (1 MPa) (Han et al. 2011). The crystal structure has been optimized by first-principles calculation (Mikami et al. 2008).

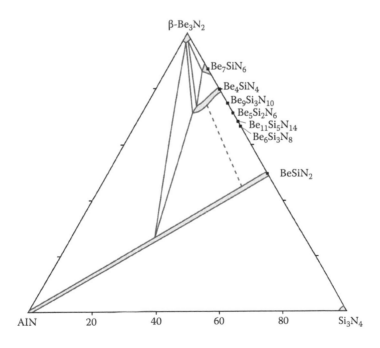

FIGURE 4.1 Isothermal section of the AlN–Be₃N₂–Si₃N₄ quasiternary system at 1760°C. (Reprinted from *Ceramurgia Int.*, 5(3), Schneider, G. et al., Phase equilibria in the Si, Al, Be/C, N system, 101–104, Copyright (1979), with permission from Elsevier.)

4.11 ALUMINUM–MAGNESIUM–OXYGEN–NITROGEN

AlN–MgO–Al$_2$O$_3$. At 1150°C, all γ-Al$_2$O$_3$ has been transformed into α-Al$_2$O$_3$, MgO has disappeared in favor of MgAl$_2$O$_4$, and AlN is still observed (Granon et al. 1994). Until 1350°C, the same phases were observed but the content of MgAl$_2$O$_4$ increases while Al$_2$O$_3$ decreases. At 1400°C, solid solution with spinel structure and α-Al$_2$O$_3$ were present but after 1 h duration alumina disappeared, and only the solid solution was found.

The isothermal sections of the AlN–MgO–Al$_2$O$_3$ quasiternary system at 1750°C and at 1800°C were constructed by Weiss et al. (1982) and Jack (1976), respectively. The spinel structure may be described by the formula Mg$_y$Al$_{3-y-1/3x}\square_{2/3x}O_{3+x+y}N_{1-x-y}$ with the temperature-dependent solubility limits ($\square$ – vacancy; $0 \le x \le 1$; $0 \le y \le 1$; $x + y \le 1$) (Weiss et al. 1982). AlN, Al$_2$O$_3$, MgO, and Mg$_3$N$_2$ were weighed, milled for 1 h in an agate jar with agate balls, and dried. Mixtures containing Mg$_3$N$_2$ were hot-pressed immediately to avoid hydrolysis. Hot-pressing was done under N$_2$ pressure in BN-coated graphite dies at 39 MPa. The isothermal section of the 2AlN–3MgO–Al$_2$O$_3$ quasiternary system at 1800°C is presented in Figure 4.2 (Jack 1976). The isothermal section at this temperature was also presented by Sun et al. (1990). An extensive solid solution of Mg–Al–oxynitride spinel and three spinel-containing two-phase region was determined.

According to the data of Granon et al. (1994), the starting materials were mixed for 72 h by ball-milling with Al$_2$O$_3$ balls in anhydrous EtOH. Pellets were obtained by isostatic pressing in plastic moulds at 400 MPa and were embedded in a powder for firing in a

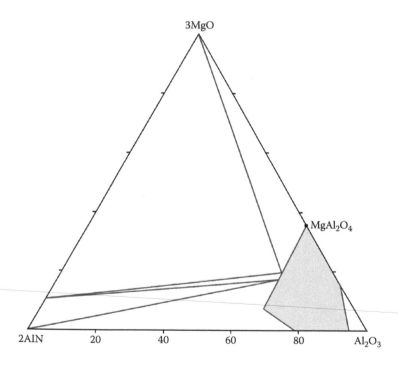

FIGURE 4.2   Isothermal section of the 2AlN–3MgO–Al$_2$O$_3$ quasiternary system at 1800°C. (With kind permission from Springer Science+Business Media: *J. Mater. Sci.*, Sialons and related nitrogen ceramics, 11(6), 1976, 1135–1158, Jack, K.H.)

graphite furnace under N_2 atmosphere. Some samples were prepared by hot-pressing in graphite dies at 20 MPa to simulate a closed thermodynamic system.

The solubility limit of Mg in aluminum oxynitride was measured using wavelength-dispersive spectroscopy mounted on a scanning electron microscope (SEM) from the samples quickly cooled from 1870°C (Miller and Kaplan 2008). It was found to be more than 4000 ppm.

AIN–MgAl$_2$O$_4$. Two nonstoichiometric spinels have been studied ($Mg_{0.980}Al_{2.013}\square_{0.0056}O_4$ and $Mg_{0.870}Al_{2.086}\square_{0.043}O_4$, where $\square$ is vacancy) (Granon et al. 1994). Until 1200°C, no reaction occurs; at 1700°C, samples were single phase in both cases with the same lattice parameter ($a = 799.4 \pm 0.5$ pm). A difference was observed at 1550°C: two phases were present (AIN + solid solution with spinel structure), but a higher nitrogen concentration was allowed by a higher cationic vacancy concentration in the initial phase (for $\square = 0.043$, [N] = 2.3 mass%; for $\square = 0.0056$, [N] = 1.8 mass%). From 1200°C to 1550°C, a severe mass loss was observed.

AIN–MgO. No reaction occurs until 1200°C at mixing 60.4 mol% MgO and 39.6 mol% AIN (Granon et al. 1994). Then a severe mass loss linked to AIN disappearing was observed. At 1550°C, three phases, AIN, MgO (traces), and $MgAl_2O_4$, were noted. At 1700°C for 1 h, a single-phase solid solution was obtained, but the nitrogen content was only 3 mass%, while its initial content was 13.3 mass%.

The phase diagram of the AIN–MgO system at 0.1 MPa was calculated by Kaufman (1979) and is shown in Figure 4.3. The **5AIN·MgO** (**Al$_5$MgON$_5$**) quaternary compound is formed in this system according to the calculations.

According to the data of Weiss et al. (1982), the **6AIN·MgO** (**Al$_6$MgON$_6$**) quaternary compound, which crystallizes in the hexagonal structure with the lattice parameters $a = 299.8$ and $c = 532.8$ pm, is formed in the AIN–MgO system.

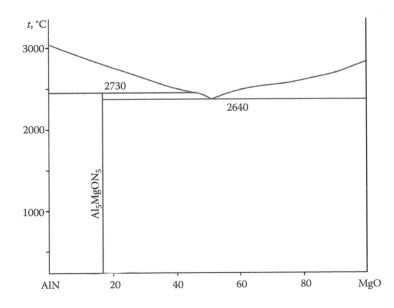

FIGURE 4.3 Calculated phase diagram of the AIN–MgO system at 0.1 MPa. (Reprinted from *CALPHAD*, 3(4), Kaufman, L., Calculation of quasibinary and quasiternary oxynitride systems—III, 275–291, Copyright (1979), with permission from Elsevier.)

4.12 ALUMINUM–CALCIUM–SILICON–NITROGEN

The **CaAlSiN₃** quaternary compound, which crystallizes in the orthorhombic structure with the lattice parameters $a = 973.1 \pm 0.1$, $b = 564.0 \pm 0.8$, and $c = 504.1 \pm 0.7$ pm and a calculated density of 3.7919 g·cm⁻³, is formed in the Al–Ca–Si–N system (Piao et al. 2007) [$a = 980.20 \pm 0.04$, $b = 565.06 \pm 0.02$, and $c = 506.33 \pm 0.02$ pm (Uheda et al. 2006a, 2006b); $a = 988.69$, $b = 566.45$, and $c = 508.40$ pm (Mikami et al. 2006, 2008; $a = 978.13 \pm 0.03$, $b = 566.26 \pm 0.03$, and $c = 505.70 \pm 0.02$ pm (Li et al. 2007); $a = 976.76 \pm 0.08$, $b = 564.40 \pm 0.01$, and $c = 505.29 \pm 0.01$ pm (Li et al. 2008a); $a = 975.58 \pm 0.02$, $b = 564.73 \pm 0.01$, and $c = 505.24 \pm 0.01$ pm and a calculated density of 3.232 g·cm⁻³ for $CaAl_{0.49}Si_{1.38}N_3$ (Li et al. 2008b)])]. To prepare this compound, AlN, Ca_3N_2, and α-Si_3N_4 were mixed in a mortar under N_2 atmosphere in a glove box (Uheda et al. 2006a, 2006b). The mixed powder or its pellet was charged in a BN crucible and then fired at 1600°C for 2 h and subsequently at 1800°C for 2 h under N_2 pressure of 1 MPa. It was also synthesized by using a self-propagating high-temperature synthesis method using CaAlSi as precursor (Piao et al. 2007). The ignition occurs at 1050°C, and further nitridation process at 1450–1550°C leads to the formation of the title compound. Its crystal structure has been optimized by first-principles calculation (Mikami et al. 2006, 2008).

The title compound could be also synthesized via the reaction of the CaAlSi alloy with ammonia at 500–800°C (Li et al. 2007, 2008a). The presence of sodium amide facilitates the production of $CaAlSiN_3$ and prevents the presence of unreacted silicon, which could be due to the formation of sodium ammonometallates as intermediates. The synthesis may be conducted even under atmospheric pressure, but the crystallinity is low. The reactions of the alloy in the presence of sodium amide in pressurized ammonia (100 MPa) at 500–800°C give poorly crystallized $CaAlSiN_3$ even after a long holding time. This is because $CaAlSiN_3$ is insoluble in the ammonia–sodium amide solution under the present conditions. The decomposition of the presynthesized sodium ammonometallates by slow heating up to 800°C leads to well-crystallized platelike and barlike $CaAlSiN_3$ nanocrystals.

Ceramics with nominal compositions **Ca$_x$Si$_{12-2x}$Al$_{2x}$N$_{16}$** ($0.2 \leq x \leq 2.6$), extending along the Si_3N_4–1/2Ca_3N_2:3AlN tie line, were prepared from AlN, Si_3N_4, and CaH_2 precursors by hot-pressing at 1800°C (Cai et al. 2007). The results show that ceramics forms continuously within the compositional range $x = 0$ to at least $x = 1.82$. The lattice parameters are well described by the linear relationships $a = (775.41 \pm 0.04) + (11.14 \pm 0.09)x$ and $c = (562.17 \pm 0.04) + (8.59 \pm 0.09)x$ pm. Such ceramics could be also obtained from a powder mixture of AlN, α-Si_3N_4, and Ca_3N_2, which was hot-pressed at 1700°C for 1 h under 20 MPa (Xie et al. 2001). The sintering behavior of the sample was found to be significantly composition dependent.

4.13 ALUMINUM–CALCIUM–OXYGEN–NITROGEN

The **CaAl₁₁O₁₆N** quaternary compound, which crystallizes in the hexagonal structure with the lattice parameters $a = 555.9 \pm 0.1$ and $c = 2190.0 \pm 0.4$ pm, is formed in the Al–Ca–O–N

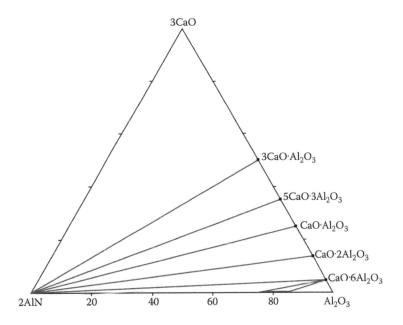

FIGURE 4.4   Subsolidus phase relationships in the 2AlN–3CaO–Al₂O₃ quasiternary system. (Reprinted from *Mater. Lett.*, 8(5), Sun, W.Y., and Yen, T.S., Phase relationships in the system Ca–Al–O–N, 150–152, Copyright (1989), with permission from Elsevier.)

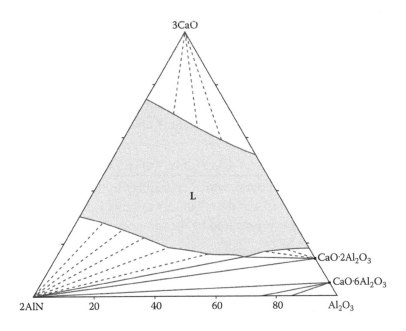

FIGURE 4.5   Isothermal section of the 2AlN–3CaO–Al₂O₃ quasiternary system at 1700°C. (Reprinted from *Mater. Lett.*, 8(5), Sun, W.Y., and Yen, T.S., Phase relationships in the system Ca–Al–O–N, 150–152, Copyright (1989), with permission from Elsevier.)

system (Jansen et al. 1997a). To prepare the title compound, $CaCO_3$, AlN, and γ-Al_2O_3 were wet-mixed in Pr^iOH for 2 h in an agate container with agate balls on a planetary mill. After mixing, Pr^iOH was evaporated. The powder was dried in a stove for one night at 160°C, and it was subsequently ground in an agate mortar. The powders were fired in a Mo crucible under a mildly flowing N_2/H_2 (9:1 volume ratio) gas mixture. Reaction was performed in a vertical high-temperature tube furnace at 1700°C for 2 h.

AlN–CaO–Al₂O₃. Phase relationships in the 2AlN–3CaO–Al_2O_3 quasiternary system at both subsolidus temperatures and 1700°C are given in Figures 4.4 and 4.5 (Sun and Yen 1989a). There is no new phase, and an extensive liquid phase region exists at 1700°C. In the subsolidus region, all the calcium aluminates join to AlN forming six AlN-containing compatibility triangles.

4.14 ALUMINUM–STRONTIUM–SILICON–NITROGEN

The **SrAlSiN₃** and **SrAlSi₄N₇** quaternary compounds are formed in the Al–Sr–Si–N system. SrAlSiN₃ crystallizes in the orthorhombic structure with lattice parameters $a = 980.87 \pm 0.01$, $b = 575.600 \pm 0.008$, and $c = 516.614 \pm 0.007$ pm and a calculated density of 4.145 g.cm^{-3} (Watanabe and Kijima 2009) [$a = 992.01$, $b = 577.92$, and $c = 520.03$ pm (Mikami et al. 2008)]. Its crystal structure has been optimized by first-principles calculation (Mikami et al. 2008).

SrAlSi₄N₇ crystallizes also in the orthorhombic structure with the lattice parameters $a = 1174.2 \pm 0.2$, $b = 2139.1 \pm 0.4$, and $c = 496.6 \pm 0.1$ pm and a calculated density of 1.731 g.cm^{-3} (Hecht et al. 2009). The title compound was prepared by high-temperature synthesis in a radio frequency furnace starting from AlN (1.0 mM), α-Si_3N_4 (0.67 mM), and Sr (0.5 mM). The starting materials were mixed in an agate mortar and filled into a W crucible under Ar atmosphere in a glove box. The reaction mixture was heated to 1630°C within 3 h. After an annealing time of 5 h at this temperature, the crucible was cooled to 1230°C in another 5 h and then quenched to room temperature by switching off the furnace. SrAlSi₄N₇ was obtained as a well crystalline colorless product.

4.15 ALUMINUM–STRONTIUM–OXYGEN–NITROGEN

The **SrAl₁₁O₁₆N** quaternary compound, which crystallizes in the hexagonal structure with the lattice parameters $a = 556.7 \pm 0.1$ and $c = 2200.1 \pm 0.2$ pm, is formed in the Al–Sr–O–N system (Jansen et al. 1997a). It was prepared by the same way as $CaAl_{11}O_{16}N$ was obtained using $SrCO_3$ instead of $CaCO_3$.

4.16 ALUMINUM–BARIUM–SILICON–NITROGEN

The **BaAl₃Si₄N₉**, **Ba₂AlSi₅N₉**, and **Ba₅Al₇Si₁₁N₂₅** quaternary compounds are formed in the Al–Ba–Si–N system. BaAl₃Si₄N₉ crystallizes in the monoclinic structure with the lattice parameters $a = 548.65 \pm 0.04$, $b = 2672.55 \pm 0.18$, and $c = 583.86 \pm 0.04$ pm and $\beta = 118.897° \pm 0.010°$ and a calculated density of 3.818 g.cm^{-3} (Hirosaki et al. 2014). It was prepared by firing appropriate amounts of AlN, Ba_3N_2, and Si_3N_4 at 1900°C for 2 h in 1.0 MPa nitrogen atmosphere using a gas pressure sintering furnace.

$Ba_2AlSi_5N_9$ crystallizes in the triclinic structure with the lattice parameters $a = 986.0 \pm 0.1$, $b = 1032.0 \pm 0.1$, and $c = 1034.6 \pm 0.1$ pm and $\alpha = 90.37° \pm 0.02°$, $\beta = 118.43° \pm 0.02°$, and $\gamma = 103.69° \pm 0.02°$ and a calculated density of 4.237 g·cm^{-3} (Kechele et al. 2009). This compound was synthesized starting from AlN, Si_3N_4, and Ba in a radio frequency furnace at temperatures about 1725°C. AlN (1.0 mM), Si_3N_4 (0.7 mM), and Ba (0.8 mM) were mixed in an agate mortar and filled into a W crucible under Ar atmosphere in a glove box. Under purified N_2, the crucible was heated to 800°C with a rate of about 78°C·min^{-1} before the temperature was increased to 1725°C during 2 h. This maximum temperature was kept for 5 h. Subsequently, the crucible was cooled down to 1300°C with a rate of about 85°C·h^{-1} before quenching to room temperature by switching off the furnace. The samples contain $Ba_2AlSi_5N_9$ as colorless, air- and water-resistant crystals. This compound is stable up to about 1515°C under He atmosphere.

$Ba_5Al_7Si_{11}N_{25}$ crystallizes in the orthorhombic structure with the lattice parameters $a = 959.23 \pm 0.02$, $b = 2139.91 \pm 0.05$, and $c = 588.89 \pm 0.02$ pm and a calculated density of 4.226 g·cm^{-3} (Hirosaki et al. 2014). It was also prepared by firing appropriate amounts of AlN, Ba_3N_2, and Si_3N_4 at 1900°C for 2 h in 1.0 MPa nitrogen atmosphere using a gas pressure sintering furnace.

4.17 ALUMINUM–BARIUM–OXYGEN–NITROGEN

AlN–BaAl$_2$O$_4$–Al$_2$O$_3$. The isothermal section of this quasiternary system at 1650°C is presented in Figure 4.6 (Jansen et al. 1996, 1997b). The **BaAl$_{11}$O$_{16}$N** quaternary compound, which crystallizes in the hexagonal structure with the lattice parameters

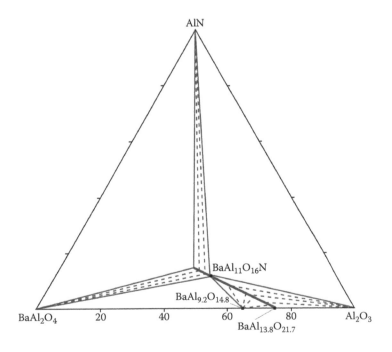

FIGURE 4.6   Isothermal section of the AlN–BaAl$_2$O$_4$–Al$_2$O$_3$ quasiternary system at 1650°C. (Reprinted from *J. Solid State Chem.*, 129(1), Jansen, S.R. et al., Phase relations in the BaO–Al$_2$O$_3$–AlN system: Materials with the β-alumina structure, 66–73, Copyright (1997), with permission from Elsevier.)

$a = 560.19 \pm 0.02$ and $c = 2265.79 \pm 0.08$ pm (Jansen et al. 1997b) [$a = 560.1 \pm 0.1$ and $c = 2266.5 \pm 0.1$ pm (Jansen et al. 1997a); $a = 559.97 \pm 0.03$ and $c = 2267.09 \pm 0.10$ pm for **$BaAl_{11.54}O_{17.11}N_{0.80}$** composition (Jansen et al. 1996)], is formed in this system. It is seen that at this temperature a solid solution exists between $BaAl_{13.8}O_{21.7}$ and $BaAl_{11}O_{16}N$. The lattice parameter a increases whereas c decreases for increasing nitrogen concentration in this solid solution. $BaAl_{11}O_{16}N$ is in equilibrium with $BaAl_{9.2}O_{14.8}$, α-Al_2O_3, and AlN. The phase equilibria at 1400°C and 1500°C are similar; only the extent of the solid solution line decreases with decreasing temperature (Jansen et al. 1997b).

$BaAl_{11}O_{16}N$ was prepared by the same way as $CaAl_{11}O_{16}N$ was obtained using $BaCO_3$ instead of $CaCO_3$ (Jansen et al. 1996, 1997a).

4.18 ALUMINUM–GALLIUM–INDIUM–NITROGEN

AlN–GaN–InN. The unstable mixing region in the wurtzite structure of the $Al_xGa_yIn_{1-x-y}N$ solid solution was calculated from the free energy of mixing on the basis of the strictly regular solution model (Matsuoka 1997; Vigdorovich and Sveshnikov 2000; Takayama et al. 2001). The interaction parameter used in this calculation was obtained using the delta lattice parameter model. The calculated spinodal isotherms for up to 3000°C are given in Figure 4.7 (Takayama et al. 2001). The spinodal isotherms for this ternary system at temperatures of 600–1000°C were also calculated by Matsuoka (1998) with the same results.

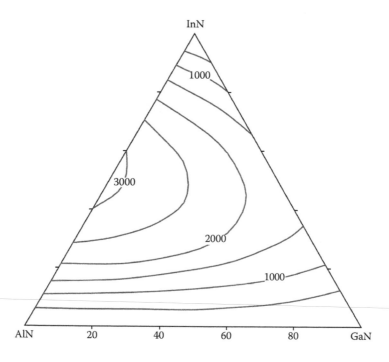

FIGURE 4.7   Calculated mixing region for the AlN–GaN–InN quasiternary system. (Reprinted from *J. Cryst. Growth*, 222(1–2), Takayama, T. et al., Analysis of phase-separation region in wurtzite group III nitride quaternary material system using modifed valence force field model, 29–37, Copyright (2001), with permission from Elsevier.)

The $Al_xGa_yIn_{1-x-y}N$ solid solutions have been grown by metalorganic vapor phase epitaxy on sapphire substrates at 800°C (Matsuoka et al. 1992). The Al, Ga, and In contents were estimated to be 74.3, 22.5, and 2.2 at%, respectively.

4.19 ALUMINUM–YTTRIUM–OXYGEN–NITROGEN

$2AlN + Y_2O_3 \Leftrightarrow Al_2O_3 + 2YN$. The subsolidus phase relations and the isothermal section at 1850°C (Figure 4.8) in this ternary mutual system have been determined by Sun et al. (1991a). No quaternary compound occurs in this system. In the Y_2O_3–YN, Al_2O_3–Y_2O_3, and AlN–Al_2O_3 subsystems, there exist Y_3O_3N, $Y_4Al_2O_9$ (YAM) $YAlO_3$ (YAP), $Y_3Al_5O_{12}$ (YAG), and aluminum oxynitride spinel.

To investigate the AlN–Al_2O_3–Y_2O_3 subsystem, the mixtures of AlN, Al_2O_3, and Y_2O_3 were ground under absolute alcohol for 1.5 h in an agate mortar (Sun et al. 1991a). Pellets of the mixtures were then isostatically pressed under 200 MPa and fired in a furnace. The temperatures used for the determination of the subsolidus phase relationships were in the range 1600–1800°C. AlN, YN, and Y_2O_3 were used for the determination of the phase relationships in the AlN–YN–Y_2O_3 subsystem. The mixtures without YN were ground under 2-propanol in an agate mortar. YN was added to the mixtures previously mentioned and then further dry-mixed in a dry box under N_2 flowing prior to firing. The samples were fired at 1900°C for 2 h under N_2 in a furnace which was evacuated to 0.1 Pa before heating to 1100–1200°C.

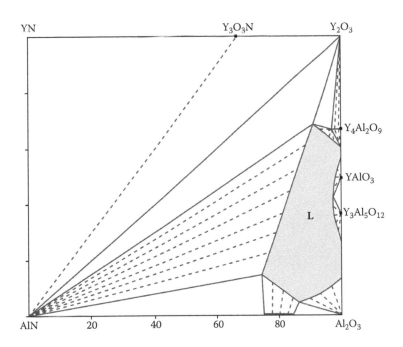

FIGURE 4.8 Isothermal section of the $2AlN + Y_2O_3 \Leftrightarrow Al_2O_3 + 2YN$ ternary mutual system at 1850°C. (Reprinted from *Mater. Lett.*, 11(3–4), Sun, W.-Y. et al., Phase relationships in the system Y–Al–O–N, 67–69, Copyright (1991), with permission from Elsevier.)

AlN–Al$_2$O$_3$–Y$_2$O$_3$. Liquidus surface of the AlN–Al$_2$O$_3$–Y$_2$O$_3$ system was calculated by Fabrichnaya et al. (2013) and compared with experimental results on primary crystallization fields (Figure 4.9). Calculated temperatures of invariant reactions were in agreement with differential thermal analysis (DTA) results. Vertical sections of this system were also calculated and compared with experimental data. The data yielded the lowest eutectic temperature in the Al$_2$O$_3$-rich corner at 1791°C (Neher et al. 2013). With increasing Y$_2$O$_3$/Al$_2$O$_3$ ratio, the temperatures of eutectics increased, reaching 1870°C for melting of the three-phase assemblage composed of AlN, Y$_2$O$_3$, and YAM. During AlN sintering, Al$_2$O$_3$ could react with AlN to form the gas species Al$_2$O and N$_2$ and thus change the compositions of the sintering additives. Due to this process, the fraction of liquid phase could decrease or the liquid phase could even disappear due to the associated increase in the eutectic temperature. The oxide-rich melt showed a high degree of undercooling resulting in large crystals which have an effect on the properties of the ceramic.

The powder samples were densified via spark plasma sintering for 5 min at 1450°C in a vacuum with a pressure of 50 MPa (Neher et al. 2013). Then they were heat-treated at 1700°C for 20 h and investigated through DTA/thermogravimetry, XRD, and SEM.

According to the data of Medraj et al. (2003b), five ternary eutectic points occur in this quasiternary system in the temperature range of 1776–1861°C. They crystallize at 1776°C, 1783°C, 1846°C, 1850°C, and 1861°C.

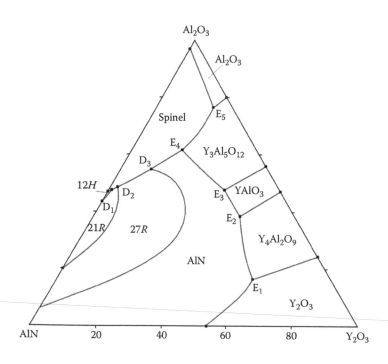

FIGURE 4.9   Calculated liquidus surface of the AlN–Al$_2$O$_3$–Y$_2$O$_3$ quasiternary system. (Reprinted from *J. Eur. Ceram. Soc.*, 33(13–14), Fabrichnaya, O. et al., Liquid phase formation in the system AlN–Al$_2$O$_3$–Y$_2$O$_3$: Part II. Thermodynamic assessment, 2457–2463, Copyright (2013), with permission from Elsevier.)

The isothermal sections of the AlN–Al$_2$O$_3$–Y$_2$O$_3$ system were developed by Gibbs energy minimization using interpolation procedures based on modeling the binary subsystem and are presented in Figures 4.10 and 4.11 (Medraj et al. 2003a, 2003b, 2005). The results of in situ high-temperature neutron diffractometry agree with the thermodynamic calculations.

At a very high temperature prior to solidification, the whole triangle would be composed of a homogeneous melt with no phase boundaries. However, 2500°C is below the primary crystallization of AlN but still above the melting point or the other two components (Figure 4.10). This isothermal section shows that besides the region of homogeneous melt, heterogeneous regions of the primary crystallization of AlN and AlN-based polytypes exist in equilibrium with the liquid phase. In addition, there are two small three-phase areas L + AlN + Al$_9$N$_7$O$_3$ (27R) and L + Al$_7$N$_5$O$_3$ (21R) + 27R.

By cooling from 2500°C to 2000°C (Figure 4.11a), further primary solidification of different phases takes place and instead of three, there are five regions of primary solidifications in which the crystalline types of AlN, Y$_2$O$_3$, 27R, γ-spinel, and Al$_2$O$_3$ occur in equilibrium with liquid phase.

At 1900°C (Figure 4.11b), most of the binary eutectic points of the system appeared. There are seven areas of binary eutectic crystallization (i.e., L + two crystalline phases). Also, a new primary solidification for the ternary compounds YAM, YAP, and YAG occurred at this temperature.

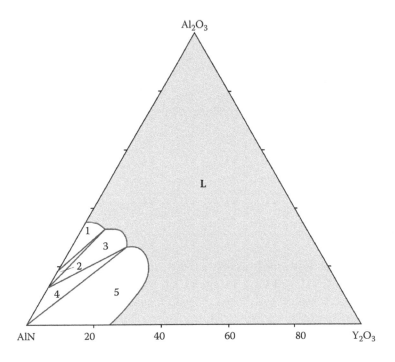

FIGURE 4.10   Calculated isothermal section of the AlN–Al$_2$O$_3$–Y$_2$O$_3$ quasiternary system at 2500°C: 1, L + 21R; 2, L + 21R + 27R; 3, L + 27R; 4, L + 27R + AlN; 5, L + AlN. (From Medraj, M. et al., *Canad. Metallurg. Quart.*, 42(4), 495–507, 2003. With permission.)

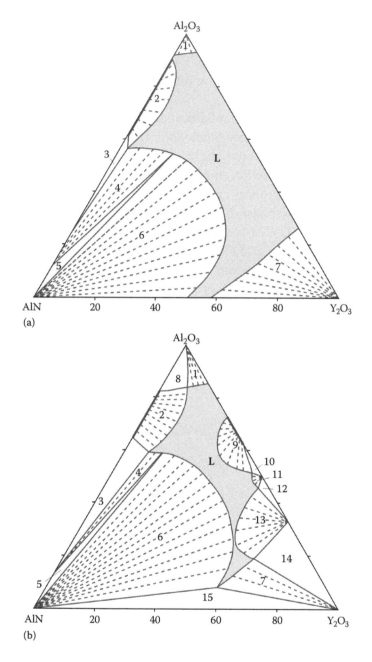

FIGURE 4.11 Calculated isothermal section of the AlN–Al$_2$O$_3$–Y$_2$O$_3$ quasiternary system at (a) 2000°C and (b) 1900°C: 1, L + Al$_2$O$_3$; 2, L + spinel; 3, L + spinel + 27R; 4, L + 27R; 5, L + 27R + AlN; 6, L + AlN; 7, L + Y$_2$O$_3$; 8, L + spinel + Al$_2$O$_3$; 9, L + Y$_3$Al$_5$O$_{12}$; 10, L + Y$_3$Al$_5$O$_{12}$ + YAlO$_3$; 11, L + YAlO$_3$; 12, L + YAlO$_3$ + Y$_4$Al$_2$O$_9$; 13, L + Y$_4$Al$_2$O$_9$; 14, L + Y$_2$O$_3$ + Y$_4$Al$_2$O$_9$; 15, L + AlN + Y$_2$O$_3$; 16, L + Al$_2$O$_3$ + Y$_3$Al$_5$O$_{12}$; 17, L + AlN + spinel; 18, L + AlN + Y$_3$Al$_5$O$_{12}$; 19, AlN + Y$_3$Al$_5$O$_{12}$ + YAlO$_3$, 20, AlN + YAlO$_3$ + Y$_4$Al$_2$O$_9$; 21, AlN + Y$_4$Al$_2$O$_9$ + Y$_2$O$_3$; 22, L + Al$_2$O$_3$ + Y$_3$Al$_5$O$_{12}$; 23, Al$_2$O$_3$ + spinel + Y$_3$Al$_5$O$_{12}$; 24, spinel + Y$_3$Al$_5$O$_{12}$; 25, spinel + AlN + Y$_3$Al$_5$O$_{12}$.   (*Continued*)

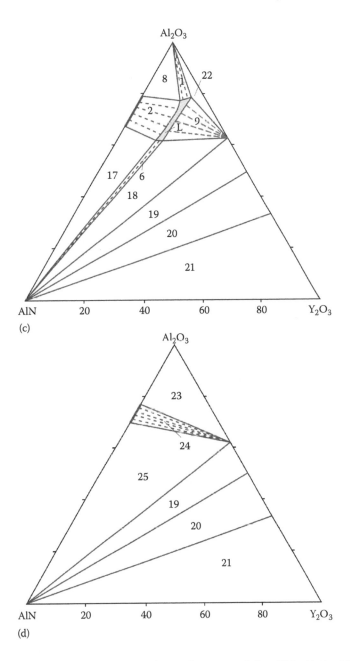

FIGURE 4.11 (CONTINUED) Calculated isothermal section of the AlN–Al$_2$O$_3$–Y$_2$O$_3$ quasiternary system at (c) 1800°C and (d) 1700°C: 1, L + Al$_2$O$_3$; 2, L + spinel; 3, L + spinel + 27R; 4, L + 27R; 5, L + 27R + AlN; 6, L + AlN; 7, L + Y$_2$O$_3$; 8, L + spinel + Al$_2$O$_3$; 9, L + Y$_3$Al$_5$O$_{12}$; 10, L + Y$_3$Al$_5$O$_{12}$ + YAlO$_3$; 11, L + YAlO$_3$; 12, L + YAlO$_3$+ Y$_4$Al$_2$O$_9$; 13, L + Y$_4$Al$_2$O$_9$; 14, L + Y$_2$O$_3$ + Y$_4$Al$_2$O$_9$; 15, L + AlN + Y$_2$O$_3$; 16, L + Al$_2$O$_3$ + Y$_3$Al$_5$O$_{12}$; 17, L + AlN + spinel; 18, L + AlN + Y$_3$Al$_5$O$_{12}$; 19, AlN + Y$_3$Al$_5$O$_{12}$ + YAlO$_3$; 20, AlN + YAlO$_3$ + Y$_4$Al$_2$O$_9$; 21, AlN + Y$_4$Al$_2$O$_9$ + Y$_2$O$_3$; 22, L + Al$_2$O$_3$ + Y$_3$Al$_5$O$_{12}$; 23, Al$_2$O$_3$ + spinel + Y$_3$Al$_5$O$_{12}$; 24, spinel + Y$_3$Al$_5$O$_{12}$; 25, spinel + AlN + Y$_3$Al$_5$O$_{12}$. (Reprinted from *J. Mater. Proc. Technol.*, 161(3), Medraj, M. et al., Understanding AlN sintering through computational thermodynamics combined with experimental investigation, 415–422, Copyright (2005), with permission from Elsevier.)

At 1800°C (Figure 4.11c), the melt region has shrunk very significantly. However, some liquid is in equilibrium with other phases and is contained in the two- and three-phase regions. Moreover, there are still regions of primary solidifications of AlN, YAG, γ-spinel, and Al_2O_3 in equilibrium with their respective melt.

A later stage of solidification (1700°C) is shown in Figure 4.11d, where there is no residual liquid. At this temperature, there are five three-phase regions and one two-phase region.

The isothermal sections of the $AlN-Al_2O_3-Y_2O_3$ system at 1600°C and 1800°C were also calculated by Fabrichnaya et al. (2013). The obtained results practically coincide with the date of Medraj et al. (2003a, 2003b, 2005).

Two new AlN polytypoid phases, $24H$ and $39R$, were identified in the $AlN-Al_2O_3$ quasibinary system where Y_2O_3 was used as a sintering additive (Yu et al. 2004). The $39R$ phase existed as single grains while $24H$ was observed only between subblocks of $33R$ and $39R$. Lattice parameters for this phases were determined to be $a = 310$ and $c = 654$ pm for hexagonal $24H$ and $a = 310$ and $c = 1020$ pm (in hexagonal setting) for rhombohedral $39R$.

The effect of Al_2O_3 addition on the phase reaction of the $AlN-Y_2O_3$ system at high temperatures was investigated by Kim et al. (1996a, 1996b). The amounts of Al_2O_3 and Y_2O_3 were adjusted in the range of 0–40 and 0–10 mass%, respectively. The samples were sintered at 1600–2000°C under 0.6 MPa of N_2 for 2–6 h. It was shown that the density of the sintered AlN specimens increased monotonically with firing temperature, but the maximum value was limited to 95%. All the specimens containing Y_2O_3, regardless of their Al_2O_3 content, achieved full density by sintering at 1800°C. However, at and above 1900°C, the density decreases with increased Al_2O_3 content.

In the $AlN-Al_2O_3-Y_2O_3$ system, YAG phase formed in the first stage at a temperature below 1600°C, after which aluminum oxynitrides appeared (Kim et al. 1996a, 1996b). In the specimens sintered at 1950°C or higher, $Al_{23}N_5O_{27}$ (γ-phase) and $27R$ phases coexisted.

AlN–AlN polytypoid composite materials were prepared in situ using pressureless sintering of $AlN-Al_2O_3$ mixtures (3.7–16.6 mol% Al_2O_3) with Y_2O_3 (1.4–1.5 mass%) as a sintering additive (Tangen et al. 2004). Materials fired at 1950°C consisted of elongated grains of AlN polytypoids embedded in equiaxed AlN grains. The Al_2O_3 content in the polytypoids varied systematically with the overall Al_2O_3 content, but equilibrium phase composition was not established because of slow nucleation rate and rapid grain growth of the polytypoid grains. The polytypoids $24H$ and $39R$ were identified using high-resolution transmission electron microscopy. Solid solution of Y_2O_3 in the polytypoids was demonstrated, and Y_2O_3 was shown to influence the stability of the AlN polytypoids. A solid solution of Y_2O_3 (about 1 mass%) in the polytypoids was observed.

The solubility limit of Y in aluminum oxynitride was measured using wavelength-dispersive spectroscopy mounted on a SEM from the samples quickly cooled from 1870°C (Miller and Kaplan 2008). It was found to be 1775 ± 128 ppm.

AlN–Y_2O_3. The calculated phase diagram of this quasibinary system is a eutectic type (Figure 4.12) (Medraj et al. 2003b; Fabrichnaya et al. 2013). The eutectic contains 46.7 mol% AlN and crystallizes at 2024°C [47 mol% AlN and crystallizes at 1942°C (Medraj et al. 2003b); crystallizes at 1965°C (Neher et al. 2013)].

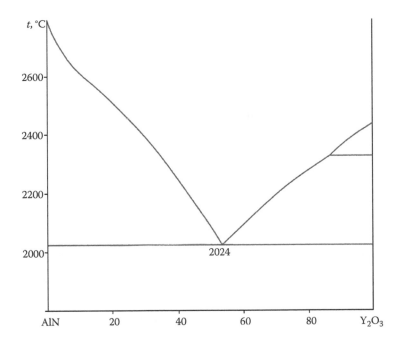

FIGURE 4.12 Calculated phase diagram of the AlN–Y_2O_3 quasibinary system. (Reprinted from *J. Eur. Ceram. Soc.*, 33(13–14), Fabrichnaya, O. et al., Liquid phase formation in the system AlN–Al_2O_3–Y_2O_3: Part II. Thermodynamic assessment, 2457–2463, Copyright (2013), with permission from Elsevier.)

4.20 ALUMINUM–LANTHANUM–SILICON–NITROGEN

The **$La_{17}Si_9Al_4N_{33}$** quaternary compound, which crystallizes in the cubic structure with the lattice parameter $a = 1542.79 \pm 0.06$ pm, is formed in the Al–La–Si–N system (Pilet et al. 2006). Powder samples of this compound were prepared at 1500–1650°C in a graphite furnace under N_2, from La, Si_3N_4, and AlN as starting materials. The powders were mixed, ground in a dry box under Ar, and then placed in Nb tubes that were sealed with parafilm during the transfer to the furnace. In a typical run, the sample was heated to the synthesis temperature over 2 h, kept there for 12 h, then cooled to 1200°C over 4 h, whereupon the furnace was switched off. Yellow-to-orange single crystals were found in preparations at 1650°C.

4.21 ALUMINUM–LANTHANUM–OXYGEN–NITROGEN

The **La_2AlO_3N** and **$LaAl_{12}O_{18}N$** quaternary compounds are formed in the Al–La–O–N system. La_2AlO_3N crystallizes in the tetragonal structure with the lattice parameters $a = 378.9 \pm 0.1$ and $c = 1282.7 \pm 0.3$ pm (Marchand 1976). It was prepared by heating a mixture of AlN and La_2O_3 at 1350°C. Yellow powder of the title compound was obtained, which is stable in ambient air. According to the data of Huang et al. (1990), such compound was not found in the AlN–La_2O_3 system.

$LaAl_{12}O_{18}N$ crystallizes in the hexagonal structure with the lattice parameters $a = 557.0$ and $c = 2202.5$ pm (Wang et al. 1988). It was synthesized by solid-state reactions of the mixtures of La_2O_3, Al_2O_3, and AlN (molar ratio 1:11:2) at 1850°C.

The solubility limit of La in aluminum oxynitride was measured using wavelength-dispersive spectroscopy mounted on a SEM from the samples quickly cooled from 1870°C (Miller and Kaplan 2008). It was found to be 498 ± 82 ppm.

4.22 ALUMINUM–CERIUM–OXYGEN–NITROGEN

The subsolidus phase relationship in the $2AlN–Al_2O_3–Ce_2O_3$ quasiternary system was determined by Sun et al. (1991a, 1991b) and is presented in Figure 4.13. It is seen that this relationship contains three two-phase and five three-phase regions. Two quaternary compounds, **Ce_2AlO_3N (τ_1)** and **$CeAl_{12-x}O_{18}N_{1-x}$ ($0 < x < 1$) (τ_2)**, are formed in this system (Sun et al. 1991a, 1991b; Wang et al. 1988). A second compound crystallizes in the hexagonal structure with the lattice parameters $a = 556.6$ and $c = 2199.4$ pm for $x = 0$ (Wang et al. 1988). It was synthesized by solid-state reactions at high temperature.

4.23 ALUMINUM–PRASEODYMIUM–OXYGEN–NITROGEN

The subsolidus phase relationship in the $2AlN–Al_2O_3–Pr_2O_3$ quasiternary system was determined by Sun et al. (1991a, 1991b), and it is the same as for the $2AlN–Al_2O_3–Ce_2O_3$ quasiternary system (Figure 4.13). Two quaternary compounds, **Pr_2AlO_3N (τ_1)** and **$PrAl_{12-x}O_{18}N_{1-x}$ ($0 < x < 1$) (τ_2)**, are formed in this system (Huang et al. 1990; Sun et al. 1991a, 1991b; Wang et al. 1988). The first of them crystallizes in the tetragonal structure with the lattice parameters $a = 371.5$ and $c = 1260$ pm (Huang et al. 1990). The second compound crystallizes in the hexagonal structure with the lattice parameters $a = 556.8$ and $c = 2199.4$ pm for $x = 0$ (Wang et al. 1988). It was synthesized by solid-state reactions at high temperature.

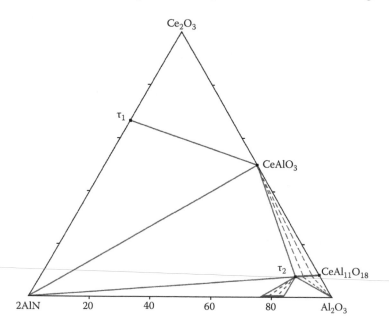

FIGURE 4.13 Subsolidus phase relationships in the $2AlN–Al_2O_3–Ce_2O_3$ quasiternary system. (Reprinted from *J. Solid State Chem.*, 95(2), Sun, W.Y. et al., Subsolidus phase relationships in the systems Re–Al–O–N (where Re = rare earth elements), 424–429, Copyright (1991), with permission from Elsevier.)

4.24 ALUMINUM–NEODYMIUM–OXYGEN–NITROGEN

The subsolidus phase relationship in the $2AlN-Al_2O_3-Nd_2O_3$ quasiternary system at both subsolidus temperatures and at 1700°C were determined by Sun and Yen (1989b) and Sun et al. (1991a, 1991b) and is presented in Figures 4.14 and 4.15. It is seen that the subsolidus relationship contains one two-phase and six three-phase regions (Figure 4.14). At 1700°C, an extensive liquid phase region occurs in the triangle Nd_2O_3–**Nd_2AlO_3N** (τ_1)–$NdAlO_3$ (Figure 4.15). Two quaternary compounds, τ_1 and **$NdAl_{12}O_{18}N$** (τ_2), are formed in this system. Nd_2AlO_3N melts congruently at 1750°C (Huang et al. 1990) and crystallizes in the tetragonal structure with the lattice parameters $a = 372.0 \pm 0.1$ and $c = 1252.0 \pm 0.5$ pm and the calculated and experimental densities of 7.23 and 7.05 g·cm^{-3}, respectively (Marchand 1976) [$a = 370.4$ and $c = 1250.5$ pm (Huang et al. 1990)]. Orange powder of the title compound was prepared by heating a mixture of Al_2O_3 and Nd_2O_3 mixtures at 1250°C using the carbothermal reduction and nitridation route: $Al_2O_3 + 2Nd_2O_3 + 3C + N_2 = 2Nd_2AlO_3N + 3CO$ (Cheviré et al. 2011). It could be also obtained by heating a mixture of AlN and Nd_2O_3 at 1350°C (Marchand 1976).

$NdAl_{12}O_{18}N$ crystallizes in the hexagonal structure with the lattice parameters $a = 556.4$ and $c = 2192.0$ pm (Wang et al. 1988) [$a = 556.4$ and $c = 2200$ pm (Sun and Yen 1989b; Sun et al. 1991a, 1991b)]. This compound is stable at least up to 1850°C under N_2 atmosphere (Sun and Yen 1989b). It was synthesized by solid-state reactions at high temperature (Wang et al. 1988).

$AlN-Nd_2O_3$. The melting behavior of this system was determined by visual observation and by DTA (Huang et al. 1990). The resulting tentative phase diagram is presented in Figure 4.16.

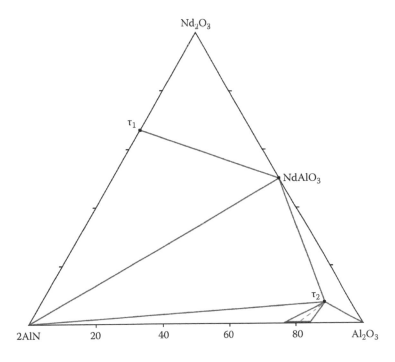

FIGURE 4.14 Subsolidus phase relationships in the $2AlN-Al_2O_3-Nd_2O_3$ quasiternary system. (Reprinted from *Mater. Lett.*, 8(5), Sun, W.Y., and Yen, T.S., Phase relationships in the system Nd–Al–O–lN, 145–149, Copyright (1989), with permission from Elsevier.)

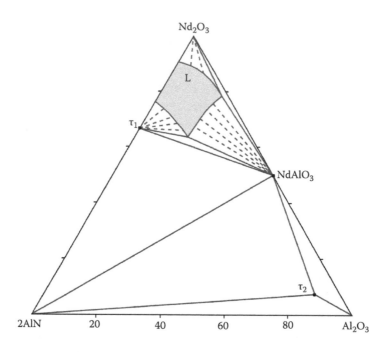

FIGURE 4.15 Isothermal section of the 2AlN–Al_2O_3–Nd_2O_3 quasiternary system at 1700°C. (Reprinted from *Mater. Lett.*, 8(5), Sun, W.Y., and Yen, T.S., Phase relationships in the system Nd–Al–O–N, 145–149, Copyright (1989), with permission from Elsevier.)

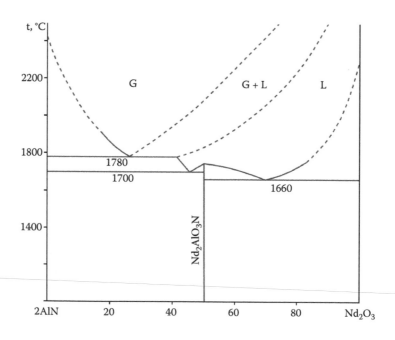

FIGURE 4.16 Tentative phase diagram of the AlN–Nd_2O_3 quasibinary system. (Reprinted from *J. Solid State Chem.*, 85(1)., Huang, Z.K. et al., Compound formation and melting behavior in the *AB* compound and rare earth oxide systems, 51–55, Copyright (1990), with permission from Elsevier.)

4.25 ALUMINUM–SAMARIUM–OXYGEN–NITROGEN

The subsolidus phase relationship in the $2AlN-Al_2O_3-Sm_2O_3$ quasiternary system was determined by Sun et al. (1991a, 1991b), and it is the same as for the $2AlN-Al_2O_3-Nd_2O_3$ quasiternary system (Figure 4.14). Two quaternary compounds, **Sm_2AlO_3N** (τ_1) and **$SmAl_{12}O_{18}N_1$** (τ_2), are formed in this system. Sm_2AlO_3N crystallizes in the tetragonal structure with the lattice parameters $a = 369.0 \pm 0.1$ and $c = 1237.1 \pm 0.5$ pm (Marchand 1976). Thermal analyses carried under air on this compound show that oxidation starts at 600°C and is rapid from 700°C to 800°C (Cheviré et al. 2011). The product of the oxidation is a mixture of $Sm_4Al_2O_9$ with small amounts of $SmAlO_3$ and Sm_2O_3. Yellow powder of the title compound was prepared by heating a mixture of Al_2O_3 and Sm_2O_3 at 1250°C using the carbothermal reduction and nitridation route: $Al_2O_3 + 2Sm_2O_3 + 3C + N_2 = 2Sm_2AlO_3N + 3CO$ (Cheviré et al. 2011). It could be also obtained by heating a mixture of AlN and Sm_2O_3 at 1350°C (Marchand 1976).

$SmAl_{12}O_{18}N$ crystallizes in the hexagonal structure with the lattice parameters $a = 556.7$ and $c = 2194.8$ pm (Wang et al. 1988) [$a = 556.4$ and $c = 2200$ pm (Sun et al. 1991a, 1991b)]. It was synthesized by solid-state reactions at high temperature (Wang et al. 1988).

4.26 ALUMINUM–EUROPIUM–OXYGEN–NITROGEN

The **$EuAl_{12}O_{18}N$** quaternary compound, which crystallizes in the hexagonal structure with the lattice parameters $a = 556.8$ and $c = 2200.9$ pm, is formed in the Al–Eu–O–N system (Wang et al. 1988). It was synthesized by solid-state reactions at high temperature.

$AlN-Eu_2O_3$. The melting behavior of this system was determined by visual observation and by DTA (Huang et al. 1990). The resulting tentative phase diagram is presented in Figure 4.17. **Eu_2AlO_3N** quaternary compound, which melts incongruently at 1400°C and crystallizes in the tetragonal structure with the lattice parameters $a = 368.2$ and $c = 1238$ pm, is formed in this system (Huang et al. 1990).

4.27 ALUMINUM–GADOLINIUM–OXYGEN–NITROGEN

The **$GdAl_{12}O_{18}N$** quaternary compound, which crystallizes in the hexagonal structure with the lattice parameters $a = 556.1$ and $c = 2179.3$ pm, is formed in the Al–Gd–O–N system (Wang et al. 1988). It was synthesized by solid-state reactions at high temperature.

4.28 ALUMINUM–CARBON–SILICON–NITROGEN

$AlN-Al_4C_3-SiC$. The thermodynamic properties of this quasiternary system were assessed using the Calculation of Phase Diagrams (CALPHAD) approach (Pavlyuchkov et al. 2012), and the isothermal section at 1860°C was constructed (Figure 4.18). The isothermal section of the $AlN-Al_4C_3-SiC$ system at 1860°C was also constructed by Schneider et al. (1979), and Oden and McCune (1990) constructed the isothermal sections of the **$4AlN + 3SiC$** ⇔ **$Al_4C_3 + Si_3N_4$** ternary mutual system at 1760°C and 1860°C. **Al_5C_4SiN** (τ_1) and **Al_9C_7SiN** (τ_2) quaternary compounds are formed in the $AlN-Al_4C_3-SiC$ system. The first of them decomposes at about 2265°C (Oden and McCune 1990) and crystallizes in the

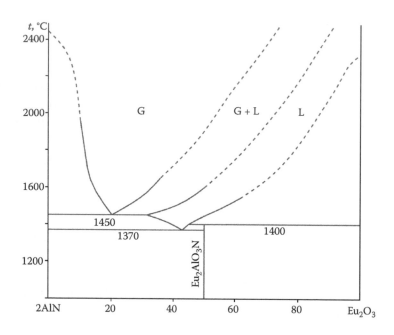

FIGURE 4.17 Tentative phase diagram of the $AlN–Eu_2O_3$ quasibinary. (Reprinted from *J. Solid State Chem.*, 85(1)., Huang, Z.K. et al., Compound formation and melting behavior in the AB compound and rare earth oxide systems, 51–55, Copyright (1990), with permission from Elsevier.)

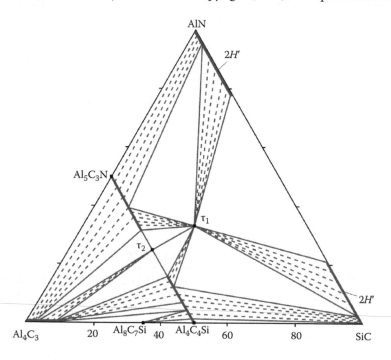

FIGURE 4.18 Calculated isothermal section of the $AlN–Al_4C_3–SiC$ quasiternary system at 1860°C. (With kind permission from Springer Science+Business Media: *J. Phase Equilibr. Dif.*, Thermodynamic assessments of the $Al_2O_3–Al_4C_3–AlN$ and $Al_4C_37–AlN–SiC$ systems, 33(5), 2012, 357–368, Pavlyuchkov, D. et al.)

hexagonal structure with the lattice parameters $a = 325$ and $c = 4017.7$ pm (Schneider et al. 1979). The homogeneity region of τ_1 is rather small (Schneider et al. 1979).

AlN–SiC. Phase relationships in this system were determined by analytical electron microscopy of local equilibria among adjacent phases in hot-pressed samples and in diffusion couples (Zangvil and Ruh 1988). The tentative phase diagram was constructed and is presented in Figure 4.19. At 2100–2300°C, a 4H–2H equilibrium exists, the 4H field extending from ca. 2 mol% AlN to an upper limit of 11 and 14 mol% AlN. The wurtzite-type 2H solid solution extends from an impurity-sensitive lower limit of 17–24 mol% AlN up to 100 mol% AlN. The existence of a miscibility gap below 1950°C was confirmed, but its limits were not determined accurately. A faulted metastable cubic phase (β′) exists below ca. 2000°C and contains up to ≈4 mol% AlN. The transformation into the stable α structure occurs through diffusion-controlled rearrangements. The calculation of Pavlyuchkov et al. (2012) indicated that the separation of the 2H solid solution starts at 2150°C.

According to the data of Rafaevich et al. (1990), homogenization of the 2H solid solution takes place by the diffusion of SiC in AlN. Mechanical activation of AlN and SiC by quasihydrostatic compression promotes the formation of the 2H solid solutions in AlN–SiC system (Ordan'yan et al. 1997).

High-density SiC–AlN compositions were fabricated from powder mixtures by hot-pressing in the 1700–2300°C temperature range (Ruh and Zangwil 1982; Zangvil and Ruh 1985). At 2100°C, a 2H solid solution was found from ≈35 to 100 mass% AlN. Lattice parameters closely followed Vegard's law (Ruh and Zangwil 1982). For compositions with

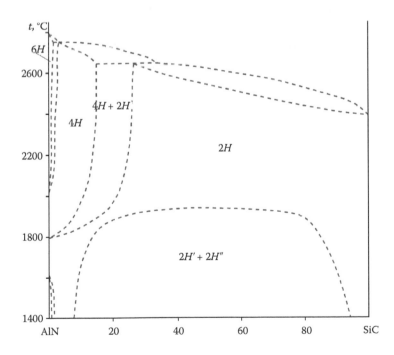

FIGURE 4.19 Tentative phase diagram of the AlN–SiC quasibinary system. (Zangvil, A., and Ruh, R.: Phase relationships in the silicon carbide–aluminum nitride system. *J. Amer. Ceram. Soc.* 1988. 71 (10). 884–890. Copyright Wiley-VCH Verlag GmbH & Co. KGaA. Reproduced with permission.)

<35 mass% AlN, multiphase assemblages were found. Approximate diffusion depths range from 0.5 µm at 1950°C to 15 µm at 2300°C (both for 1 h), and the corresponding diffusion coefficients are of the order of from $1.5 \cdot 10^{-12}$ to $6 \cdot 10^{-10}$ cm$^2 \cdot$s^{-1} (Zangvil and Ruh 1985). It has been suggested that low diffusion coefficients not allow the formation of the solid solutions by heating a mixture of the powdered solid components (Cutler et al. 1978; Rafaniello et al. 1981). In contrast, it was found that pressing compositions of ≥35 mol% AlN at ≥2100°C resulted in $2H$ solid solutions and multiphase solid-solution assemblages were found for compositions of <35 mol% AlN. However, all samples featured residual inhomogeneities in both grain size and composition and were evidently far from having reached equilibrium.

Cutler et al. (1978) obtained $2H$ solid solutions with 2–100 mol% AlN through the carbothermal reduction at 1600°C of fine SiO$_2$ and Al$_2$O$_3$ with a carbon source and under N$_2$ atmosphere. The hot-pressing of these powders also resulted in $2H$ solid solutions through the full range of compositions (Rafaniello et al. 1981), but a later report of Rafaniello et al. (1983) indicated that the phase separation occurred upon annealing at 1700°C, and it suggested the existence of a miscibility gap at the low-temperature range.

$2H$ solid solutions decompose into modulated structures (Kuo et al. 1987; Kuo and Virkar 1990). The precipitates are lenticular in shape with a circular cross section in the basal plane. These solid solutions were fabricated by hot-pressing powder mixtures in graphite dies. The samples were annealed between 1600°C and 2300°C for up to 1000 h.

According to the data of Tsukuma et al. (1982), **Al$_4$C$_3$Si$_3$N$_4$** and **Al$_5$C$_3$Si$_3$N$_5$** were obtained by reacting AlN, Al$_4$C$_3$, and Si$_3$N$_4$ at 1800°C for 1 h under Ar at 10 MPa. Both compounds crystallize in the hexagonal structure with the lattice parameters a = 310.7 and c = 500.8 pm for the first compound and a = 301.8 and c = 500.5 pm for the second one. These compounds were later interpreted to be the compositions in the continuous solid-solution range with 57.1 and 62.5 mol% AlN, respectively (Zangvil and Ruh 1984).

High-purity $2H$ solid solutions with wurtzite-type structure were also obtained by a reaction of powdered AlN, Si, and C at 1900–2000°C (Avignon-Poquillon et al. 1990; Bu et al. 2002). Heat capacities of these solid solutions were measured at constant pressure between 40°C and 260°C, and also the enthalpy changes between 25°C and 180°C (Avignon-Poquillon et al. 1990). It was observed that addition of small amounts of SiC to AlN resulted in c_p and ΔH values close to those of SiC.

The study of polytypism of $2H$ solid solutions showed that the concentration of AlN significantly affects their polytypic structure (Safaraliev et al. 1991, 1993).

The microstructure and toughening of AlN–SiC ceramics from powder mixtures of β-SiC, AlN, and α-SiC by hot-pressing were studied in the 1870–2030°C temperature range by Lim (2010). The reaction of AlN and β-SiC powders causing transformation to the $2H$ (wurtzite) structure appeared to depend on the hot-pressing temperature and the seeding with α-SiC. For the compositions of 49 mass% AlN + 49 mass% SiC with 2 mass% α-SiC and 47.5 mass% AlN + 47.5 mass% SiC with 5 mass% α-SiC heated at 2030°C for 1 h, complete solid solutions with a single phase of $2H$ could be obtained. An appreciable amount of α-SiC seeding could develop columnar intergrains of the $4H$ phase and the stable $2H$ phase with a relatively uniform composition and grain size distribution. The seeding effect of α-SiC on the present phases and compositional microstructures with columnar intergrains was investigated in detail. The mechanical properties of the hot-pressed solid solutions were evaluated in terms of fracture toughness and Vickers hardness.

The effect of AlN on the stabilization of the $2H$ structure of SiC was investigated by furnace heating under vacuum and in field-activated combustion synthesis of powder compacts of $Si_3N_4 + 4Al + 3C$ and by furnace heating compacts of $xAlN + Si + C$, where x was either 0, 10, or 57 mol% (Carrillo-Heian et al. 2000). Stoichiometric mixtures of Si and C with AlN additives were heated at 1500°C for 20 min. In the absence of AlN, the product was cubic β-SiC. With 10 mol% AlN added, the reaction did not go to completion and the resulting SiC had the cubic structure. With 57 mol% AlN as an additive, the reaction to form SiC was even less complete. Heating of compacts of $Si_3N_4 + 4Al + 3C$ for 30 min at temperatures of 1130°C and 1400°C resulted in incomplete conversion to AlN and cubic β-SiC. At 1600°C, the conversion was nearly complete and the product contained the same two phases, with a trace of h-SiC. When heating was done at 1650°C, the product contained two regions, an outer layer which was analyzed to be β-SiC and an inner region which contained the cubic phase of SiC and h-AlN. At this temperature, the previously formed AlN phase decomposed and evaporated. When composites and solid solutions were synthesized by a field-activated self-propagating combustion, the products were AlN and β-SiC, or a solid solution of the two phases, depending on the electric field applied. When these samples were heated under vacuum at 1700°C, the end product contained only β-SiC. Field-activated combustion synthesis seems to be a very successful technique for making h-SiC in the presence of AlN, compared with furnace heating under vacuum.

4.29 ALUMINUM–CARBON–TITANIUM–NITROGEN

This quaternary system was investigated using powder methods and XRD by Pietzka and Schuster (1996) and reviewed by Raghavan (2006a). Phase equilibria at 1375°C were presented in an isothermal network for alloys up to 50 at% Ti. In the composition range studied, the following four-phase spaces were identified: (Al) + Ti(C,N) + AlN + Al_4C_3, (Al) + Ti(C,N) + AlN + $TiAl_3$, $TiAl_3$ + Ti_5Al_{11} + Ti_3AlC_2 + $Ti_2AlC_xN_{1-x}$, and Ti(C,N) + Ti_4AlN_{3-x} + Ti_3AlC_2 + $Ti_2AlC_xN_{1-x}$. A tentative network of three-phase and four-phase spaces for the composition region studied at 1375°C was proposed. In the vertical section Ti_2AlC_{1-x}–Ti_2AlN_{1-x}, a complete series of the solid solutions exist at 1495°C, but a wide miscibility gap occurs at 1375°C. The vertical section Ti_3AlC_{1-x}–Ti_3AlN_{1-x} is more complex because of the occurrence of the quaternary tetragonally distorted phase $Ti_3Al(C,N)_{1-x}$ ($a = 411.35 \pm 0.04$ and $c = 413.66 \pm 0.05$ pm) and the transformation of the perovskite-type Ti_3AlN_{1-x} into filled Re_3B-type Ti_3AlN_{1-x} below 1200°C.

The cold-pressed pellets of the appropriate mixtures of the Al–C–Ti–N system were sintered in H_2 atmosphere at 1075°C for 333 h, 1175°C for 144 h, 1275°C for 62 h, 1375°C for 20 h, and 1485°C for 15 h (Pietzka and Schuster 1996). To obtain $Ti_2AlC_{0.5}N_{0.5}$ solid solution, AlN, Al_4C_3, Ti, and C powders were mixed, cold-pressed, and sealed under vacuum on Pyrex glass tubes (Barsoum et al. 2000). The latter were placed in a hot isostatic press and heated at 10°C·min^{-1} to 850°C and held at this temperature for 30 min. The chamber was pressurized with Ar to 25 MPa, and the heating was resumed with initial rate to the final processing temperature (1300°C) at which time the pressure increased to ca. 40 MPa for 15 h. This solid solution crystallizes in the hexagonal structure with the lattice parameters $a = 302.1$ and $c = 1361.0$ pm (Barsoum et al. 2000) [$a = 302.3$ and $c = 1361.0$ pm and an experimental density of 4.2 g·cm^{-3} (Radovic et al. 2008)].

According to the data of Yeh et al. (2010), the preparation of Ti$_2$AlC$_{0.5}$N$_{0.5}$ solid solution from Al$_4$C$_3$-containing powder compacts was conducted by the self-propagating high-temperature synthesis under N$_2$ pressure of 0.42–1.82 MPa. The molar proportion of three reactant powders was formulated as Ti/Al$_4$C$_3$/Al (or AlN) = 2:1/6:1/3. The increase of N$_2$ pressure augments the combustion temperature and thus accelerates the reaction front.

Manoun et al. (2007) noted that **Ti$_3$AlCN** quaternary compound, which is stable to pressure ca. 50 GPa and crystallizes in the hexagonal structure with the lattice parameters $a = 304.4 \pm 0.8$ and $c = 1841.4 \pm 0.6$ pm, is formed in the Al–C–Ti–N system. The experimental density of this compound is equal to 4.5 g·cm^{-3} (Radovic et al. 2008). Bulk polycrystalline samples of the title compound were fabricated by mixing AlN, Ti, and graphite powders in the proper stoichiometric composition. The mixtures were ball-milled for ≈30 min and cold-pressed using a pressure of ≈600 MPa. The cylindrical pellets were presintered in a vacuum hot press which was heated at a rate of circa 5°C·min^{-1} under vacuum at 525°C, held at that temperature for 2 h, then heated at 4°C·min^{-1} to 625°C and held there for 10 h. The presintered pellets were then placed in a hot isostatic press and heated at 5°C·min^{-1} to 850°C, then at 2°C·min^{-1} to 1000°C, at which time the chamber was pressurized with Ar to ≈80 MPa. The heating was resumed at a rate of 10°C·min^{-1} to 1400°C and held at that temperature for 10 h before the furnace cooling. The chamber pressure at 1400°C was ≈100 MPa.

4.30 ALUMINUM–CARBON–OXYGEN–NITROGEN

AlN–Al$_4$C$_3$–Al$_2$O$_3$. The thermodynamic properties of this quasiternary system were assessed by modeling the Gibbs energy of the various phases (Qiu and Metselaar 1997; Pavlyuchkov et al. 2012). The projection of the liquidus surface according to the thermodynamic calculations is given in Figure 4.20 (Qiu and Metselaar 1997). On each liquidus surface, the liquid is in equilibrium with only one solid phase; thus, only one solid phase is marked in each area, except the area with liquid + carbon (i.e., aluminum-rich liquid and graphite) that is caused by the decomposition of Al$_4$C$_3$. That solid phase will precipitate first during solidification when the sample composition falls in this area. The liquidus of the $2H$ phase notably covers a wide area; this is because the $2H$ [(AlN)$_x$(Al$_2$OC)$_{1-x}$] phase has been modeled as a complete solid solution between Al$_2$OC and AlN, and its liquidus extends from the Al$_2$O$_3$–Al$_4$C$_3$ side to the AlN–Al$_4$C$_3$ side. Hence, the liquidus of the $2H$ phase has contact with all other liquidi. All of the liquid has solidified after the eutectic reaction L ⇔ Al$_2$O$_3$ + $2H$ + Al$_4$O$_4$C at 1838°C. The evaluation of the AlN–Al$_4$C$_3$ liquid was based on very limited information. Therefore, the calculated liquidus projection should be regarded as an approximation.

A series of isothermal sections have been also calculated by Qiu and Metselaar (1997) at temperatures in the range of 1000–2100°C and are presented in Figure 4.21. The obtained isothermal sections show that AlN extends in the quasiternary system to form

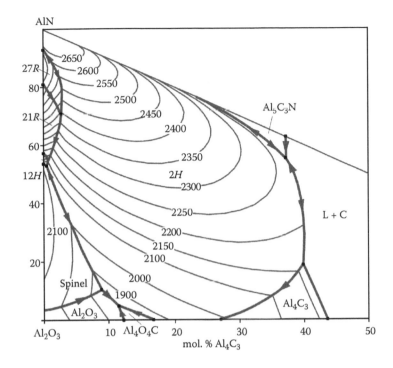

FIGURE 4.20 Calculated liquidus surface of the AlN–Al$_4$C$_3$–Al$_2$O$_3$ quasiternary system. (Qiu, C., and Metselaar, R.: Phase relations in the aluminum carbide–aluminum nitride–aluminum oxide system. *J. Amer. Ceram. Soc.* 1997. 80(8). 2013–2020. Copyright Wiley-VCH Verlag GmbH & Co. KGaA. Reproduced with permission.)

the solid-solution phase 2*H*, and the solubility of Al$_2$OC in the 2*H* phase increases as the temperature increases until the complete solution forms between AlN and Al$_2$OC in the range of 1710–1990°C, where Al$_2$OC becomes stable. After Al$_2$OC melts at 1990°C, the liquid becomes a dominant phase in the Al$_2$O$_3$-rich corner at higher temperatures. The calculated section at 1800°C shows that Al$_5$C$_3$N exists only in the three-phase Al$_5$C$_3$N + 2*H* +Al$_4$C$_3$ and two-phase Al$_5$C$_3$N + 2*H* regions. These regions occupy only a small part of the entire composition triangle.

The existence of the homogeneous solid solution in the AlN–Al$_2$OC system at elevated temperatures was also determined by Kuo and Virkar (1989, 1990) and Chen et al. (1993). Below about 1800°C, this solid solution is unstable and decomposes into two solid solutions (Kuo and Virkar 1989). Samples of near equimolar composition form disk-shaped precipitates upon annealing at temperatures below 1900°C (Chen et al. 1993). According to the data of Kuo and Virkar (1989), modulated structures occur in this system, and the modulations exist in the [001] direction. The precipitates are lenticular in shape with a circular cross section in the basal plane.

A thermodynamic dataset for the AlN–Al$_4$C$_3$–Al$_2$O$_3$ was reassessed based on available literature data using the CALPHAD approach by Pavlyuchkov et al. (2012), and the isothermal sections at 1600°C, 2000°C, and 2100°C were reconstructed with some corrections.

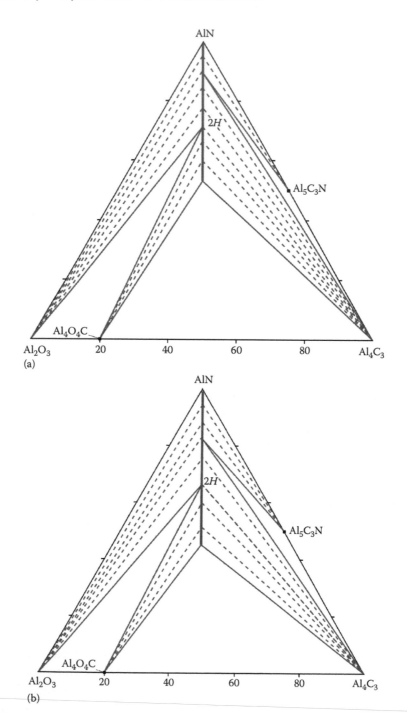

FIGURE 4.21 Calculated isothermal section of the AlN–Al$_4$C$_3$–Al$_2$O$_3$ quasiternary system at (a) 1000°C and (b) 1200°C. (*Continued*)

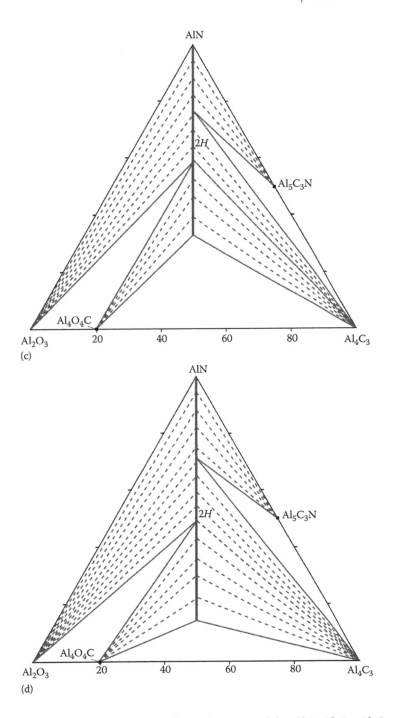

FIGURE 4.21 (CONTINUED) Calculated isothermal section of the AlN–Al$_4$C$_3$–Al$_2$O$_3$ quasiternary system at (c) 1400°C and (d) 1600°C. *(Continued)*

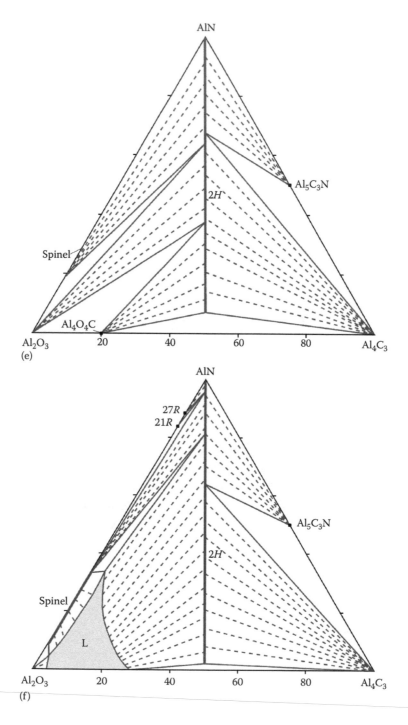

FIGURE 4.21 (CONTINUED)   Calculated isothermal section of the AlN–Al$_4$C$_3$–Al$_2$O$_3$ quasiternary system at (e) 1800°C and (f) 2000°C. *(Continued)*

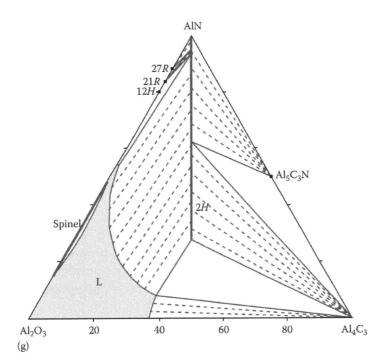

FIGURE 4.21 (CONTINUED)   Calculated isothermal section of the AlN–Al$_4$C$_3$–Al$_2$O$_3$ quasiternary system at (g) 2100°C. (Qiu, C., and Metselaar, R.: Phase relations in the aluminum carbide aluminum nitride–aluminum oxide system. *J. Amer. Ceram. Soc.* 1997. 80(8). 2013–2020. Copyright Wiley-VCH Verlag GmbH & Co. KGaA. Reproduced with permission.)

Some quaternary compounds were obtained in the AlN–Al$_4$C$_3$–Al$_2$O$_3$ quasiternary system. The **"AlCON"** phase was determined by Tabary et al. (2000). Its melting temperature is higher than 1950°C. The molar composition of this phase linearly varies between 25% Al$_2$O$_3$ + 75% AlN and 60% Al$_2$O$_3$ + 20% AlN + 20% Al$_4$C$_3$. It crystallizes in the hexagonal structure with the lattice parameters a = 315.3 and c = 498.0 pm.

Al$_4$CN$_3$O quaternary compound crystallizes in the orthorhombic structure with the lattice parameters a = 574.31 ± 0.05, b = 852.8 ± 0.1, and c = 909.4 ± 0.1 pm (Zhang and Tanaka 2003). To prepare the title compound, the mixed powder of AlN and Al was pressed into pellet. The pellet was put in a BN crucible and then heated with a graphite susceptor in a radio frequency furnace. The mixture was first heated to about 1500°C in flowing N$_2$ atmosphere, kept there for 8–12 h, then cooled slowly (about 40°C·h^{-1}) down to 500°C. Finally, the power was switched off and the melt allowed to cool down to room temperature. Single crystals with needle shape were obtained on the surface of the Al matrix and on the wall of the crucible. The crystals were transparent to light red in color.

An aluminum oxycarbonitride, **Al$_5$(O$_x$C$_y$N$_{4-x-y}$)** (x ~ 1.4 and y ~ 2.1), has been synthesized and characterized by XRD, TEM, and electron energy loss spectroscopy (Inuzuka et al. 2010). It crystallizes in the hexagonal structure with the lattice parameters a = 328.455 ± 0.006 and c = 2159.98 pm and the calculated density of 3.12 g·cm^{-3}. This compound was synthesized as follows. AlN and Al$_4$C$_3$ were mixed in molar ratio 1:1.5. The well-mixed chemicals were pressed into pellets, loaded into an open carbon crucible, and heated at 1900°C for 1 h in an Ar atmosphere with the next cooling to ambient temperature by cutting furnace power. The procedures of mixing and heating were repeated twice. The reaction product was a slightly sintered polycrystalline material.

$Al_{28}C_6O_{21}N_6$ quaternary compound crystallizes in the rhombohedral structure with the lattice parameters $a = 589.7 \pm 0.2$ and $\alpha = 55.15° \pm 0.04°$ or $a = 545.9 \pm 0.2$, $c = 1495.2 \pm 0.6$ pm (in hexagonal setting) and the calculated and experimental densities of 3.13 and 3.11 ± 0.05 g·cm^{-3}, respectively (Groen et al. 1995a, 1995b). For obtaining this compound, AlN, Al_2O_3, and Al_4C_3 were mixed in the appropriate ratios in an agate mortar. From the resultant powder mixture, pellets were pressed at 5 MPa. These pellets were heated in a furnace under Ar flow. All experiments were done using an h-BN crucible. The heating rate was 20°C·min^{-1} to the set point temperature minus 100°C. A rate of 10°C·min^{-1} was used between the aforementioned point and the set point to prevent overshoot. The cooling rate was approximately 20°C·min^{-1} down to 1000°C. The powder was prepared in two heating cycles with intermediate grinding. In the first cycle, the powder was heated for 2 h at 1600°C, and in the second cycle for 4 h at 1650°C.

4.31 ALUMINUM–SILICON–TITANIUM–NITROGEN

The phase diagram of the Al–Si–Ti–N quaternary system was calculated at 630°C (Bhansali et al. 1990). Interior tie lines exist between AlN and titanium silicides and titanium aluminum silicides. Additionally, interior tie lines emanate from Ti_5Si_3 to the titanium aluminum nitrides. It is apparent from the quaternary phase diagram that both AlN and TiN are stable phases in this system. Tie planes indicate AlN–TiN–TiSi$_x$, AlN–TiN–Si$_3$N$_4$, and AlN–TiN–Si three-phase equilibria. Although TiN is stable with the titanium silicides and Si$_3$N$_4$, Al is not stable with any of these phases.

4.32 ALUMINUM–SILICON–OXYGEN–NITROGEN

4AlN + 3SiO$_2$ ⇔ 2Al$_2$O$_3$ + Si$_3$N$_4$. The liquidus surface of this ternary mutual system was calculated for the first time by Hillert and Jonsson (1992). Then, two sets of self-consistent data for this system were favored: one with zero reciprocal reaction energy and the other with nonzero energy for the β-SiAlON phase (Mao and Selleby 2007). According to these datasets, two sets of isothermal sections, liquidus projections and the thermochemical properties of the β-SiAlON phase, were presented. The agreement with the experimental data is good. The liquidus surface calculated with nonzero energy for the β-SiAlON phase is presented in Figure 4.22.

The isothermal sections of the 4AlN + 3SiO$_2$ ⇔ 2Al$_2$O$_3$ + Si$_3$N$_4$ ternary mutual system at 1650°C, 1700°C, 1750°C, and 1800°C calculated with nonzero energy for the β-SiAlON phase is given in Figure 4.23 (Mao and Selleby 2007).

The phase relations in a series of isothermal sections of this system were also determined by Gauckler et al. (1975) (at 1760°C), Jack (1976) and Naik et al. (1978) (at 1750°C), Kaufman (1979) (at 1930°C), Dörner et al. (1981) (at 1630°C and 1830°C), Hillert and Jonsson (1992) (at 1600°C, 1650°C, 1700°C, 1750°C, 1800°C, 1900°C, 2000°C, and 2100°C),

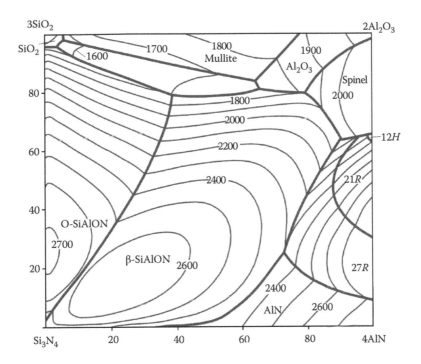

FIGURE 4.22 Calculated liquidus surface of the $4AlN + 3SiO_2 \Leftrightarrow 2Al_2O_3 + Si_3N_4$ ternary mutual system. (Reprinted from *CALPHAD: Comput. Coupling Phase Diagr. and Thermochem.*, 31(2), Mao, H., and Selleby, M., Thermodynamic reassessment of the Si_3N_4–AlN–Al_2O_3–SiO_2 system – Modeling of the SiAlON and liquid phase, 269–280, Copyright (2007), with permission from Elsevier.)

and Dumitrescu and Sundman (1995) (at 1600°C, 1650°C, 1700°C, 1750°C, and 1900°C). The region of the solid solution in the $4AlN + 3SiO_2 \Leftrightarrow 2Al_2O_3 + Si_3N_4$ system at 1730°C was investigated by Oyama (1972, 1974). The boundary between the α- and β-SiAlONs at various temperatures was determined in Sun et al. (1987).

Some compounds and phases exist in this system (Metselaar 1998). **$Al_xSi_{2-x}N_{2-x}O_{1+x}$** (0.04 ≤ x ≤ 0.40), O-SiAlON, crystallizes in the orthorhombic structure with the lattice parameters, which depend on the composition, $a = (888.07 \pm 0.13)$–(892.54 ± 0.05), $b = (549.65 \pm 0.04)$–(549.88 ± 0.04), and $c = (485.50 \pm 0.06)$–(485.96 ± 0.02) pm and a calculated density of 2.81 ± 0.01 g·cm^{-3} (Lindqvist et al. 1991) [$a = 550.0 \pm 0.3$, $b = 890.4 \pm 0.5$, and $c = 486.1 \pm 0.3$ pm for $x = 0.2$ (Trigg and Jack 1987); $a = 550.0$, $b = 890.5$, and $c = 486.1$ pm for $x = 0.2$ (Bergman et al. 1991)]. $Al_xSi_{2-x}N_{2-x}O_{1+x}$ transforms to high-temperature phase at 1350°C without a significant mass change and without the appearance or disappearance of a secondary phase (Bowden et al. 1998).

The samples of the title phase were prepared by mixing α-Si_3N_4, SiO_2, and Al_2O_3 (Lindqvist et al. 1991). Mixing/milling was carried out in an agate ball mill with ethanol and pressing agents. The dried powder mixture was then formed and cold isostatic-pressed at 275 MPa, and the remaining ethanol and pressing agents were driven off at 300°C. The powder compacts were fired at 1820°C for 30 min under a protective powder (preoxidized Si_3N_4). The sintered samples were crushed to powder and etched in hot 1 M NaOH and hot

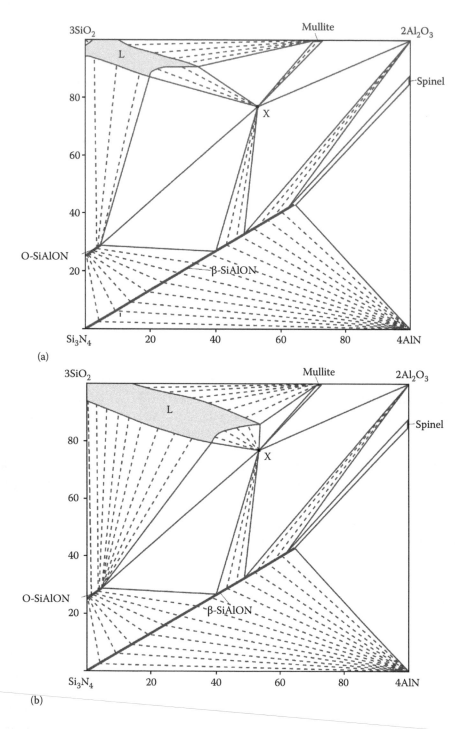

FIGURE 4.23 Calculated isothermal sections of the $4AlN + 3SiO_2 \Leftrightarrow 2Al_2O_3 + Si_3N_4$ ternary mutual system at (a) 1650°C, (b) 1700°C. *(Continued)*

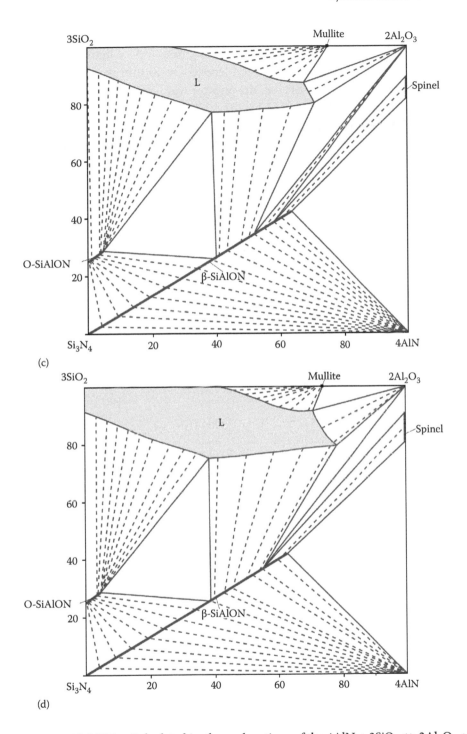

FIGURE 4.23 (CONTINUED)   Calculated isothermal sections of the $4AlN + 3SiO_2 \Leftrightarrow 2Al_2O_3 + Si_3N_4$ ternary mutual system at (c) 1750°C, and (d) 1800°C. (Reprinted from *CALPHAD: Comput. Coupling Phase Diagr. and Thermochem.*, 31(2), Mao, H., and Selleby, M., Thermodynamic reassessment of the Si_3N_4–AlN–Al_2O_3–SiO_2 system – Modeling of the SiAlON and liquid phase, 269–280, Copyright (2007), with permission from Elsevier.)

aqua regia to remove remaining glass and metal impurities and finally washed in hot deionized water. They could be also obtained using mixtures of light kaolin, Si, and SiO_2 (Bowden et al. 1998). These mixtures were blended for 20 min in a ball mill using Si_3N_4 media and hexane solvent. After drying, the powder was lightly pressed (8 MPa) into disks and heated under flowing H_2 (10%) in N_2 for 8 h at 1270°C.

$Al_xSi_{6-x}N_{8-x}O_x$ ($x = 1$–4) (β-SiAlON) [according to the data of Jack and Wilson (1972), the composition of this phase is $Al_{0.67x}Si_{6-0.75x}N_{8-x}O_x$ at $0 < x < 5$] crystallizes in the hexagonal structure with the lattice parameters, which show a linear dependence of both a and c on x: a (pm) $= 760.3 + 2.74x$ and c (pm) $= 290.7 + 2.45x$ (Liang et al. 1999) [a (pm) $= (760.3 \pm 0.6) + (2.96 + 0.04)x$ and c (pm) $= (290.7 \pm 0.8) + (2.55 \pm 0.06)x$ (Ekström et al. 1989); a (pm) $= (760.58 \pm 0.24) + (2.694 + 0.010)x$ and c (pm) $= (290.82 \pm 0.10) + (2.44 \pm 0.04)x$ (Loong et al. 1996)]. The lattice parameters of this phase for $x = 1$–4 are presented in Table 4.1.

TABLE 4.1   Lattice Parameters of the $Al_xSi_{6-x}N_{8-x}O_x$ Phase

x	a (pm)	c (pm)	Calculated Density (g·cm⁻³)	References
1	763.6 ± 0.1	293.70 ± 0.05	3.151	Mitomo et al. (1978)
2	766.3 ± 0.1	296.30 ± 0.05	3.112	
3	768.5 ± 0.1	299.50 ± 0.05	3.070	
4	771.6 ± 0.1	300.50 ± 0.05	3.045	
1	761.8 ± 0.1	295.0 ± 0.1	3.152	Mitomo and Kuramoto (1979)
2	765.8 ± 0.1	295.6 ± 0.1	3.123	
3	767.7 ± 0.1	299.2 ± 0.1	3.080	
4	771.5 ± 0.1	300.5 ± 0.1	3.046	
2	764.9 ± 0.3	295.0 ± 0.3		Gillott et al. (1981)
	764.3 ± 0.4[a]	294.5 ± 0.3[a]		
2.9	766.8 ± 0.3	296.6 ± 0.3		
	766.4 ± 0.3[a]	296.3 ± 0.3[a]		
4	769.5 ± 0.4	299.5 ± 0.4		
	769.2 ± 0.3[a]	299.0 ± 0.3[a]		
1	763.3	292.4		Takase et al. (1982)
2	765.7	295.4		
3	768.4	297.4		
4	770.4	300.2		
1	763.0	292.9		Takase and Tani (1984)
2	766.1	295.6		
3	769.1	298.2		
4	770.5	299.8		
0.3	761.29 ± 0.03	291.40 ± 0.02		Shen et al. (1999)

[a] At 4 K.

XRD of β-SiAlON shows that β-Si_3N_4 can take into solid solution relatively large amounts of Al and oxygen (up to about $x = 4.2$) without evidence of any structural changes (Wild et al. 1978). The only observable change is a general increase in the a and c dimensions of the hexagonal cell with increasing x. However, the results of infrared studies indicate that structural changes are occurring.

Analysis of the results of high-temperature (1300–1500°C) synthesis of β-SiAlON showed that different compositions correspond to different morphological forms—fibers and spheroids (Vorobyev and Shveikin 2000). For the fibers $x = 3.58$ and for the spheroids $x = 2.60$.

Enthalpies of formation were determined for β-SiAlON phases ($x = 0.46$ to 3.6) by high-temperature oxidative drop solution calorimetry using an alkali-metal borate (52 mass% $LiBO_2$ + 48 mass% $NaBO_2$) solvent (Liang et al. 1999). Oxygen gas was bubbled through the melt to accelerate oxidation of the oxynitride samples during dissolution. β-SiAlON near $x = 2$ appears less stable energetically than ones with higher or lower nitrogen content. This phase was prepared as follows (Yi et al. 2010). First, powders of $Al_xSi_{6-x}N_{8-x}O_x$ were combustion-synthesized from well-mixed powders of Si, Al, and SiO_2 under N_2 pressure of 1 MPa using $Al_{1.14}Si_{4.86}N_{6.86}O_{1.14}$ as the diluent. Then, the synthesized powders were subjected to spark plasma sintering in vacuum for 12 min under a stress of 50 MPa and a temperature of 1600°C, without any sintering additives. XRD revealed that a small amount of unreacted phase was found in the combustion-synthesized powders; however, pure β-SiAlON compacts were successfully obtained after spark plasma sintering. It could be also prepared if the powders of AlN, Al_2O_3, and Si_3N_4 were mixed and the mixture was hot-pressed at 1750–1850°C for 1 h under a pressure of 0.2–0.3 GPa (Mitomo et al. 1978; Mitomo and Kuramoto 1979; Takase and Tani 1984; Ekström et al. 1989). The pressure was applied from room temperature. The rate of temperature increasing was kept to 30°C·min^{-1}. Microstructural studies revealed that β-SiAlON with $x = 4$ was the most homogeneous.

$Al_xSi_{3-x}N_{4-x}O_x$ ($0 \leq x \leq 5$) solid solutions, which crystallize in the cubic structure with the lattice parameter $a = 782.34 \pm 0.03$ pm and a calculated density of 3.92 g·cm^{-3} for $x = 1$ (Schwarz et al. 2002). To prepare $AlSi_2N_3O$, a powder mixture of Si_3N_4 (66.61 mass%), AlN (11.40 mass%), and Al_2O_3 (21.99 mass%) was prepared. This powder was attrition-milled in 2-propanol for 2 h with high-purity Si_3N_4 milling medium in a Teflon-coated jar. The slurry was dried in a polyethylene beaker under a halogen lamp while being stirred. Finally, charges of the powder were uniaxially hot-pressed (30 MPa) in a graphite furnace (1775°C, 2 h, N_2 atmosphere).

β-$AlSi_5N_7O$ was combustion-synthesized by slowly heating the raw materials of Al, Si, and SiO_2 under a constant N_2 pressure of 1 MPa (Yi et al. 2013). The temperature record showed several exothermic and endothermic reactions during the entire process and two main exothermic reactions at ~800°C and ~1350°C. The measured ignition temperature was approximately 1350°C. The total enthalpy of reaction estimated from the two exothermic reactions was 166.53 kJ·mol^{-1}.

$Al_{2.83}Si_{3.17}N_{5.17}O_{2.83}$ crystallizes in the hexagonal structure with the lattice parameters $a = 297.69 \pm 0.01$ and $c = 767.85 \pm 0.01$ pm (Navrotsky et al. 1997). Thermochemical measurements gave the standard enthalpy of formation: -67.8 ± 15.5 kJ·mol^{-1} from the compounds and -2768.3 ± 8.2 kJ·mol^{-1} from the elements. To prepare this SiAlON, starting powders of Si_3N_4 and Al_2O_3 and a low oxygen content AlN were placed in a graphite die. The synthesis involved a resistance heating to 1800°C (6 min) to form a dense disk.

$Al_{1+x}Si_{3-x}N_{5-x}O_x$ ($x \approx 2.2$) (8H-SiAlON) also crystallizes in the hexagonal structure with the lattice parameters $a = 298.877 \pm 0.008$ and $c = 230.872 \pm 0.005$ pm and a calculated density of 3.407 g·cm^{-3} (Banno et al. 2014b). It was prepared as follows. Si_3N_4, Al_2O_3, and AlN were mixed in the molar ratio of 1:2:8. The well-mixed chemicals were heated under a N_2 pressure of 0.1 MPa at 1750°C for 1 h, followed by cooling to ambient temperature by cutting furnace power. The reaction product was a slightly sintered polycrystalline material consisting exclusively of 15R-SiAlON, which was subsequently ground and heated again under the same condition. The resulting sintered specimen was composed mainly of 8H-SiAlON with small amounts of α-Al_2O_3, 15R-SiAlON, and β-SiAlON.

$Al_4SiN_4O_2$ (15R-SiAlON) also crystallizes in the hexagonal structure with the lattice parameters $a = 301.332 \pm 0.003$ and $c = 4186.16 \pm 0.04$ pm and a calculated density of 3.390 g·cm^{-3} (Banno et al. 2014c) [in the rhombohedral structure with the lattice parameters $a = 301$ and $c = 4180$ pm (in the hexagonal setting) (Bando et al. 1986)]. It was obtained at the first step of 8H-SiAlON preparation. In this SiAlON polytype, two-layered precipitations with a structure of α-Al_2O_3 are formed (Gilev 2001). 15R-SiAlON could be described as the combination of three layers of AlN and a double layer with an average composition of $AlSiO_2N$ in the block.

$Al_5SiN_5O_2$ (12H-SiAlON) also crystallizes in the hexagonal structure with the lattice parameters $a = 303.153 \pm 0.003$ and $c = 3281.53 \pm 0.03$ pm and a calculated density of 3.370 g·cm^{-3} (Bando et al. 1986; Banno et al. 2014d). To prepare $Al_5SiN_5O_2$, the starting powder with mixture of AlN, Al_2O_3, and Si_3N_4 (molar ratio 11:2:1) was hot-pressed in a BN-coated carbon die at 1750°C for 1 h in a nitrogen atmosphere (0.1 MPa) and annealed at 1925°C for 1 h in the same atmosphere (1 MPa). The reaction product was a slightly sintered polycrystalline material consisting mainly of 12H-SiAlON with a small amount of 15R-SiAlON.

According to the data of Metselaar (1998), $Al_6SiN_6O_2$ (21R-SiAlON) also exists in the Al, Si$\|$N, and O system.

The composition of J-SiAlON, which melts at the temperature higher than 1800°C, lies between $Al_6Si_3N_2O_{12}$ and $Al_{10}Si_6N_4O_{21}$ (Land et al. 1978).

$Al_{6+x}Si_{3-x}N_{10-x}O_x$ ($x = 1.9 \pm 0.1$) (27R-SiAlON) [according to the data of Metselaar (1998), the composition of this compound is $Al_8SiN_8O_2$] crystallizes in the trigonal structure with the lattice parameters $a = 305.991 \pm 0.004$ and $c = 714.54 \pm 0.01$ pm and a calculated density of 3.336 g·cm^{-3} (Banno et al. 2014a). To obtain it, Si_3N_4, Al_2O_3, AlN, and $SrCO_3$ (molar ratio 1:2:8:1.375) were mixed and heated under a N_2 pressure of 1.1 MPa at 1750°C for 1 h, followed by cooling to ambient temperature by cutting furnace powder.

$Al_9Si_2N_7O_7$ or $Al_{12}Si_3N_{10}O_9$ (Θ-phase) also exists in the Al, Si$\|$N, O system (Land et al. 1978) [according to the data of Gauckler et al. (1975), the composition of this compound is $Al_{10}Si_2N_8O_7$].

$Al_{18}Si_{12}N_8O_{39}$ (X-phase) exists in a narrow region (Bergman et al. 1991; Anya and Hendry 1992) and crystallizes in the triclinic structure with the lattice parameters $a = 968$, $b = 855$, $c = 1121$ pm, $\alpha = 91.4°$, $\beta = 124.4°$, and $\gamma = 99.2°$ from the XRD pattern (Anya and Hendry 1992) [$a = 990$, $b = 970$, $c = 950$ pm, $\alpha = 109°$, $\beta = 95°$, and $\gamma = 95°$ (Drew and Lewis 1974); $a = 856$, $b = 985$, $c = 969$ pm, $\alpha = 70.0 \pm 0.5°$, $\beta = 81.0 \pm 0.5°$, and $\gamma = 81.0 \pm 0.5°$ (Zangvil 1978); $a = 1120 \pm 4$, $b = 968.5 \pm 3.0$, $c = 954.5 \pm 3.0$ pm, $\alpha = 99.2 \pm 0.1°$, $\beta = 90.1 \pm 0.1°$, and $\gamma = 124.30 \pm 0.05°$ and a calculated density of 3.00 ± 0.05 g·cm^{-3} (Zangvil 1980)]. However, in electron diffraction, the patterns were indexed as an orthorhombic structure with the lattice parameters $a = 966$, $b = 284$, and $c = 1117$ pm (Anya and Hendry 1992). According to the data of Jack (1976), X-phase crystallizes in the monoclinic structure with the lattice parameters $a = 972.8$, $b = 840.4$, $c = 957.2$ pm, and $\beta = 108.96°$.

Stable X-phase is formable by simultaneous carbothermal reduction and nitridation route (Anya and Hendry 1992).

AlN–SiO$_2$. This system is a nonquasibinary section of the $4AlN + 3SiO_2 \Leftrightarrow 2Al_2O_3 + Si_3N_4$ ternary mutual system (Bhansali et al. 1990). The mixture of AlN and SiO$_2$ reacts at 1820°C and 1950°C to form 21R and 27R polytypoids (Cannard et al. 1991).

4.33 ALUMINUM–TIN–TITANIUM–NITROGEN

Phase equilibria in this quaternary system at 900°C were investigated within the region AlN–TiN–(Al,Sn)$_{liq.}$–Ti$_3$(Al,Sn) using XRD and metallography (Pietzka and Schuster 1997). The alloys were annealed at 900°C for 240 h. No quaternary phase was observed. In the quaternary system, Ti_2AlN_{1-x} shows a high solubility for Sn, which enlarges the lattice parameters to $a = 305.85 \pm 0.09$ and $c = 1372.6 \pm 0.4$ pm. No other quaternary solid solubility was found. The following four-phase equilibria were observed: Ti$_2$AlN + Ti$_5$Sn$_3$ + TiAl$_2$ +TiAl, Ti$_2$AlN + Ti$_6$Sn$_5$ + Ti$_5$Sn$_3$ + TiAl$_2$, Ti$_2$AlN + Ti$_6$Sn$_5$ + TiAl$_3$ +TiAl$_2$, (Sn, Al)$_{liq.}$ + Ti$_2$AlN + Ti$_6$Sn$_5$ + TiAl$_3$, (Sn,Al)$_{liq.}$ + TiN + Ti$_2$AlN + TiAl$_3$, TiN + Ti$_2$AlN + Ti$_5$Sn$_3$ + Ti$_3$(Sn,Al), and (Sn,Al)$_{liq.}$ + Ti$_2$AlN + Ti$_6$Sn$_5$ + TiN.

4.34 ALUMINUM–TITANIUM–OXYGEN–NITROGEN

The phase diagram of the Al–Ti–O–N quaternary system, valid in the range of temperatures between 450°C and 550°C, has been calculated by Caron et al. (1996). For this diagram, the composition range of each of the ternary constituents was not considered, but the binary compounds homogeneity ranges were considered at a temperature around 500°C in order to predict which phase is stable for a given concentration. The quaternary phase diagram of this system predicts that the reaction which involves Al, Ti, oxygen, and nitrogen in the form of TiN, TiO$_2$, and Al results in the formation of Al$_2$O$_3$, AlN, and TiAl$_3$.

Chemical and microstructural changes occurring when AlN and TiO$_2$ are put in contact at 1400–1600°C in a N$_2$ atmosphere have been studied by Mocellin and Bayer (1985). It was shown that solid TiO$_2$ oxidizes AlN into Al$_2$O$_3$ with corresponding evolution of nitrogen and formation of titanium suboxides. The end products of the reaction are Al$_2$O$_3$ and TiN.

Al$_2$O$_3$–TiN. Al$_2$O$_3$–TiN powder mixtures containing 60 and 66.66 mol% TiN may be prepared by in situ reaction between AlN and TiO$_2$ (Mukerji and Biswas 1990). The powder mixture can be pressure less sintered at 1600°C in a N$_2$ atmosphere to 94–97% theoretical dense product. Al_2O_3 + TiN composite (molar ratio 1:2) resists oxidation up to 700°C with the formation of an adherent layer of rutile. Above 820°C, the linear oxidation kinetics was observed after an initial diffusion-controlled reaction.

4.35 ALUMINUM–ZIRCONIUM–OXYGEN–NITROGEN

AlN–ZrO₂. To investigate this system, homogeneous mixtures between AlN and ZrO_2 were prepared by milling for 24 h with 2-propanol in an alumina ball mill (Stoto et al. 1989). After the mixture was dried and screened, it was vacuum hot-pressed into pellets in graphite dies under a nominal 45 MPa stress at 1400°C for 1 h. A quaternary zirconium aluminum oxynitride with unknown composition was present in the samples with AlN, ZrO_2, ZrN, and Al_2O_3. It crystallizes in the cubic structure with the lattice parameter $a = 1830$ pm.

The displacement reaction between AlN and ZrO_2 is $2AlN + 1.5ZrO_2 = 1.5ZrN + Al_2O_3 + 0.25N_2$ (Bayer and Mocellin 1986). This reaction is found not to be complete, and, at this composition, ZrO_2 is found to be present in both tetragonal and monoclinic forms with Al_2O_3 and ZrN. These data were confirmed by Mukerji (1989) and Toy and Savrun (1998). AlN–ZrO_2 compositions containing AlN varying from 2 to 70 mol% were sintered at 1500°C for 1 h in a N_2 atmosphere (Mukerji 1989). XRD analysis showed that in all the samples containing up to 60 mol% AlN, no free AlN was present. It clearly indicated that AlN, up to 60 mol%, completely reacted with ZrO_2. Cubic ZrO_2 is stabilized at room temperature by the addition of AlN. The percentage of c-ZrO_2 in the mixture of c-ZrO_2 and monoclinic ZrO_2 increases linearly with addition of up to 20 mol% AlN. The above-mentioned displacement reaction (Bayer and Mocellin 1986) begins above 20 mol% AlN and is complete at 50 mol% AlN.

According to the data of Toy and Savrun (1998), the solid-state displacement reactions in the AlN–ZrO_2 system yield a ZrAlON phase, which crystallizes in the cubic structure with the lattice parameter $a = 1834$ pm. The amount of this phase increases with increasing AlN/ZrO_2 molar ratio and with increasing reaction temperature. AlN additions were found to stabilize the (cubic + tetragonal) ZrO_2 structure in composites. The relative amount of tetragonal phase was less than 30 vol%. The amount of ZrO_2 was determined to decrease with increasing sintering temperature and increasing AlN/ZrO_2 molar ratio.

The granulated powder mixtures were consolidated by hot-pressing in a graphite die (Toy and Savrun 1998). The samples were hot-pressed under pressure of 45 MPa at 1500°C and 1600°C for 1 h under flowing N_2 and vacuum conditions.

4.36 ALUMINUM–HAFNIUM–OXYGEN–NITROGEN

AlN–HfO₂. No reaction occurred in this system (Toy and Savrun 1998), contrary to what is reported by Bayer and Mocellin (1986). Earlier it was claimed (Bayer and Mocellin 1986) that the AlN–HfO_2 system displays the next solid-state reaction: $2AlN + 1.5HfO_2 = 1.5HfN + Al_2O_3 + 0.25N_2$, but no data were presented to support this claim. XRD of the hot-pressed samples indicated that there was no reaction between AlN and HfO_2 after hot-pressing at 1500°C under N_2 atmosphere, as thermodynamic calculations predict (Toy and Savrun 1998). The only phases present are the major orthorhombic and monoclinic forms of HfO_2 and AlN.

The granulated powder mixtures were consolidated by hot-pressing in a graphite die (Toy and Savrun 1998). The samples were hot-pressed under pressure of 45 MPa at 1500°C and 1600°C for 1 h under flowing N_2 and vacuum conditions.

4.37 ALUMINUM–PHOSPHORUS–CHLORINE–NITROGEN

The [$Cl_2AlNPCl_3$]$_2$ quaternary compound, which melts at 192°C and crystallizes in the triclinic structure with lattice parameters $a = 719.1 \pm 0.2$, $b = 843.5 \pm 0.2$, and $c = 857.4 \pm 0.2$ pm and $a = 64.23 \pm 0.01°$, $\beta = 76.62 \pm 0.01°$, and $\gamma = 68.21 \pm 0.01°$ and a calculated density of 1.910 g·cm^{-3}, is formed in the Al–P–Cl–N (Jäschke and Jansen 2002b). The thermal treatment of [$Me_3SiNPCl_3 \cdot AlCl_3$] and simultaneously removing the by-product Me_3SiCl results in the formation of [$Cl_2AlNPCl_3$]$_2$. Its crystallization from CH_2Cl_2 yields colorless crystals of the title compound.

CHAPTER 5

Systems Based on AlP

5.1 ALUMINUM–HYDROGEN–OXYGEN–PHOSPHORUS

Some quaternary compounds are formed in the Al–H–O–P system. **AlPO$_4$·2H$_2$O** is stable at 25°C (Jameson and Salmon 1954; Martin 1960; Sveshnikova 1960). This compound also exists as minerals variscite and metavariscite (Kniep and Mootz 1973; Bass and Sclar 1979; Duggan et al. 1990). Variscite crystallizes in the orthorhombic structure with the lattice parameters $a = 982.3 \pm 0.4$, $b = 856.2 \pm 0.9$, and $c = 962.0 \pm 0.5$ pm (Onac et al. 2004) [$a = 985$, $b = 955$, and $c = 850$ pm and experimental and calculated densities of 2.5 and 2.61 g·cm^{-3}, respectively (McConnell 1940); $a = 983.1$ and 990.2, $b = 963.5$ and 965.9, and $c = 855.5$ and 1718.0 pm and calculated densities of 2.58 and 2.50 g·cm^{-3} for two types of variscite (Salvador and Fayos 1972); $a = 982.2 \pm 0.3$, $b = 856.1 \pm 0.3$, and $c = 963.0 \pm 0.3$ pm and a calculated density of 2.59 g·cm^{-3} (Kniep et al. 1977)].

Metavariscite crystallizes in the monoclinic structure with the lattice parameters $a = 517.8 \pm 0.2$, $b = 951.4 \pm 0.2$, $c = 845.4 \pm 0.2$ pm, and $\beta = 90.35 \pm 0.02°$ and experimental and calculated densities of 2.54 and 2.535 g·cm^{-3}, respectively (Kniep and Mootz 1973) [$a = 515$, $b = 945$, $c = 845$ pm, and $\beta = 90°$ and experimental and calculated densities of 2.54 and 2.53 g·cm^{-3}, respectively (McConnell 1940); $a = 845 \pm 1$, $b = 952 \pm 2$, $c = 517 \pm 1$ pm, and $\beta = 89.5 \pm 0.2°$ (Bass and Sclar 1979)].

According to the data of Brosheer et al. (1954), **AlPO$_4$·xH$_2$O** ($x \leq 3$ and decreases with increasing in temperature) exists in this quaternary system.

Al$_2$PO$_4$(OH)$_3$ [mineral augelite (Duggan et al. 1990)] crystallizes in the monoclinic structure with the lattice parameters $a = 1309 \pm 1$, $b = 798.6 \pm 0.3$, $c = 506.6 \pm 0.3$ pm, and $\beta = 112°20' \pm 3'$ and a calculated density of 2.711 g·cm^{-3} (Wise and Loh 1976) [$a = 1310$, $b = 796$, $c = 506$ pm, and $\beta = 112°26.5'$ and experimental and calculated densities of 2.696 and 2.704 g·cm^{-3}, respectively (Peacock and Moddle 1941); $a = 1312.4 \pm 0.6$, $b = 798.8 \pm 0.5$, $c = 506.6 \pm 0.3$ pm, and $\beta = 112.25 \pm 0.02°$ (Araki et al. 1968); $a = 1312.6 \pm 0.5$, $b = 799.6 \pm 0.3$, $c = 507.5 \pm 0.2$ pm, and $\beta = 112.26 \pm 0.02°$ (Bass and Sclar 1979)]. Al$_2$PO$_4$(OH)$_3$ was synthesized in sealed gold tubes at water pressure of 50 MPa to 0.3 GPa in standard, externally heated, cold-seal pressure vessels (Wise and Loh 1976). Starting materials were mixtures of Al(OH)$_3$ or Al$_2$O$_3$ with AlPO$_4$.

$Al_2(PO_4)(OH)_3 \cdot H_2O$ (mineral senegalite) crystallizes in the orthorhombic structure with the lattice parameters $a = 967.8 \pm 0.2$, $b = 759.7 \pm 0.2$, and $c = 766.8 \pm 0.2$ pm and experimental and calculated densities of 2.552 ± 0.007 and 2.551 g·cm^{-3}, respectively (Johan 1976; Fleischer et al. 1977) [$a = 767.5 \pm 0.4$, $b = 971.1 \pm 0.4$, and $c = 763.5 \pm 0.4$ pm (Keegan et al. 1979)]. The final product of senegalite dehydration is $AlPO_4$ (Johan 1976; Fleischer et al. 1977).

$Al_2(PO_4)(OH)_3 \cdot (4-5)H_2O$ (mineral bolivarite) is an amorphous mineral with an experimental density of $1.97-2.05$ g·cm^{-3} (van Wambeke 1971; García-Guinea et al. 1995).

$Al_3(PO_4)(OH)_6 \cdot 6H_2O$ (mineral evansite) is also amorphous mineral with an experimental density of $1.8-2.2$ g·cm^{-3} (García-Guinea et al. 1995).

$Al_3(PO_4)_2(OH)_3 \cdot (4.5-5)H_2O$ (mineral wavellite) crystallizes in the orthorhombic structure with the lattice parameters $a = 962.1 \pm 0.2$, $b = 1736.3 \pm 0.4$, and $c = 699.4 \pm 0.3$ pm (Araki and Zoltai 1968). According to the data by Gordon (1950), the formula of this mineral is $Al_3(PO_4)_2(OH,F)_3 \cdot 5H_2O$.

$Al_4(PO_4)_2(OH)_3$ [mineral trolleite (Duggan et al. 1990)] crystallizes in the monoclinic structure with the lattice parameters $a = 1885.7 \pm 0.5$, $b = 711.0 \pm 0.4$, $c = 717.8 \pm 0.4$ pm, and $\beta = 100°29' \pm 2'$ and a calculated density of 3.115 g·cm^{-3} (Wise and Loh 1976) [$a = 1889.4 \pm 0.5$, $b = 716.1 \pm 0.1$, $c = 716.2 \pm 0.2$ pm, and $\beta = 99.99 \pm 0.02°$ and experimental and calculated densities of 3.09 and 3.08 g·cm^{-3}, respectively (Moore and Araki 1974); $a = 1889 \pm 1$, $b = 716.2 \pm 0.1$, $c = 714.2 \pm 0.4$ pm and $\beta = 99.9 \pm 0.7°$ (Bass and Sclar 1979)]. The entropy and the Gibbs free energy of formation of $Al_4(PO_4)_2(OH)_3$ from the elements were calculated to be 281.7 J·K^{-1}·mol^{-1} and -6504.2 kJ·mol^{-1}, respectively (Brunet et al. 2004) [270.9 ± 25.1 J·K^{-1}·mol^{-1} and -6081.83 ± 14.65 kJ·mol^{-1}, respectively (Bass and Sclar 1979)]. The title compound was synthesized in sealed gold tubes at water pressure of 50 MPa to 0.3 GPa in standard, externally heated, cold-seal pressure vessels (Wise and Loh 1976). Starting materials were mixtures of $Al(OH)_3$ or Al_2O_3 with $AlPO_4$.

$Al_4(PO_4)_3(OH)_3 \cdot 9H_2O$ (mineral vantasselite) crystallizes in the orthorhombic structure with the lattice parameters $a = 1052.8 \pm 0.4$, $b = 1654.1 \pm 0.3$, and $c = 2037.3 \pm 0.6$ pm and experimental and calculated densities of 2.30 and 2.312 g·cm^{-3}, respectively (Fransolet 1987; Jambor et al. 1988).

$Al_6(PO_4)_4(OH)_6 \cdot 11H_2O$ (mineral kobokoboite) crystallizes in the triclinic structure with the lattice parameters $a = 746.0 \pm 0.1$, $b = 737.7 \pm 0.1$, and $c = 1238.5 \pm 0.5$ pm and $\alpha = 102.79 \pm 0.02°$, $\beta = 90.20 \pm 0.03°$, and $\gamma = 116.33 \pm 0.02°$ and experimental and calculated densities of 2.21 ± 0.03 and 2.287 g·cm^{-3}, respectively (Mills et al. 2010; Cámara et al. 2014).

$Al_{11}(PO_4)_9(OH)_6 \cdot 38H_2O$ (mineral vashegyite) crystallizes in the orthorhombic structure with the lattice parameters $a = 1075.4 \pm 0.3$, 1077.3 ± 0.3, 1076.4 ± 0.3, and 1075.7 ± 0.3; $b = 1497.1 \pm 0.5$, 1497.1 ± 0.5, 1495.4 ± 0.5, and 1499.1 ± 0.5; and $c = 2267.5 \pm 0.6$, 2062.6 ± 0.6, 2052.6 ± 0.6, and 2060.4 ± 0.6 pm and experimental and calculated densities of $(1.930 \pm 0.001)-(1.994 \pm 0.002)$ and $1.934-2.005$ g·cm^{-3}, respectively (Johan et al. 1983) [$a = 2092$, $b = 1982$, and $c = 1448$ pm and a calculated density of 1.98 g·cm^{-3} (McConnell 1974)].

Earlier investigations also indicated the existence of the next compounds in the Al–H–O–P quaternary system: $AlPO_4 \cdot H_3PO_4 \cdot 3H_2O$ and $AlPO_4 \cdot 2H_3PO_4$ (Brosheer et al. 1954); $Al_2O_3 \cdot P_2O_5 \cdot 7H_2O$, $2Al_2O_3 \cdot 3P_2O_5 \cdot 10H_2O$, and $Al_2O_3 \cdot 3P_2O_5 \cdot 6H_2O$ or $Al_2(H_2PO_4)_3$ (Jameson and Salmon 1954; Martin 1960); $AlH_3(PO_4)_2 \cdot H_2O$ and $AlH_3(PO_4)_2 \cdot 3H_2O$ (Martin 1960); and $2Al_2(H_2PO_4)_2 \cdot 5H_2O$ (Sveshnikova 1960).

5.2 ALUMINUM–LITHIUM–SODIUM–PHOSPHORUS

The $LiNa_2AlP_2$ quaternary compound, which crystallizes in the orthorhombic structure with the lattice parameters $a = 1156.6 \pm 0.3$, $b = 1359.2 \pm 0.3$, and $c = 580.1 \pm 0.1$ pm, is formed in the Al–Li–Na–P system (Somer et al. 1995d). This compound was synthesized from a stoichiometric mixture of Na, LiAl, and P in a sealed steel ampoule at 1000°C.

5.3 ALUMINUM–LITHIUM–POTASSIUM–PHOSPHORUS

The LiK_2AlP_2 quaternary compound, which crystallizes in the orthorhombic structure with the lattice parameters $a = 615.34 \pm 0.08$, $b = 1465.6 \pm 0.2$, and $c = 610.12 \pm 0.09$ pm, is formed in the Al–Li–K–P system (Somer et al. 1995a). This compound was synthesized from a stoichiometric mixture of K, LiP, and AlP in a sealed steel ampoule at 650°C.

5.4 ALUMINUM–LITHIUM–OXYGEN–PHOSPHORUS

The $LiAl(PO_3)_4$ and $Li_9Al_3(P_2O_7)_3(PO_4)_2$ quaternary compounds are formed in the Al–Li–O–P system. $LiGa(PO_3)_4$ crystallizes in the orthorhombic structure with the lattice parameters $a = 1244.54 \pm 0.01$, $b = 823.40 \pm 0.01$, and $c = 892.46 \pm 0.01$ pm and a calculated density of 2.541 g·cm^{-3} (Brühne and Jansen 1994). Single crystals of the title compound have been obtained by reaction of Li_2CO_3, Al_2O_3, and H_3PO_4 (molar ratio 1:1:16) at 350°C.

$Li_9Al_3(P_2O_7)_3(PO_4)_2$ incongruently melts at 790°C and crystallizes in the trigonal structure with the lattice parameters $a = 955.3 \pm 0.1$ and $c = 1349.2 \pm 0.2$ pm and a calculated density of 2.666 g·cm^{-3} (Poisson et al. 1998). It was prepared by crystallization in a flux of Li_3PO_4. Starting materials were Li_3PO_4, Al_2O_3, and $NH_4H_2PO_4$ mixed in molar ratios of Li/Al/P = 55.5:4:40.5. The mixture was gradually heated up to 900°C in a Pt crucible and cooled at a rate of 50°C·h^{-1} down to 850°C, at a rate of 2°C·h^{-1} between 850°C and 600°C, and finally at a rate of ca. 200°C·h^{-1} down to room temperature. In order to isolate the title compound, the mixture was treated during 5 days by an aqueous 1 M solution of CH_3COOH, which dissolved $Li_4P_2O_7$. A saturated solution of NaCl was then added to the remaining solid to eliminate $LiPO_3$. Colorless platelike crystals of $Li_9Al_3(P_2O_7)_3(PO_4)_2$ were obtained.

5.5 ALUMINUM–LITHIUM–SULFUR–PHOSPHORUS

The $LiAlP_2S_6$ quaternary compound, which crystallizes in the monoclinic structure with the lattice parameters $a = 678.3 \pm 0.3$, $b = 1036.5 \pm 0.4$, $c = 1177.6 \pm 0.4$ pm, and $\beta = 94.399 \pm 0.005°$ and a calculated density of 2.20 g·cm^{-3}, is formed in the Al–Li–S–P system (Kuhn et al. 2013). It was prepared via a typical solid-state synthesis carried out in evacuated quartz tubes. Li_2S, Al, P, and S served as starting materials, and all manipulations were carried out under Ar atmosphere. The quartz tube with the mixture was heated in a furnace up to 750°C (100°C·h^{-1}) and kept at that temperature for 24 h, followed by cooling down to room temperature at a rate of 10°C·h^{-1}. The obtain compound is air and water sensitive.

5.6 ALUMINUM–SODIUM–POTASSIUM–PHOSPHORUS

The **NaK$_2$AlP$_2$** quaternary compound, which crystallizes in the orthorhombic structure with the lattice parameters $a = 1451.9 \pm 0.4$, $b = 660.8 \pm 0.2$, and $c = 628.6 \pm 0.2$ pm and a calculated density of 2.10 g·cm^{-3}, is formed in the Al–Na–K–P system (Ohse et al. 1993). It was prepared from stoichiometric mixture of the elements or from Na, NaP, K, and AlP at 730°C.

5.7 ALUMINUM–SODIUM–CALCIUM–PHOSPHORUS

Na$_3$Ca$_3$AlP$_4$ quaternary compound, which crystallizes in the hexagonal structure with the lattice parameters $a = 916.55 \pm 0.09$ and $c = 702.66 \pm 0.13$ pm and a calculated density of 2.21 g·cm^{-3} at 200 K is formed in the Al–Na–Ca–P system (Wang et al. 2017a). The title compound was prepared from the elements. The starting materials were weighed in the glove box and loaded into Nb tube, which was then closed by arc-welding under high purity argon gas. The sealed Nb container was subsequently jacketed within evacuated fused silica tube. The reaction mixture was heated in a muffle furnace to 900°C at a rate of 100°C·h^{-1} and equilibrated at this temperature for 24 h. After that, the sample were cooled to 500°C at a rate of 5°C·h^{-1}, at which point, the cooling rate to room temperature was changed to 40°C·h^{-1}. All manipulations were carried out inside an argon-field glove box with controlled atmosphere or under vacuum.

5.8 ALUMINUM–SODIUM–EUROPIUM–PHOSPHORUS

The **Na$_{3.12\pm0.01}$Eu$_{2.88}$AlP$_4$** quaternary compound, which crystallizes in the hexagonal structure with the lattice parameters $a = 932.22 \pm 0.04$ and $c = 728.80 \pm 0.06$ pm and a calculated density of 4.00 g·cm^{-3} at 200 K, is formed in the Al–Na–Eu–P system (Wang et al. 2017a). The title compound was synthesized from the elements. The starting materials were weighed in the glove box and loaded into Nb tube, which was then closed by arc-welding under high purity argon gas. The sealed Nb container was subsequently jacketed within evacuated fused silica tube. The reaction mixture was heated in a muffle furnace to 900°C at a rate of 100°C·h^{-1} and equilibrated at this temperature for 24 h. After that, the sample were cooled to 500°C at a rate of 5°C·h^{-1}, at which point, the cooling rate to room temperature was changed to 40°C·h^{-1}. All manipulations were carried out inside an argon-field glove box with controlled atmosphere or under vacuum.

5.9 ALUMINUM–SODIUM–OXYGEN–PHOSPHORUS

Some quaternary compounds are formed in the Al–Na–O–P system. **NaAlP$_2$O$_7$** congruently melts at 850°C (Ust'yantsev and Zholobova 1977) and crystallizes in the monoclinic structure with the lattice parameters $a = 719.76$, $b = 769.57$, $c = 931.20$ pm, and $\beta = 111.736°$ and calculated density of 3.103 g·cm^{-3} (Gamondès et al. 1971). The title compound was obtained either by thermolysis of a triphosphate $NaAlP_3O_{10}$ or by reaction in the solid state at ca. 800°C between 1 mole of Na_2CO_3 and 1 mole of $AlH_2P_3O_{10}$ or of another phosphate with an Al/P ratio of 1:3.

$NaAl_2PO_6$ congruently melts at 882°C (Berul' and Voskresenskaya 1968). It was obtained by melting Al_2O_3 and $NaPO_3$.

$Na_3Al_2(PO_4)_3$ incongruently melts at 680°C (Gusarov et al. 2002) [congruently at 760°C (Berul' and Voskresenskaya 1968); incongruently at 700°C forming $AlPO_4$ and liquid (Ust'yantsev and Zholobova 1977)]. It was also obtained by melting Al_2O_3 and $NaPO_3$ (Berul' and Voskresenskaya 1968).

$Na_6Al_2(P_2O_7)_3$ congruently melts at 750°C [at 780°C (Berul' and Voskresenskaya 1968)] and has polymorphic transformation at 580°C (Gusarov et al. 2002). It was also obtained by melting Al_2O_3 and $NaPO_3$ (Berul' and Voskresenskaya 1968).

$Na_9Al_2(P_3O_{10})_3$ congruently melts at 832°C (Berul' and Voskresenskaya 1968). It was obtained by melting Al_2O_3 and $NaPO_3$. According to the data by Gusarov et al. (2002), this compound was not obtained in the Al–Na–O–P system.

5.10 ALUMINUM–SODIUM–SULFUR–PHOSPHORUS

The $NaAlP_2S_6$ quaternary compound, which crystallizes in the orthorhombic structure with the lattice parameters $a = 804.00 \pm 0.02$, $b = 1094.52 \pm 0.02$, and $c = 2088.01 \pm 0.04$ pm and a calculated density of 2.32 g·cm^{-3}, is formed in the Al–Na–S–P system (Kuhn et al. 2013). It was prepared by reaction stoichiometric amounts of the corresponding elements. All manipulations were carried out under Ar atmosphere. The quartz tube with the mixture was heated in a furnace up to 750°C (100°C·h^{-1}) and kept at that temperature for 24 h, followed by cooling down to room temperature at a rate of 10°C·h^{-1}. The obtain compound is air and water sensitive.

5.11 ALUMINUM–POTASSIUM–OXYGEN–PHOSPHORUS

The $KAlP_2O_7$ quaternary compound, which crystallizes in the monoclinic structure with the lattice parameters $a = 730.8 \pm 0.8$, $b = 966.2 \pm 0.6$, $c = 802.5 \pm 0.4$ pm, and $\beta = 106.69 \pm 0.07°$ and a calculated density of 2.927 g·cm^{-3} (Ng and Calvo 1973) [$a = 732$, $b = 976$, $c = 911$ pm, and $\beta = 122.5°$ and a calculated density of 2.93 g·cm^{-3} (Klyucharov and Skoblo 1964); $a = 732.91$, $b = 965.41$, $c = 804.53$ pm, and $\beta = 106.934°$ and experimental and calculated densities of 2.923 and 2.927 g·cm^{-3}, respectively (Gamondès et al. 1971)], is formed in the Al–K–O–P system. This compound has no phase transformations from 25°C to 500°C (Ng and Calvo 1973). The crystals of $KAlP_2O_7$ were prepared by heating a mixture of $AlPO_4$ and KH_2PO_4 (molar ratio 1:4) in a Pt crucible (Ng and Calvo 1973). The mixture was melted, and the temperature was held at 1100°C for 6 h. It was then cooled at a rate of 6°C·h^{-1} to 600°C and quenched to room temperature. Clear and colorless parallelepipeds of the title compound could be separated from the powder matrix by washing the product in boiling H_2O.

KAlP$_2$O$_7$ could be also prepared either by thermolysis of a triphosphate **KAlP$_3$O$_{10}$** or by reaction in the solid state at ca. 800°C between 1 mole of K$_2$CO$_3$ and 1 mole of AlH$_2$P$_3$O$_{10}$ or of another phosphate with an Al/P ratio of 1:3 (Gamondès et al. 1971).

5.12 ALUMINUM–CESIUM–OXYGEN–PHOSPHORUS

The **Cs$_2$AlP$_3$O$_{10}$** and **Cs$_2$Al$_2$P$_2$O$_9$** quaternary compounds are formed in the Al–Cs–O–P system. The first of them melts at 780°C (Lutsko et al. 1986) and crystallizes in the monoclinic structure with the lattice parameters $a = 942.0 \pm 0.4$, $b = 901.6 \pm 0.3$, $c = 1225.8 \pm 0.2$ pm, and $\beta = 94.88 \pm 0.03°$ (Devi and Vidyasagar 2000). It was synthesized in polycrystalline form by solid state reactions from stoichiometric mixture of CsNO$_3$, NH$_4$H$_2$PO$_4$, and Al(OH)$_3$ (Devi and Vidyasagar 2000). This mixture was heated in open air initially at 400°C for 12 h, and the temperature was raised in steps of 100°C to the maximum value of 700°C, at which temperature the compound was heated for 12 h. Single crystals of Cs$_2$AlP$_3$O$_{10}$ were grown by the flux method using CsPO$_3$ as the flux. The mixture of CsNO$_3$, NH$_4$H$_2$PO$_4$, and Al(OH)$_3$ taken in the 6:1 mass ratio of CsPO$_3$ flux to Al was heated at 700°C for one day and then cooled to 550°C at a rate of 3°C·h^{-1}. Platelike colorless crystals were separated by washing away the flux with H$_2$O.

Cs$_2$Al$_2$P$_2$O$_9$ crystallizes in the triclinic structure with the lattice parameters $a = 492.5 \pm 0.2$, $b = 712.1 \pm 0.2$, and $c = 806.6 \pm 0.2$ pm and $\alpha = 96.51 \pm 0.02°$, $\beta = 107.12 \pm 0.02°$, and $\gamma = 108.68 \pm 0.02°$ and a calculated density of 3.502 g·cm^{-3} (Huang and Hwu 1999). Crystals of this compound were grown by high-temperature, solid-state methods using CsCl flux. Al (2.0 mM), KO$_2$ (2.0 mM), CuO (1.0 mM), and P$_2$O$_5$ (1.0 mM) were ground and mixed with the flux CsCl. The charge to flux ratio was 1:2 by mass. The reaction was carried out in a carbon-coated quartz ampoule. The mixture was heated up to 500°C over 24 h and isothermed for 6 h, followed by heating to 800°C and isothermal treatment for an additional 4 days. The reaction was cooled slowly to 500°C over 72 h, followed by furnace cooling to room temperature. Colorless columnar crystals of the title compound, as well as some other unidentified black and white polycrystalline phases, were retrieved from the flux by washing the products with deionized water using suction filtration.

5.13 ALUMINUM–SILVER–SULFUR–PHOSPHORUS

The **AgAlP$_2$S$_6$** quaternary compound, which crystallizes in the orthorhombic structure with the lattice parameters $a = 771.1 \pm 0.1$, $b = 1083.8 \pm 0.1$, and $c = 2086.3 \pm 0.2$ pm, is formed in the Al–Ag–S–P system (Pfitzner et al. 2004). It was obtained by solid-state reactions from the stoichiometric quantities of the elements in evacuated silica ampoule. This compound is air and moisture sensitive.

5.14 ALUMINUM–SILVER–SELENIUM–PHOSPHORUS

The **AgAlP$_2$Se$_6$** quaternary compound, which congruently melts at 588°C (Pfeiff and Kniep 1992) and crystallizes in the monoclinic structure with the lattice parameters $a = 635.3 \pm 0.1$, $b = 1100.7 \pm 0.1$, $c = 1358.8 \pm 0.1$ pm, and $\beta = 98.46 \pm 0.01°$, is formed in the

Al–Ag–Se–P system (Pfitzner et al. 2004) [$a = 634.8 \pm 0.5$, $b = 1098.9 \pm 0.3$, $c = 702.8 \pm 0.4$ pm, $\beta = 107.2 \pm 0.5°$, and energy gap 2.42 eV (Pfeiff and Kniep 1992)]. It was obtained by solid-state reactions from the mixture of $AgAlSe_2$, P, and Se in evacuated silica ampoule (Pfitzner et al. 2004) or from the stoichiometric mixture of the elements in evacuated silica ampoule by heating up to 750°C (10 h) followed by cooling from the melt ($30°C \cdot h^{-1}$) (Pfeiff and Kniep 1992). After quenching the melt to liquid nitrogen, the yellow sample was annealed at 550°C for 21 days.

5.15 ALUMINUM–MAGNESIUM–OXYGEN–PHOSPHORUS

The $Mg_xAl_{1-x}PO_4$ (0.4 < x < 0.5) and $MgAlPO_5$ quaternary phases are formed in the Al–Mg–O–P system. $Mg_xAl_{1-x}PO_4$ (0.4 < x < 0.5) exists in three polymorphic modifications (Bu et al. 1997). First modification crystallizes in the hexagonal structure with the lattice parameters $a = 1770.0$ and $c = 2735.2$ pm, second modification crystallizes in the tetragonal structure with the lattice parameters $a = 1909.0$ and $c = 2748.6$ pm, and third modification crystallizes in the rhombohedral structure with the lattice parameters $a = 1766.1$ and $c = 4160.8$ pm (in hexagonal setting). This material was hydrothermally made at 180°C with the reaction times between 3 and 5 days. Starting materials included $(Me_2CHO)_3Al$, $MgHPO_4 \cdot 3H_2O$, and 85% H_3PO_4.

$MgAlPO_5$ has two polymorphic modifications (Schmid-Beurmann et al. 2007). α-$MgAlPO_5$ crystallizes in the monoclinic structure with the lattice parameters $a = 711.69 \pm 0.09$, $b = 1037.0 \pm 0.1$, $c = 544.98 \pm 0.07$ pm, and $\beta = 98.34 \pm 0.02°$ and a calculated density of 2.708 $g \cdot cm^{-3}$ at 25°C (Schmid-Beurmann et al. 2007) [$a = 711.1 \pm 0.2$, $b = 1036.2 \pm 0.3$, $c = 545.5 \pm 0.3$ pm, and $\beta = 98.38 \pm 0.05°$ and a calculated density of 2.71 $g \cdot cm^{-3}$ (Hesse and Čemič 1994b)]. Room-temperature modification was found to undergo an isosymmetric and reversible phase transition toward a nonquenchible β-$MgAlPO_5$ at 485°C (Schmid-Beurmann et al. 2007). The enthalpy of this transformation is equal to 650 $J \cdot mol^{-1}$. The standard enthalpy of formation of this modification is -2394 ± 5 $kJ \cdot mol^{-1}$. β-$MgAlPO_5$ also crystallizes in the monoclinic structure with the lattice parameters $a = 714.5 \pm 0.2$, $b = 1041.9 \pm 0.3$, $c = 527.9 \pm 0.1$ pm, and $\beta = 91.12 \pm 0.03°$ and a calculated density of 2.743 $g \cdot cm^{-3}$ at 550°C (Schmid-Beurmann et al. 2007). The standard enthalpy and entropy of formation of $MgAlPO_5$ are -2405 ± 9 $kJ \cdot mol^{-1}$ and 90 $J \cdot mol^{-1}K^{-1}$, respectively (Čemič and Schmid-Beurmann 1995).

The colorless crystals of the title compound were produced in the course of an investigation of the p-T-stability of mineral lazulite, $MgAl_2(PO_4)_2(OH)_2$. They formed as decomposition products between 0.1 MPa and 0.7 GPa water pressure and at temperatures above 500°C (Hesse and Čemič 1994b). Single crystals of the title compound were synthetically prepared. It could be also synthesized using hydrothermal as well as dry conditions (Schmid-Beurmann et al. 2007). For hydrothermal synthesis stoichiometric mixtures of $AlPO_4$ (berlinite) and MgO, heated beforehand at 1000°C, were reacted in Au capsules in the presence of water at 0.2 GPa and 800°C for 3 days. Under dry conditions $MgAlPO_5$ was synthesized from mixtures of $AlPO_4$ (tridymite form) and MgO pressed to pellets and heated at 1300°C in a Pt crucible in air.

5.16 ALUMINUM–CALCIUM–OXYGEN–PHOSPHORUS

The $Ca_9Al(PO_4)_7$ quaternary compound, which has two polymorphic modifications, one of which crystallizes in the trigonal structure with the lattice parameters $a = 1031.40 \pm 0.02$ and $c = 3725.3 \pm 0.1$ pm (Watras et al. 2016) [1031.9 ± 0.2 and $c = 3720 \pm 2$ pm (Golubev et al. 1990)], is formed in the Al–Ca–O–P system. The phase transition of this compound takes place at the temperature higher than 1450°C (Golubev et al. 1990). The title compound was synthesized from the solid-state reaction of $(NH_4)_2HPO_4$ and $AlPO_4{\cdot}3H_2O$ at 900–1200°C for 30–50 h.

5.17 ALUMINUM–CALCIUM–PLATINUM–ANTIMONY

The $Ca_2Pt_{\sim7}Al_{\sim1}P_{\sim3}$ quaternary compound, which crystallizes in the tetragonal structure with the lattice parameters $a = 400.0 \pm 0.1$ and $c = 2683.8 \pm 0.3$ pm, is formed in the Al–Ca–Pt–P system (Lux et al. 1991). It was prepared by heating the respective elements in a stoichiometric ratio corresponding to the formula (corundum crucible, melted quartz glass vials, Ar atmosphere). The reaction temperature was 1050°C, and the reaction times was 2–3 weeks, the annealing being interrupted several times in order to grind still inhomogeneous preparations in the Ar glove box.

5.18 ALUMINUM–ZINC–OXYGEN–PHOSPHORUS

The $Zn_xAl_{1-x}PO_4$ (0.4 < x < 0.5) quaternary phase, which exists in three polymorphic modifications, is formed in the Al–Zn–O–P system (Bu et al. 1997). First modification crystallizes in the hexagonal structure with the lattice parameters $a = 1768.8$ and $c = 2739.6$ pm, second modification crystallizes in the tetragonal structure with the lattice parameters $a = 1899.9$ and $c = 2795.8$ pm, and third modification crystallizes in the rhombohedral structure with the lattice parameters $a = 1762.5$ and $c = 4158.1$ pm (in hexagonal setting). This material was made hydrothermally at 180°C with the reaction times between 3 and 5 days. Starting materials included $(Me_2CHO)_3Al$, $Zn(NO_3)_2{\cdot}6H_2O$, and 85% H_3PO_4.

5.19 ALUMINUM–ZINC–SULFUR–PHOSPHORUS

AlP–ZnS. The solubility of AlP in ZnS is not higher than 1 mol% (Addamiano 1960). This system was investigated through X-ray diffraction (XRD). The ingots were obtained by the sintering of powderlike ZnS and AlP at 900°C for 24 h in the argon atmosphere.

5.20 ALUMINUM–ZINC–IRON–PHOSPHORUS

The isothermal section of the Al–Zn–Fe–P quaternary system at 450°C and at 93 at% Zn (Figure 5.1) was constructed by Wang et al. (2012) and included in the review of Raghavan (2013). The ternary phase $Al_6Zn_{86}Fe_8$ (τ) is present at this temperature. Six four-phase regions exist in this section. All four-phase regions, except L + ζ + Fe₂P + FeP, were found experimentally. The solubility of P in FeAl₃, Fe₂Al₅, ζ, and τ phases is negligible. The AlP

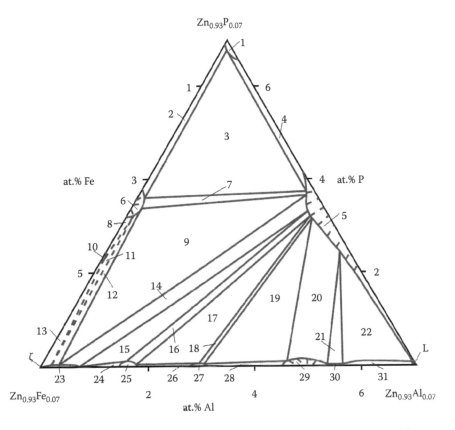

FIGURE 5.1 Isothermal section of the Al–Zn–Fe–P quaternary system at 450°C and at 93 at.% Zn: 1, L + α-Zn_3P_2; 2, L + FeP + α-Zn_3P_2; 3, L + α-Zn_3P_2 + AlP + FeP; 4, L + α-Zn_3P_2 + AlP; 5, L + AlP; 6, L + FeP; 7, L + FeP + AlP; 8, L + FeP + Fe_2P; 9, L + ζ + AlP + FeP; 10, L + Fe_2P; 11, L + ζ + Fe_2P + FeP; 12, L + ζ + FeP; 13, L + Fe_2P + ζ; 14, L + ζ + AlP; 15, L + δ + ζ + AlP; 16, L + δ + AlP; 17, L + δ + τ + AlP; 18, L + τ + AlP, 19, L + Fe_2Al_5 + τ + AlP; 20,, L + Fe_2Al_5 + AlP; 21, L + Fe_2Al_5 + $FeAl_3$ + AlP; 22, L + $FeAl_3$ + AlP; 23, L + ζ; 24, L + ζ + δ; 25, L + δ; 26, L + δ + τ; 27, L + τ; 28, L + Fe_2Al_5 + τ; 29, L + Fe_2Al_5; 30, L + Fe_2Al_5 + $FeAl_3$; 31, L + $FeAl_3$. (Reprinted from *CALPHAD: Comput. Coupling Phase Diagr. and Thermochem.*, 38, Wang, J. et al., The zinc-rich corner of the Zn–Fe–Al–P quaternary system at 450°C, 122–126, Copyright (2012), with permission from Elsevier.)

and L_{Zn} phases both are in equilibrium with all other phases in this section, respectively, except Fe_2P phase. Zn can be dissolved in all compounds existed in the equilibrium alloys. The alloys, which had a constant Zn content of 93 at%, were annealed at 450°C for 60 days and quenched in water. The phase equilibria were studied with scanning electron microscopy equipped with energy-dispersive X-ray spectroscopy and XRD.

5.21 ALUMINUM–GALLIUM–INDIUM–PHOSPHORUS

AlP–GaP–InP. The calculated liquidus and solidus surfaces for this system are given in Figure 5.2, and the miscibility gap is presented in Figure 5.3 (Ishida et al. 1989). The thermodynamics

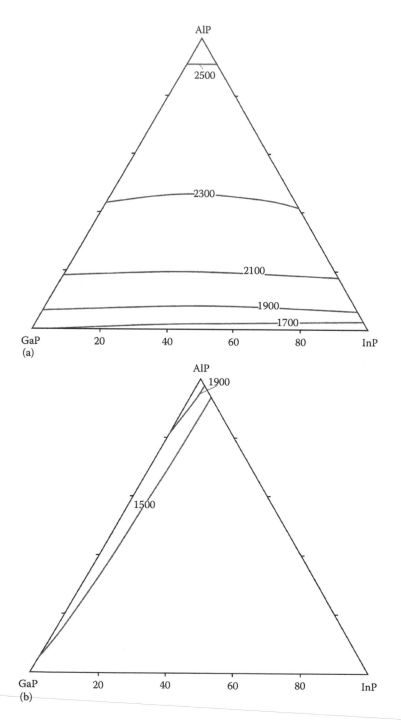

FIGURE 5.2 Liquidus (a) and solidus (b) surfaces of the AlP–GaP–InP quasiternary system. (Reprinted from *J. Cryst. Growth*, 98(1–2), Ishida, K. et al., Data base for calculating phase diagrams of III-V alloy semiconductors, 140–147, Copyright (1989), with permission from Elsevier.)

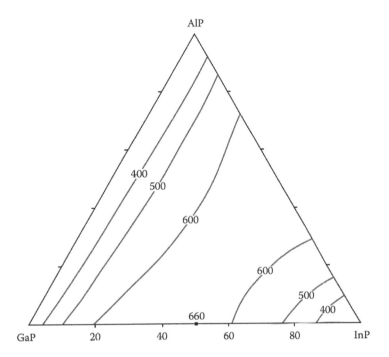

FIGURE 5.3 Miscibility gaps in the AlP–GaP–InP quasiternary system at 400–600°C. (Reprinted from *J. Cryst. Growth*, 98(1–2), Ishida, K. et al., Data base for calculating phase diagrams of III-V alloy semiconductors, 140 147, Copyright (1989), with permission from Elsevier.)

of spinodal decomposition in the AlP–GaP–InP quasiternary system have been developed by Stringfellow (1983). Based on the delta-lattice parameter solution model of the free energy of mixing of semiconductor alloys, an analysis has been developed for the calculation of the spinodal surface and the critical temperature for solid alloys. Solid–solid isotherms were presented for this system. Concepts are also developed for the thermodynamic analysis of spinodal decomposition in this quaternary alloys, including the effect of the coherency strain energy. This addition to the free energy of the inhomogeneous solid is shown to completely stabilize the alloys of interest even at temperatures below room temperature. The calculated critical point for this system is 500°C (Stringfellow 1983) [660°C (Ishida et al. 1989)].

5.22 ALUMINUM–GALLIUM–ARSENIC–PHOSPHORUS

AlP + GaAs ⇔ AlAs + GaP. The calculated liquidus and solidus surfaces for this system are given in Figure 5.4 (Ishida et al. 1989). Experimental liquidus and solidus data along several isoconcentration curves on the 900°C and 1000°C isotherm surfaces were also presented for this system by Ilegems and Panish (1974). The liquidus data were obtained by liquid observation techniques and solidus compositions determined on self-nucleated

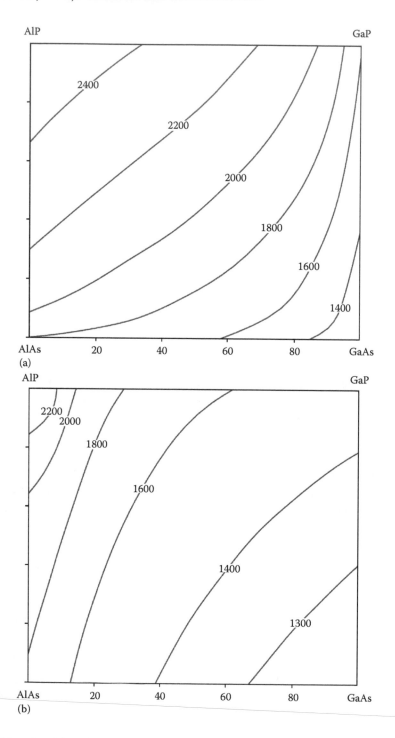

FIGURE 5.4   Liquidus (a) and solidus (b) surfaces of the AlP + GaAs ⇔ AlAs + GaP ternary mutual system. (Reprinted from *J. Cryst. Growth*, 98(1–2), Ishida, K. et al., Data base for calculating phase diagrams of III-V alloy semiconductors, 140–147, Copyright (1989), with permission from Elsevier.)

crystals and on mixed crystal layers grown on GaP substrates. The experimental results are consistent with the calculated phase diagram for this system.

The solid–liquid equilibria in the AlP + GaAs ⇔ AlAs + GaP ternary mutual system were thermodynamically modeled at 900°C and 1000°C by Acharya and Hajra (2005). The use of multiparameter function for the modeling complies with the specific requirement of more exact compositional description of the liquid alloys in the system.

5.23 ALUMINUM–GALLIUM–ANTIMONY–PHOSPHORUS

AlP + GaSb ⇔ AlSb + GaP. The calculated liquidus surface for this system is given in Figure 5.5, and the miscibility gap is presented in Figure 5.6 (Ishida et al. 1989). According to the data by Stringfellow (1982b), the calculated critical point for this system is 1772°C [2276°C (Ishida et al. 1989)].

5.24 ALUMINUM–GALLIUM–CHROMIUM–PHOSPHORUS

The temperature dependence of the GaP solubility in the gallium solutions containing 0.31 and 2.1 at% Al with 0.5–3 at% Cr was determined by Saidov et al. (1979).

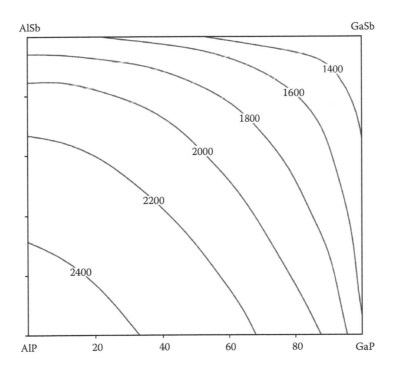

FIGURE 5.5 Liquidus surfaces of the AlP + GaSb ⇔ AlSb + GaP ternary mutual system. (Reprinted from *J. Cryst. Growth*, 98(1–2), Ishida, K. et al., Data base for calculating phase diagrams of III-V alloy semiconductors, 140–147, Copyright (1989), with permission from Elsevier.)

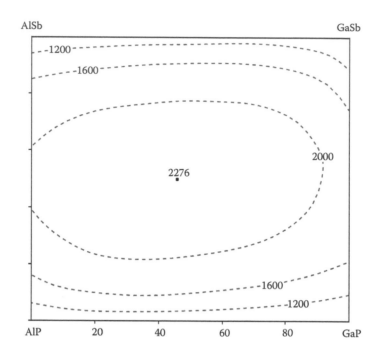

FIGURE 5.6 Miscibility gaps in the AlP + GaSb ⇔ AlSb + GaP ternary mutual system at 1200–2000°C. (Reprinted from *J. Cryst. Growth*, 98(1–2), Ishida, K. et al., Data base for calculating phase diagrams of III-V alloy semiconductors, 140–147, Copyright (1989), with permission from Elsevier.)

5.25 ALUMINUM–INDIUM–ARSENIC–PHOSPHORUS

AlP + InAs ⇔ AlAs + InP. The calculated liquidus and solidus surfaces for this system are given in Figure 5.7, and the miscibility gap is presented in Figure 5.8 (Ishida et al. 1989). According to the data by Stringfellow (1982b), the calculated critical point for this system is 746°C [847°C (Ishida et al. 1989)].

5.26 ALUMINUM–INDIUM–ANTIMONY–PHOSPHORUS

AlP + InSb ⇔ AlSb + InP. The calculated liquidus surface for this system is given in Figure 5.9, and the miscibility gap is presented in Figure 5.10 (Ishida et al. 1989). According to the data by Stringfellow (1982b), the calculated critical point for this system is 2346°C [2276°C (Ishida et al. 1989)].

5.27 ALUMINUM–EUROPIUM–PLATINUM–ANTIMONY

$Eu_2Pt_7AlP_{2.95}$ quaternary compound, which crystallizes in the tetragonal structure with the lattice parameters $a = 404.6 \pm 0.1$ and $c = 2685.0 \pm 0.2$ pm and experimental and calculated densities of 13.21 and 13.52 g·cm^{-3}, respectively, is formed in the Al–Eu–Pt–P system (Lux et al. 1991). It was prepared by heating the respective elements in a stoichiometric ratio

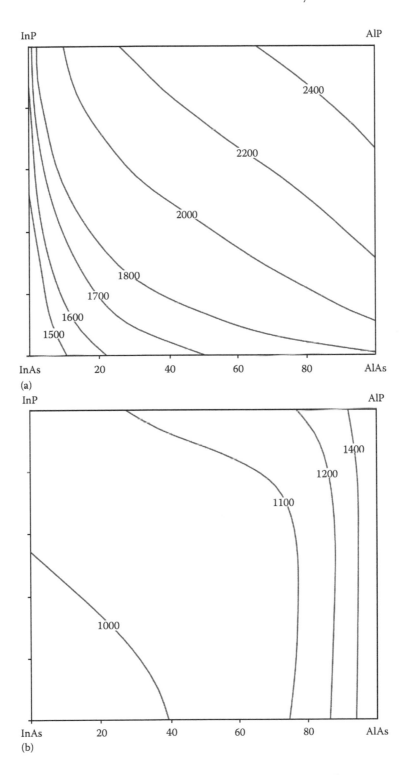

FIGURE 5.7   Liquidus (a) and solidus (b) surfaces of the AlP + InAs ⇔ AlAs + InP ternary mutual system. (Reprinted from *J. Cryst. Growth*, 98(1–2), Ishida, K. et al., Data base for calculating phase diagrams of III-V alloy semiconductors, 140–147, Copyright (1989), with permission from Elsevier.)

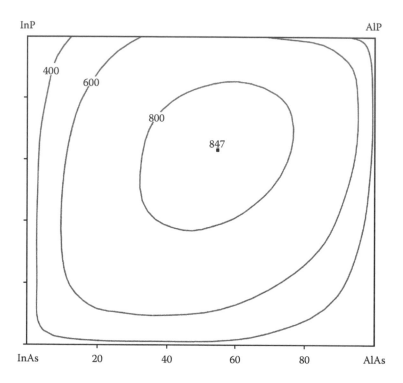

FIGURE 5.8   Miscibility gaps in the AlP + InAs ⇔ AlAs + InP ternary mutual system at 400–800°C. (Reprinted from *J. Cryst. Growth*, 98(1–2), Ishida, K. et al., Data base for calculating phase diagrams of III-V alloy semiconductors, 140–147, Copyright (1989), with permission from Elsevier.)

corresponding to the formula (corundum crucible, melted quartz glass vials, Ar atmosphere). The reaction temperature was 1050°C, and the reaction times was 2–3 weeks, the annealing being interrupted several times in order to grind still inhomogeneous preparation in the Ar glove box.

5.28 ALUMINUM–ARSENIC–ANTIMONY–PHOSPHORUS

AlP–AlAs–AlSb. The calculated liquidus and solidus surfaces for this system are given in Figure 5.11, and the miscibility gap is presented in Figure 5.12 (Ishida et al. 1989). The calculated critical point for this system is 2033°C.

5.29 ALUMINUM–OXYGEN–MANGANESE–PHOSPHORUS

$Mn_xAl_{1-x}PO_4$ (0.4 < x < 0.5) quaternary phase, which exists in two polymorphic modifications, is formed in the Al–O–Mn–P system (Bu et al. 1997). First modification crystallizes in the hexagonal structure with the lattice parameters $a = 1781.7$ and $c = 2753.1$ pm and the second one crystallizes in the tetragonal structure with the lattice parameters $a = 1939.0$ and $c = 2706.7$ pm. This material was hydrothermally made at 180°C with the reaction

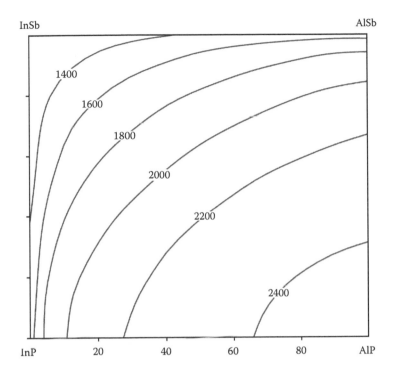

FIGURE 5.9 Liquidus surface of the AlP + InSb ⇔ AlSb + InP ternary mutual system. (Reprinted from *J. Cryst. Growth*, 98(1–2), Ishida, K. et al., Data base for calculating phase diagrams of III-V alloy semiconductors, 140–147, Copyright (1989), with permission from Elsevier.)

times between 3 and 5 days. Starting materials included $(Me_2CHO)_3Al$, $MnCO_3$, and 85% H_3PO_4.

5.30 ALUMINUM–OXYGEN–IRON–PHOSPHORUS

FeAlPO$_5$ quaternary compound, which crystallizes in the monoclinic structure with the lattice parameters $a = 712.3 \pm 0.2$, $b = 1052.8 \pm 0.3$, $c = 584.1 \pm 0.2$ pm, and $\beta = 97.91 \pm 0.02°$ (Schmid-Beurmann et al. 1997) [$a = 714.1 \pm 0.3$, $b = 1054.9 \pm 0.4$, $c = 549.3 \pm 0.3$ pm, and $\beta = 98.19 \pm 0.09°$ and a calculated density of 3.14 g·cm^{-3}, is formed in the Al–O–Fe–P system (Hesse and Cemič 1994a)]. The title compound was formed as the decomposition product of **FeAl$_2$(PO$_4$)$_2$(OH)$_2$** at 0.2 GPa water pressure and at temperatures above 500°C, at an oxygen fugacity defined by the Ni/NiO buffer (Schmid-Beurmann et al. 1997; Hesse and Cemič 1994a). It could also be prepared from mixtures of AlPO$_4$ and FeO (Schmid-Beurmann et al. 1997). The mixtures were placed in a silica glass ampoule, together with Fe pellets to keep the oxygen fugacity low, and reacted at 800°C for 2 days.

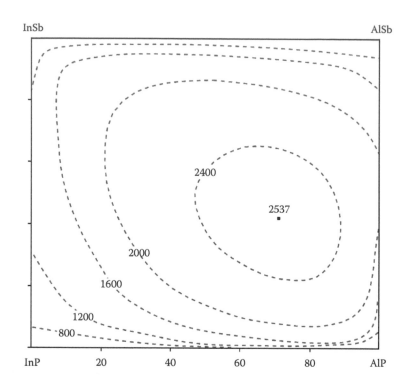

FIGURE 5.10 Miscibility gaps in the AlP + InSb ⇔ AlSb + InP ternary mutual system at 1200–2000°C. (Reprinted from *J. Cryst. Growth*, 98(1–2), Ishida, K. et al., Data base for calculating phase diagrams of III-V alloy semiconductors, 140–147, Copyright (1989), with permission from Elsevier.)

5.31 ALUMINUM–OXYGEN–COBALT–PHOSPHORUS

$Co_xAl_{1-x}PO_4$ and $Co_{1+x}Al_{1-x}P_2O_8$ quaternary phases are formed in the Al–O–Co–P system. $Co_{0.33}Al_{0.67}PO_4$ crystallizes in the cubic structure with the lattice parameter a = 2480.2 pm (Feng et al. 1997). $Co_{0.45}Al_{0.55}PO_4$ crystallizes in the hexagonal structure with the lattice parameters a = 1770.1 and c = 2734.9 pm, and $Co_xAl_{1-x}PO_4$ ($0.4 < x < 0.5$) crystallizes in two polymorphic modifications: tetragonal structure with the lattice parameters a = 1906.5 and c = 2759.4 pm and rhombohedral structure with the lattice parameters a = 1770.4 and c = 4169.0 pm (in hexagonal setting) (Bu et al. 1997). $Co_{0.8}Al_{0.2}PO_4$ crystallizes in the monoclinic structure with the lattice parameters a = 1439.2, b = 942.2, c = 1001.5 pm, and β = 132.76°, and $Co_{0.9}Al_{0.1}PO_4$ crystallizes in the trigonal structure with the lattice parameters a = 1389.0, c = 1532.9 pm (Feng et al. 1997).

$Co_{1+x}Al_{1-x}P_2O_8$ crystallizes in the orthorhombic structure with the lattice parameters a = 2014.7, b = 2051.5, and c = 1002.4 pm (Feng et al. 1997).

$Co_xAl_{1-x}PO_4$ ($0.4 < x < 0.5$) was hydrothermally made at 180°C with the reaction times between 3 and 5 days. Starting materials included $(Me_2CHO)_3Al$, $CoCO_3 \cdot xH_2O$, and 85% H_3PO_4 (Bu et al. 1997).

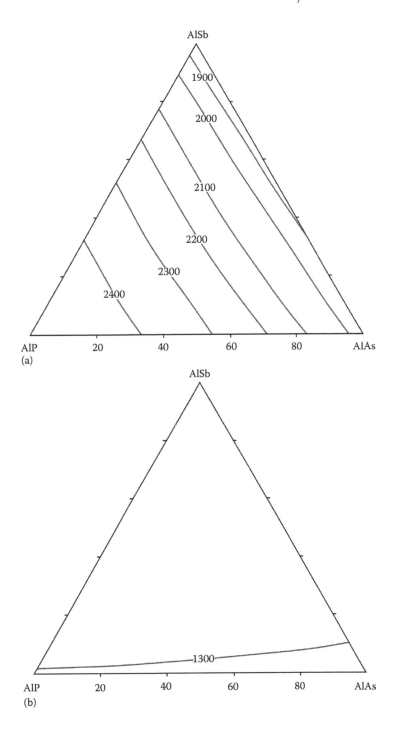

FIGURE 5.11 Liquidus (a) and solidus (b) surfaces of the AlP–AlAs–AlSb quasiternary system. (Reprinted from *J. Cryst. Growth*, 98(1–2), Ishida, K. et al., Data base for calculating phase diagrams of III-V alloy semiconductors, 140–147, Copyright (1989), with permission from Elsevier.)

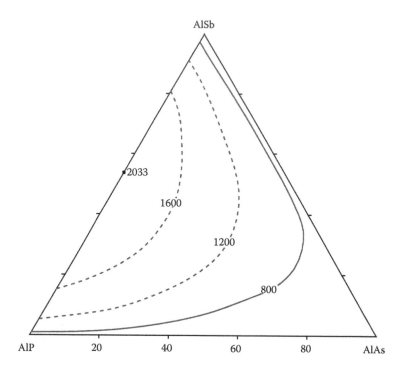

FIGURE 5.12 Miscibility gaps in the AlP–GaP–InP quasiternary system at 800–2000°C. (Reprinted from *J. Cryst. Growth*, 98(1–2), Ishida, K. et al., Data base for calculating phase diagrams of III-V alloy semiconductors, 140–147, Copyright (1989), with permission from Elsevier.)

CHAPTER 6

Systems Based on AlAs

6.1 ALUMINUM–HYDROGEN–OXYGEN–ARSENIC

Some quaternary compounds are formed in the Al–H–O–As system. **AlAsO$_4$·2H$_2$O** crystallizes in the orthorhombic structure with the lattice parameters $a = 882.18 + 0.05$, $b = 982.52 + 0.06$, and $c = 1011.63 + 0.06$ pm and experimental and calculated densities of 3.031 [for mineral mansfieldite (Allen et al. 1948)] and 3.06 g·cm^{-3}, respectively (Harrison 2000). To prepare this compound, a starting mixture of KOH (1 M, 4 mL), Al(NO$_3$)$_3$ (0.5 M, 8 mL), and H$_3$AsO$_4$ (1 M, 4 mL) was heated to 170°C in a 23 mL capacity Teflon-lined hydrothermal bomb for 4 days (Harrison 2000). Upon cooling the bomb to ambient temperature over a period of 2–3 h, a crop of small blocks and octahedra of the title compound was recovered by vacuum filtration and washing with water and acetone.

Al$_2$(AsO$_4$)(OH)$_3$·3H$_2$O (mineral bulachite) also crystallizes in the orthorhombic structure with the lattice parameters $a = 1553$, $b = 1778$, and $c = 703$ pm and experimental and calculated densities of 2.60 and 2.55 g·cm^{-3}, respectively (Walenta 1983; Dunn et al. 1985).

Al$_2$O$_3$·3As$_2$O$_5$·10H$_2$O or **Al(H$_2$AsO$_4$)$_3$·2H$_2$O** also crystallizes in the orthorhombic structure with the lattice parameters $a = 1230$, $b = 464$, and $c = 861$ pm and experimental and calculated densities of 3.21 and 3.27 g·cm^{-3}, respectively (Katz and Kedesdy 1954). In an attempt to grow large crystals of AlAsO$_4$ by the hydrothermal method, a dense polycrystalline Al(H$_2$AsO$_4$)$_3$·2H$_2$O resulted. It was in the form of fibrous crystallites in parallel arrangement which readily parted into fine, clear needles.

(H$_3$O)Al$_4$(AsO$_4$)$_3$(OH)$_4$·4-5H$_2$O (mineral hydroniumpharmacoalumite) crystallizes in the cubic structure with the lattice parameter $a = 772.69 + 0.02$ pm [$a = 773.79 + 0.01$ pm (Hochleitner et al. 2013)] and a calculated density of 2.486 g·cm^{-3} (Hochleitner et al. 2015).

Al$_6$(AsO$_4$)$_3$(OH)$_9$·16H$_2$O (mineral bettertonite) crystallizes in the monoclinic structure with the lattice parameters $a = 777.3 \pm 0.2$, $b = 2699.1 \pm 0.5$, $c = 1586.7 \pm 0.3$ pm, and $\beta = 94.22 \pm 0.03°$ and a calculated density of 2.02 cm^{-3} (Grey et al. 2015a, 2015b).

6.2 ALUMINUM–LITHIUM–OXYGEN–ARSENIC

The **LiAlAs$_2$O$_7$** and **Li$_3$Al$_2$(AsO$_4$)$_3$** quaternary compounds are formed in the Al–Li–O–As system. The first of them crystallizes in the monoclinic structure with the lattice parameters

$a = 658.3 \pm 0.1$, $b = 800.7 \pm 0.2$, $c = 463.5 + 0.1$ pm and $\beta = 104.13 \pm 0.03°$ and a calculated density of 4.146 g.cm^{-3} (Schwendtner and Kolitsch 2007). This compound was prepared by mild hydrothermal method in Teflon-lined stainless steel autoclave (220°C, slow furnace cooling). Small, colorless, hemimorphic, rounded triangular crystals were prepared (6 days, pH = 1) from a mixture of Li_2CO_3, Al_2O_3, H_3AsO_4·0.5H_2O (molar ratio ~1:1:3) and distilled water. The Teflon container was filled with distilled water to ~70-80% of its inner volume.

$Li_3Al_2(AsO_4)_3$ crystallizes in the rhombohedral structure with the lattice parameters $a = 1314.4 \pm 0.2$ and $c = 1828.2 \pm 0.3$ pm (in hexagonal setting) (d'Yvoire et al. 1986). It was obtained by ion exchange reaction.

6.3 ALUMINUM–SODIUM–POTASSIUM–ARSENIC

The NaK_2AlAs_2 quaternary compound, which crystallizes in the orthorhombic structure with the lattice parameters $a = 1486.7 \pm 0.3$, $b = 674.2 \pm 0.1$, and $c = 649.9 \pm 0.1$ pm and a calculated density of 2.85 g·cm^{-3}, is formed in the Al–Na–K–As system (Ohse et al. 1993). It was prepared from stoichiometric mixture of the elements or from Na, NaAs, K, and AlAs at 730°C.

6.4 ALUMINUM–SODIUM–CALCIUM–ARSENIC

The $Na_3Ca_3AlAs_4$ quaternary compound, which crystallizes in the hexagonal structure with the lattice parameters $a = 939.9 \pm 0.1$ and $c = 721.0 \pm 0.1$ pm, is formed in the Al–Na–Ca–As system (Somer et al. 1996c). The title compound was synthesized from a stoichiometric mixture of Na_3As, CaAs, and AlAs in a sealed Nb ampoule at 1100°C.

6.5 ALUMINUM–SODIUM–EUROPIUM–ARSENIC

The $Na_{3.10\pm0.02}Eu_{2.90}AlAs_4$ quaternary compound, which crystallizes in the hexagonal structure with the lattice parameters $a = 956.21 \pm 0.07$ and $c = 744.18 \pm 0.11$ pm and a calculated density of 4.73 g·cm^{-3} at 200 K, is formed in the Al–Na–Eu–As system (Wang et al. 2017a). The title compound was prepared from the elements. The starting materials were weighed in the glove box, loaded into the Nb tube, which was then closed by arc-welding under high-purity argon gas. The sealed Nb container was subsequently jacketed within evacuated fused silica tube. The reaction mixture was heated in a muffle furnace to 900°C at a rate of 100°C·h^{-1} and equilibrated at this temperature for 24 h. After that, the samples were cooled to 500°C at a rate of 5°C·h^{-1}, at which point the cooling rate to room temperature was changed to 40°C·h^{-1}. All manipulations were carried out inside an argon-field glove box with controlled atmosphere or under vacuum.

6.6 ALUMINUM–SODIUM–OXYGEN–ARSENIC

The $NaAlAs_2O_7$, $Na(Al_{1,2}As_{0,5})(As_2O_7)_2$, $Na_3Al_2(AsO_4)_3$, $Na_3Al_3(AsO_4)_4$, and Na_7Al_3 $(As_2O_7)_4$ quaternary compounds are formed in the Al–Na–O–As system. $NaAlAs_2O_7$ crystallizes in the monoclinic structure with lattice parameters $a = 691.14 \pm 0.07$, $b = 813.45 \pm 0.12$, $c = 954.46 \pm 0.10$ pm, and $\beta = 107.51 \pm 0.01°$ and experimental and calculated densities of 4.02 and 4.05 g·cm^{-3}, respectively (Driss and Jouini 1994). Single crystals of this compound were precipitated from mixtures with Na/As molar ratio ranging from 0.5 to 1 by the flux method and by the hydrothermal route. The best crystals were obtained by the flux method from a mixture of $NaAsO_3$ and As_2O_3 taken in a molar ratio of Na/As = 0.7 and a small quantity of $Al(OH)_3$ (molar ratio of Na/Al = 5). The mixture was melted at 650°C and left at this temperature for 3 h. The temperature was then lowered to 620°C and stabilized for 12 h. After cooling, the solid mass washed thoroughly with boiling water releases crystals in the form of thin rods with 0.5 mm in length.

$Na(Al_{1.5}As_{0.5})(As_2O_7)_2$ crystallizes in the triclinic structure with the lattice parameters $a = 772.7 \pm 0.4$, $b = 711.8 \pm 0.2$, and $c = 483.9 \pm 0.2$ pm and $\alpha = 104.43 \pm 0.03$, $\beta = 93.71 \pm 0.03°$, and $\gamma = 90.07 \pm 0.04°$ and experimental and calculated densities of 4.02 and 4.03 g·cm^{-3}, respectively (Driss and Jouini 1989).

$Na_3Al_2(AsO_4)_3$ melts congruently at 824°C (d'Yvoire et al. 1986) and undergoes cooling, near 44°C, a reversible phase transition $\alpha \Leftrightarrow \beta$ associated with a slight monoclinic distortion (d'Yvoire et al. 1986, 1988; Masquelier et al. 1995). The fused title compound can be easily quenched into a totally glassy transparent solid (d'Yvoire et al. 1986). α-$Na_3Al_2$$(AsO_4)_3$ crystallizes in the monoclinic structure with the lattice parameters $a = 1457.6 \pm 0.6$, $b = 1340.9 \pm 0.6$, $c = 972.8 \pm 0.5$ pm, and $\beta = 96.95 \pm 0.04°$ and a calculated density of 3.80 g·cm^{-3} (Masquelier et al. 1995) [$a = 1455.5 \pm 0.2$, $b = 1330.1 \pm 0.2$, $c = 978.5 \pm 0.1$ pm, and $\beta = 96.85 \pm 0.01°$ (d'Yvoire et al. 1986)]. β-$Na_3Al_2(AsO_4)_3$ crystallizes in the rhombohedral structure with the lattice parameters $a = 1328.1 \pm 0.2$ and $c = 1846.7 \pm 0.3$ pm at 50°C (in hexagonal setting) (d'Yvoire et al. 1986; Masquelier et al. 1995).

Crystals of $Na_3Al_2(AsO_4)_3$ were grown in a flux of sodium arsenate from the starting mixture of $Al(OH)_3$, $Na_4As_2O_7$, and $NaH_2AsO_4·H_2O$ (molar ratio 2:1:4) (Masquelier et al. 1995). The mixture was progressively heated and melted up to 765°C, cooled to 670°C at a rate of 2°C·min^{-1}, and then quenched. The resulting solid was mainly vitreous, but some crystals could be isolated after a prolonged immersion in H_2O.

The title compound could be also obtained as follows (d'Yvoire et al. 1986): The mixture of $Al(OH)_3$, $Na_4As_2O_7$, and As_2O_5 (molar ratio 10:5:4) was progressively heated to 760°C and cooled down in several hours. Two or three successive heating separated by grinding operations were often necessary to complete the reaction. Single crystals of $Na_3Al_2(AsO_4)_3$ were prepared by flux method (d'Yvoire et al. 1986).

$Na_3Al_3(AsO_4)_4$ is formed at the cooling of the melting $Na_3Al_2(AsO_4)_3$ (d'Yvoire et al. 1986).

$Na_7Al_3(As_2O_7)_3$ undergoes a reversible order–disorder transition at 30–36°C (Masquelier et al. 1994). XRD shows that α-$Na_7Al_3(As_2O_7)_3$ is a superstructure of the β form. This compound was prepared by crystallization in a flux of sodium arsenate. Starting materials were $Al(OH)_3$, NaH_2PO_4, and $Na_4P_2O_7$. An intimate mixture containing Na_2O, Al_2O_3, and As_2O_5 (molar ratio 51:7:42) was placed in a Pt crucible and progressively heated up to 685°C. Cooling was achieved at a rate of 2°C·h^{-1} down to 480°C and then switching off the furnace. The resulting product was washed with H_2O.

6.7 ALUMINUM–POTASSIUM–OXYGEN–ARSENIC

The $KAlAs_2O_7$ and $K_3Al_2(AsO_4)_3$ quaternary compounds are formed in the Al–K–O–As system. $KAlAs_2O_7$ crystallizes in the triclinic structure with the lattice parameters $a = 619.2 \pm 0.4$, $b = 629.7 \pm 0.3$, and $c = 810.60 \pm 0.10$ pm and $\alpha = 96.600 \pm 0.008$,

$\beta = 104.517 \pm 0.008°$, and $\gamma = 102.864 \pm 0.007°$ and a calculated density of 3.712 g.cm^{-3} (Boughzala and Jouini 1995).

$K_3Al_2(AsO_4)_3$ crystallizes in the rhombohedral structure with the lattice parameters $a = 1338.9 \pm 0.2$ and $c = 1898.1 \pm 0.6$ pm (in hexagonal setting) (d'Yvoire et al. 1986). It was obtained by ion exchange reaction.

6.8 ALUMINUM–RUBIDIUM–OXYGEN–ARSENIC

The **RbAlAs$_2$O$_7$** quaternary compound, which crystallizes in the triclinic structure with lattice parameters $a = 823.3 \pm 0.5$, $b = 634 \pm 2$, and $c = 624.1 \pm 0.5$ pm and $\alpha = 102.6 \pm 0.1$, $\beta = 103.89 \pm 0.07°$, and $\gamma = 96.7 \pm 0.1°$ and experimental and calculated densities of 3.9 $\pm$ 0.3 and 4.09 g·cm^{-3}, respectively, is formed in the Al–Rb–O–As system (Boughzala et al. 1993).

6.9 ALUMINUM–COPPER–OXYGEN–ARSENIC

The **CuAlAsO$_5$** or **CuAlO(AsO$_4$)** quaternary compound (mineral urusovite), which crystallizes in the monoclinic structure with the lattice parameters $a = 731.4 \pm 0.2$, $b = 1022.3 \pm 0.3$, $c = 557.6 \pm 0.2$ pm, and $\beta = 99.79 \pm 0.03°$ and a calculated density of 3.93 g·cm^{-3}, is formed in the Al–Cu–O–As system (Vergasova et al. 2000) [$a = 733.5 \pm 0.1$, $b = 1025.5 \pm 0.1$, $c = 559.9 \pm 0.1$ pm, and $\beta = 99.79 \pm 0.01°$ and a calculated density of 3.93 g·cm^{-3} (Krivovichev et al. 2000)].

6.10 ALUMINUM–SILVER–OXYGEN–ARSENIC

The **Ag$_3$Al$_2$(AsO$_4$)$_3$** quaternary compound, which crystallizes in the rhombohedral structure with the lattice parameters $a = 1323.2 \pm 0.3$ and $c = 1913.7 \pm 0.4$ pm (in hexagonal setting), is formed in the Al–Ag–O–As system (d'Yvoire et al. 1986). It was obtained by ion exchange reaction.

6.11 ALUMINUM–GALLIUM–INDIUM–ARSENIC

The phase diagram of the Al–Ga–In–As quaternary system has been determined experimentally for several Al isoconcentration sections by Nakajima et al. (1975). The liquidus data were obtained by differential thermal analysis (DTA), and the solidus data were determined from liquid epitaxial layers grown on GaAs substrates by using electron probe X-ray microanalysis. The phase diagram was calculated by using a simple solution model. Figure 6.1 shows the liquidus isotherms and the AlAs and InAs solidus isotherms on the entire cross section at 0.2 at% Al in the liquid.

AlAs–GaAs–InAs. The calculated liquidus and solidus surfaces for this system are given in Figure 6.2, and the miscibility gap is presented in Figure 6.3 (Ishida et al. 1989). The thermodynamics of spinodal decomposition in the AlAs–GaAs–InAs quasiternary system have been developed in Stringfellow (1983). Based on the delta-lattice parameter solution model of the free energy of mixing of semiconductor alloys, an analysis has been developed for the calculation of the spinodal surface and the critical temperature for solid alloys. Solid–solid isotherms were presented for this system. Concepts are also developed for the thermodynamic analysis of spinodal decomposition in this quaternary alloys, including the effect of the coherency strain energy. This addition to the free energy of the inhomogeneous solid is shown to completely

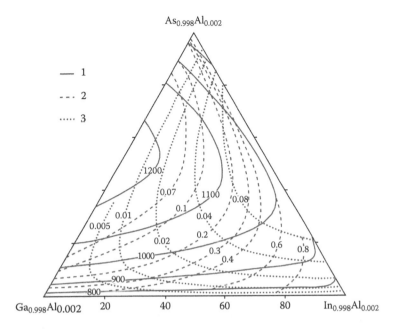

FIGURE 6.1 Calculated Al–Ga–In–As quaternary phase diagram on the entire cross section of $x^l_{Al} =$ 0.002: 1, calculated liquidus isotherms; 2, calculated AlAs solidus isotherms; 3, calculated InAs solidus isotherms. (From Nakajima, K., and Osamura, K., *J. Electrochem. Soc.*, 122(9), 1245–1248, 1975. With permission.)

stabilize the alloys of interest even at temperatures below room temperature. The calculated critical point for this system is 462°C (Stringfellow 1983) [479°C (Ishida et al. 1989)].

6.12 ALUMINUM–GALLIUM–GERMANIUM–ARSENIC

Isothermal liquidus compositions for fixed values of the Al content have been calculated at 800°C (Figure 6.4) (Ansara and Dutartre 1984). It should be observed that large amounts of Ge have only a small effect on the solubility of As at a given Al composition. In Figure 6.5, the calculated solidus lines at 800°C are shown for different Al contents in the liquid phase. The calculations indicate that at a given concentration of Al in the liquid, the mole fraction of AlAs in the solid phase decreases slightly when Ge is added to the liquid (up to 40 at% Ge). The experimental values are in agreement with this result.

6.13 ALUMINUM–GALLIUM–TIN–ARSENIC

Two techniques were used for the determination of the composition of the Al–Ga–Sn–As quaternary liquid in equilibrium with the solid solution—the "seed dissolution technique" and the "direct observation technique" (Panish 1973a). To determine the composition of the solid solution in equilibrium with a given liquidus, epitaxial layers of Sn-doped $Al_xGa_{1-x}As$ were grown onto GaAs substrates. The calculated liquidus curves at 800°C and 1000°C and the calculated solidus isotherms at 700°C, 800°C, and 1000°C for the Ga-rich part of this quaternary system for fixed values of the Al content are given in Figures 6.6 and 6.7. It is seen that a very small amount of Al in the liquid has a drastic effect upon the equilibrium compositions of both the solid and the liquid.

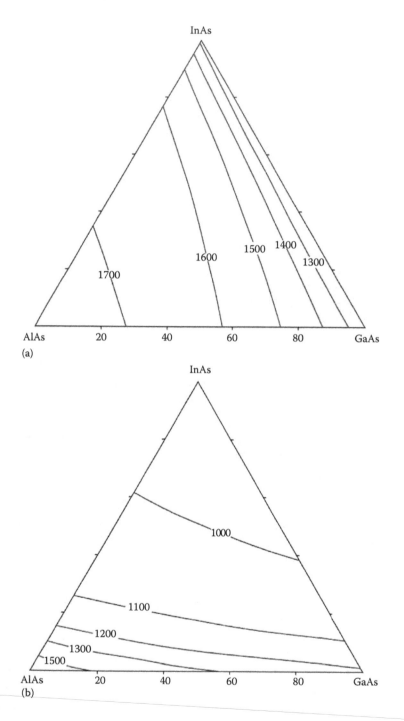

FIGURE 6.2 Liquidus (a) and solidus (b) surfaces of the AlAs–GaAs–InAs quasiternary system. (Reprinted from *J. Cryst. Growth*, 98(1–2), Ishida, K. et al., Data base for calculating phase diagrams of III-V alloy semiconductors, 140–147, Copyright (1989), with permission from Elsevier.)

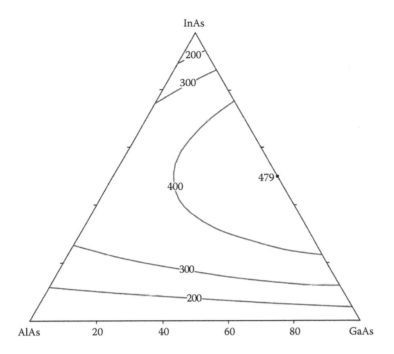

FIGURE 6.3 Miscibility gaps in the AlAs–GaAs–InAs quasiternary system at 200–400°C. (Reprinted from *J. Cryst. Growth*, 98(1–2), Ishida, K. et al., Data base for calculating phase diagrams of III-V alloy semiconductors, 140–147, Copyright (1989), with permission from Elsevier.)

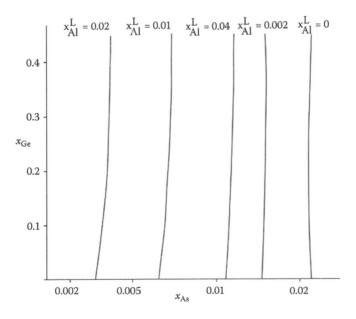

FIGURE 6.4 Calculated liquidus isotherms of the Al–Ga–Ge–As system for several fixed values of the Al content at 800°C. (Reprinted from *Calphad*, 8(4), Ansara, I., and Dutartre, D., Thermodynamic study of the Al–Ga–As–Ge system, 323–342, Copyright (1984), with permission from Elsevier.)

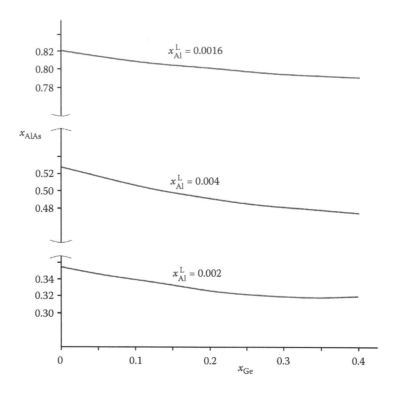

FIGURE 6.5   Calculated solidus isotherms of the Al–Ga–Ge–As system for several fixed values of the Al content at 800°C. (Reprinted from *Calphad*, 8(4), Ansara, I., and Dutartre, D., Thermodynamic study of the Al–Ga–As–Ge system, 323–342, Copyright (1984), with permission from Elsevier.)

6.14 ALUMINUM–GALLIUM–ANTIMONY–ARSENIC

AlAs + GaSb ⇔ AlSb + GaAs. The calculated liquidus and solidus surface surfaces for this system are given in Figure 6.8, and the miscibility gap is presented in Figure 6.9 (Ishida et al. 1989). According to the data of Stringfellow (1982b), the calculated critical point for this system is 700°C [1716°C (Ishida et al. 1989)]. The solidus isotherms for the AlAs + GaSb ⇔ AlSb + GaAs ternary mutual system was also constructed by Stringfellow (1982a). The extent of the solid phase miscibility gap in this system was also reported at 700°C by Pessetto and Stringfellow (1983). The bimodal data were found to be in agreement with the calculated bimodal curves. The Al distribution coefficient has been determined.

Spinodal decomposition was calculated for $Al_xGa_{1-x}As_ySb_{1-y}$ alloys lattice-matched to GaSb and InAs substrates by Asomoza et al. (2001a). Spinodal decomposition in this alloy due to mechanical and chemical origins is presented. It was shown that the spinodal decomposition temperatures are always smaller than their growth temperatures.

First-principles calculations were performed to study the structural, electronic, optical, and thermodynamic properties of technologically important $Al_xGa_{1-x}As_ySb_{1-y}$ quaternary alloys using the full potential-linearized augmented plane wave plus local orbital method within the density functional theory (El Haj Hassan et al. 2010). The investigation on the effect of composition on lattice constant, bulk modulus, and band gap for quasibinary as well as for quaternary alloys shows nonlinear dependence on the composition.

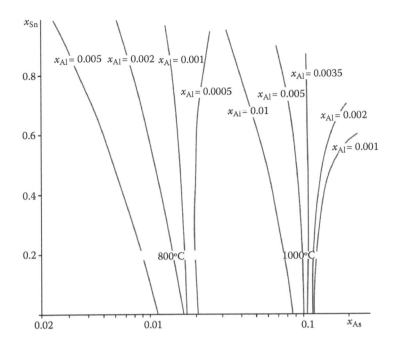

FIGURE 6.6   Calculated liquidus isotherms at 800°C and 1000°C for the Ga-rich part of the Al–Ga–Sn–As system for fixed values of the Al content. (Reprinted with permission from Panish, M.B., Phase equilibria in the system Al–Ga–As–Sn and electrical properties of Sn-doped liquid phase epitaxial $Al_xGa_{1-x}As$", J. Appl. Phys., B44(6), 2667–2675, Copyright 1973, American Institute of Physics.)

The calculation of formation energies showed that the least stable alloys are around 65 at% Al and 60 at% As.

A thermodynamic calculation of the melting diagram of this ternary mutual system was carried out using a model based on the third-order virial expansion for the average Gibbs molar energy by the number of moles of phase components (Baranov et al. 1990). The calculated data are in good agreement with the experimental data obtained at 600°C.

Layers of the $Al_xGa_{1-x}As_{1-y}Sb_y$ solid solutions were grown by liquid phase epitaxy with compositions over the range $0 < x < 0.2$ and $1 < y < 0.4$ at temperatures between 715°C and 785°C (Nahory et al. 1978). Below 715°C, a miscibility gap was observed in the solid composition of the grown layers. The concentration dependence of the energy gap for this solid solutions could be expressed by the next equation: $E_g = 1.43 - 1.9x + 1.2x^2 + 1.042(1 - y) + 0.468(1 - y)^2 + 1.0x(1 - y) - 0.1x(1 - y)^2 - 0.9\,x^2(1 - y)$. The compositions y in solid solutions versus x_{Al}^L are shown in Figure 6.10. It is seen that y is only weakly dependent on temperatures between 785°C and 745°C. Figure 6.11 shows the solidus compositions y versus x. It could be noted that x increases with decreasing temperature for a fixed value of y, and at given temperature y decreases as x increases.

6.15 ALUMINUM–GALLIUM–CHROMIUM–ARSENIC

The liquidus isotherms in the Ga corner of the GaAs–Al–Ga–Cr system were determined by the solubility method and using DTA at the temperatures between 900°C and 1050°C

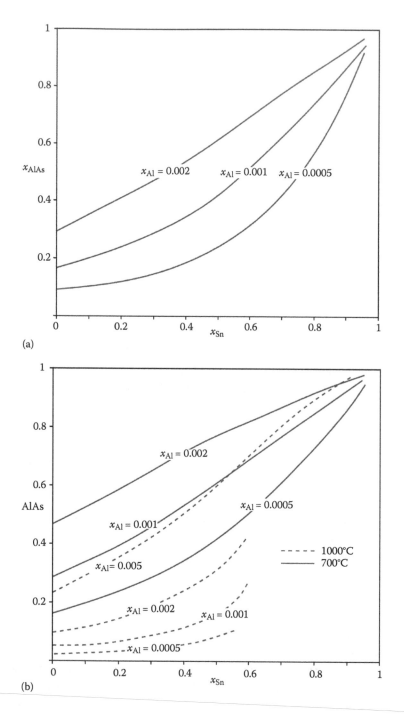

FIGURE 6.7   Calculated solidus isotherms at (a) 800°C and (b) 700°C and 1000°C for the Ga-rich part of the Al–Ga–Sn–As system for fixed values of the Al content. (Reprinted with permission from Panish, M.B., Phase equilibria in the system Al–Ga–As–Sn and electrical properties of Sn-doped liquid phase epitaxial Al$_x$Ga$_{1-x}$As", *J. Appl. Phys.*, B44(6), 2667–2675, Copyright 1973, American Institute of Physics.)

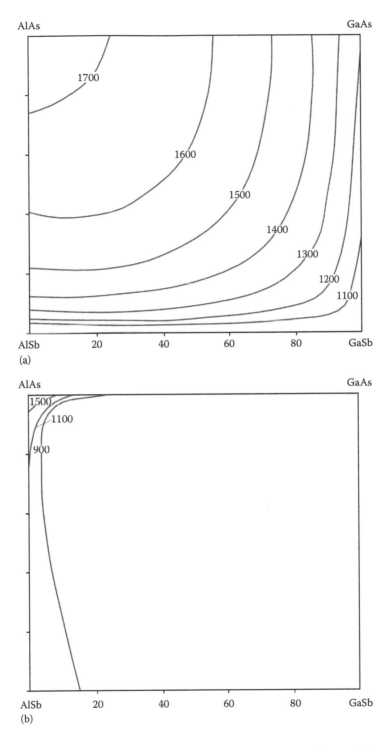

FIGURE 6.8 Liquidus (a) and solidus (b) surfaces of the AlAs + GaSb ⇔ AlSb + GaAs ternary mutual system. (Reprinted from *J. Cryst. Growth*, 98(1–2), Ishida, K. et al., Data base for calculating phase diagrams of III-V alloy semiconductors, 140–147, Copyright (1989), with permission from Elsevier.)

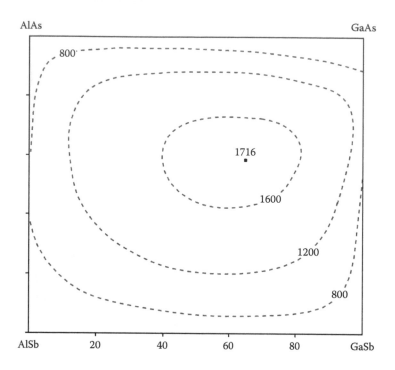

FIGURE 6.9   Miscibility gaps in the AlAs + GaSb ⇔ AlSb + GaAs ternary mutual system at 800–1600°C. (Reprinted from *J. Cryst. Growth*, 98(1–2), Ishida, K. et al., Data base for calculating phase diagrams of III-V alloy semiconductors, 140–147, Copyright (1989), with permission from Elsevier.)

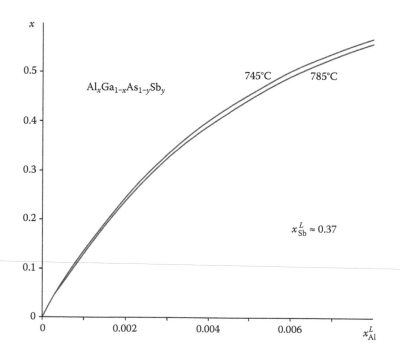

FIGURE 6.10   Compositions y in the $Al_xGa_{1-x}As_{1-y}Sb_y$ solid solutions versus x^L_{Al} at 745°C and 785°C. (From Nahory, R.E., *J. Electrochem.* Soc., 125(7), 1053–1058, 1978. With permission.)

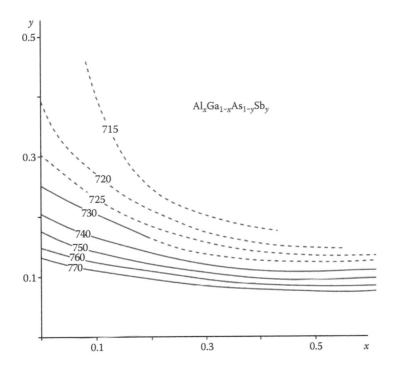

FIGURE 6.11 Solidus compositions x versus y for the $Al_xGa_{1-x}As_{1-y}Sb_y$ solid solutions at 715–770°C. (From Nahory, R.E., *J. Electrochem. Soc.*, 125(7), 1053–1058, 1978. With permission.)

(Saidov et al. 1979, 1981). It was shown that the system is susceptible to supercooling, the magnitude of which depends on the chromium content. Chromium, introduced in small amounts (up to 1.5 at%), leads to a monotonic increasing of the GaAs solubility in gallium solutions.

6.16 ALUMINUM–GALLIUM–NICKEL–ARSENIC

An approximate isothermal section at 800°C for the Al–Ga–Ni–As quaternary system at the constant composition of $Ga_{0.7}As_{0.3}$ was constructed by Députier et al. (1995) and included in the review of Raghavan (2006). It was shown that it looks like the isothermal section at this temperature for the Al–Ni–As ternary system.

6.17 ALUMINUM–INDIUM–ANTIMONY–ARSENIC

AlAs + InSb ⇔ AlSb + InAs. The calculated liquidus and solidus surfaces for this system are given in Figure 6.12, and the miscibility gap is presented in Figure 6.13 (Ishida et al. 1989). According to the data of Stringfellow (1982b), the calculated critical point for this system is 1326°C [1724°C (Ishida et al. 1989)].

6.18 ALUMINUM–SULFUR–CHLORINE–ARSENIC

The $(As_3S_5)[AlCl_4]$ quaternary compound, which crystallizes in the monoclinic structure with the lattice parameters $a = 640.48 \pm 0.04$, $b = 977.54 \pm 0.07$, $c = 1146.39 \pm 0.08$ pm and

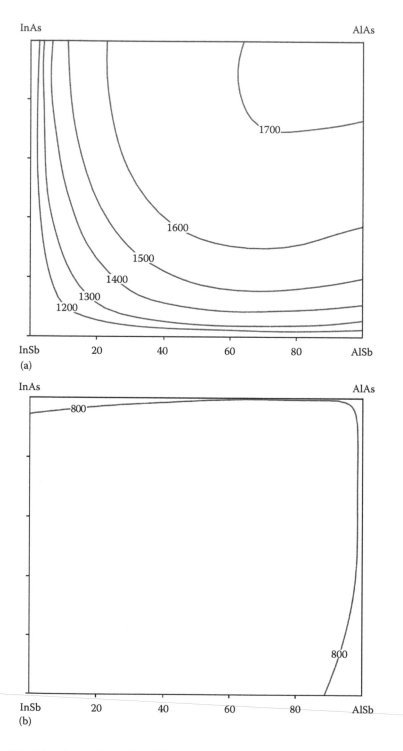

FIGURE 6.12 Liquidus (a) and solidus (b) surfaces of the AlAs + InSb ⇔ AlSb + InAs ternary mutual system. (Reprinted from *J. Cryst. Growth*, 98(1–2), Ishida, K. et al., Data base for calculating phase diagrams of III-V alloy semiconductors, 140–147, Copyright (1989), with permission from Elsevier.)

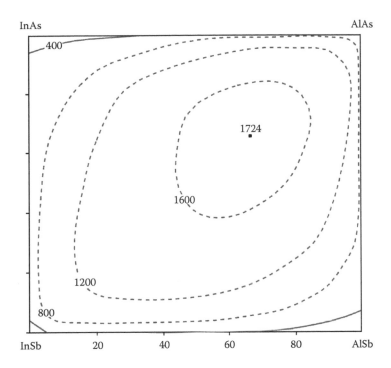

InAs   AlAs

400

1724

1600

1200

800

InSb   20   40   60   80   AlSb

FIGURE 6.13   Miscibility gaps in the AlAs + InSb ⇔ AlSb + InAs ternary mutual system at 400–1600°C. (Reprinted from *J. Cryst. Growth*, 98(1–2), Ishida, K. et al., Data base for calculating phase diagrams of III-V alloy semiconductors, 140–147, Copyright (1989), with permission from Elsevier.)

$\beta = 104.003 \pm 0.003°$ and a calculated density of 2.64 g·cm^{-3}, is formed in the Al–S–Cl–As system (Beck et al. 2005).

All manipulations including charging and opening of the reaction ampoules were performed in an argon-filled glove box or under a flux of argon. To prepare the title compound, in a glove box, As (11.25 mM), S (16.88 mM), and AlCl$_3$ (6.33 mM) were thoroughly powdered and filled in a glass ampoule. Outside the glove box, under a flow of Ar, AsCl$_3$ (3.52 mM) was added with the aid of a syringe. The content of the ampoule was frozen in a bath of dry ice/acetone, then the ampoule was evacuated and finally flame-sealed. The mixture was brought to equilibrium at a temperature of 80°C in a horizontal tube furnace for 3 days whereby a yellow melt formed. The ampoule was then cooled to 50°C within 3 days which caused the crystallization of yellow crystals. After opening of the ampoule, the small excess of AsCl$_3$ was distilled off under vacuum. The obtained compound is sensitive toward hydrolysis and oxidation by moisture and air.

6.19 ALUMINUM–SELENIUM–CHLORINE–ARSENIC

The (As$_3$Se$_5$)[AlCl$_4$] quaternary compound, which crystallizes in the monoclinic structure with the lattice parameters $a = 2080.57 \pm 0.02$, $b = 1251.41 \pm 0.01$, $c = 1367.68 \pm 0.01$ pm, and $\beta = 127.969 \pm 0.001°$ and a calculated density of 3.36 g·cm^{-3}, is formed in the Al–Se–Cl–As system (Beck et al. 2005).

All manipulations including charging and opening of the reaction ampoules were performed in an argon-filled glove box or under a flux of argon. To synthesize this compound, in a glove box, As (5.33 mM), Se (10 mM), AsCl$_3$ (2.38 mM), and AlCl$_3$ (2 mM) were filled in a glass ampoule as described previously for the obtaining of (As$_3$S$_5$)[AlCl$_4$]. A melt was formed on heating the ampoule to 80°C. On cooling to ambient temperature with 10°C per day, prism-shaped dark red crystals of (As$_3$Se$_5$)[AlCl$_4$] deposited. The ampoule was opened under a flux of Ar, and a small excess of AlCl$_3$ was distilled off in vacuum. The obtained compound is sensitive toward hydrolysis and oxidation by moisture and air.

CHAPTER 7

Systems Based on AlSb

7.1 ALUMINUM–HYDROGEN–OXYGEN–ANTIMONY

The **Al$_5$Sb$_3$O$_{14}$(OH)$_2$** quaternary compound (mineral bahianite), which crystallizes in the monoclinic structure with the lattice parameters a = 940.6 ± 0.6, b = 1154.1 ± 0.8, and c = 441.0 ± 0.3 pm and β = 90.94 ± 0.03° and calculated and experimental densities of 5.26 and 4.89–5.46 g·cm^{-3}, respectively, is formed in the Al–H–O–Sb system (Moore et al. 1978; Fleischer et al. 1979).

7.2 ALUMINUM–SODIUM–POTASSIUM–ANTIMONY

The **Na$_3$K$_6$Sb(AlSb$_3$)** quaternary compound, which crystallizes in the hexagonal structure with the lattice parameters a = 1016.8 ± 0.3 and c = 1052.7 ± 0.5 pm, is formed in the Al–Na–K–Sb system (Somer et al. 1995f). This compound was synthesized from a stoichiometric mixture of K, NaSb, and AlSb in sealed niobium ampoule at 650°C.

7.3 ALUMINUM–SODIUM–CALCIUM–ANTIMONY

Na$_{3.40±0.05}$Ca$_{2.60}$AlSb$_4$ or **Na$_3$Ca$_3$AlSb$_4$** and **Na$_{10.90±0.02}$Ca$_{2.10}$Al$_3$Sb$_8$** or **Na$_{11}$Ca$_2$Al$_3$Sb$_8$** quaternary compounds are formed in the Al–Na–Ca–Sb system. The first of them crystallizes in the hexagonal structure with the lattice parameters a = 1001.21 ± 0.05 and c = 760.45 ± 0.08 pm and a calculated density of 3.50 g·cm^{-3} at 200 K (Wang et al. 2017a). The title compound was prepared from the elements. The starting materials were weighed in the glove box and loaded into Nb tube, which was then closed by arc-welding under high-purity argon gas. The sealed Nb container was subsequently jacketed within evacuated fused silica tube. The reaction mixture was heated in a muffle furnace to 900°C at a rate of 100°C·h^{-1} and equilibrated at this temperature for 24 h. After that, the sample were cooled to 500°C at a rate of 5°C·h^{-1}, at which point the cooling rate to room temperature was changed to 40°C·h^{-1}. All manipulations were carried out inside an argon-field glove box with controlled atmosphere or under vacuum.

Na$_{10.90±0.02}$Ca$_{2.10}$Al$_3$Sb$_8$ crystallizes in the orthorhombic structure with the lattice parameters a = 1896.90 ± 0.15, b = 885.33 ± 0.07, and c = 793.04 ± 0.07 pm and a calculated density of 3.46 g·cm^{-3} at 200 K (Wang et al. 2015). Single crystals of this compound were first identified from exploratory reactions of the Na, Ca, Al, and Sb in the molar ratio of 3:3:1:4. The metals were weighed in the glove box (total weight circa 500 mg) and loaded into Nb tube, which was then closed by arc-welding under high-purity Ar gas. The sealed Nb container was subsequently jacketed within evacuated fused silica tube. The reaction mixture was heated to 900°C at a rate of 100°C·h^{-1} and equilibrated at this temperature for 24 h. Following a cooling step to 500°C at a rate of 5°C·h^{-1}, the product was brought to room temperature over 12 h. After the structure and chemical composition were established by single-crystal work, new reaction with the proper stoichiometry was loaded. It yielded this compound as major product. The produced sample contained small amounts of $Na_3Ca_3AlSb_4$, but the bulk consisted of small, shiny crystals with a dark-metallic luster that were the targeted phase. The obtained compound is brittle and air sensitive. All manipulations were carried out inside an argon-field glove box with controlled atmosphere or under vacuum.

7.4 ALUMINUM–SODIUM–STRONTIUM–ANTIMONY

The $Na_{3.18\pm0.01}Sr_{2.82}AlSb_4$ quaternary compound, which crystallizes in the hexagonal structure with the lattice parameters $a = 1015.87 \pm 0.10$ and $c = 795.09 \pm 0.15$ pm and a calculated density of 3.90 g·cm^{-3} at 200 K, is formed in the Al–Na–Sr–Sb system (Wang et al. 2017a). The title compound was synthesized from the elements. The starting materials were weighed in the glove box and loaded into Nb tube, which was then closed by arc-welding under high-purity argon gas. The sealed Nb container was subsequently jacketed within evacuated fused silica tube. The reaction mixture was heated in a muffle furnace to 900°C at a rate of 100°C·h^{-1} and equilibrated at this temperature for 24 h. After that, the sample were cooled to 500°C at a rate of 5°C·h^{-1}, at which point the cooling rate to room temperature was changed to 40°C·h^{-1}. All manipulations were carried out inside an argon-field glove box with controlled atmosphere or under vacuum.

7.5 ALUMINUM–SODIUM–EUROPIUM–ANTIMONY

The $Na_{3.41\pm0.01}Eu_{2.59}AlSb_4$ quaternary compound, which crystallizes in the hexagonal structure with the lattice parameters $a = 1012.72 \pm 0.05$ and $c = 784.53 \pm 0.08$ pm and a calculated density of 4.69 g·cm^{-3} at 200 K, is formed in the Al–Na–Eu–Sb system (Wang et al. 2017a). The title compound was obtained from the elements. The starting materials were weighed in the glove box and loaded into Nb tube, which was then closed by arc-welding under high-purity argon gas. The sealed Nb container was subsequently jacketed within evacuated fused silica tube. The reaction mixture was heated in a muffle furnace to 900°C at a rate of 100°C·h^{-1} and equilibrated at this temperature for 24 h. After that, the sample were cooled to 500°C at a rate of 5°C·h^{-1}, at which point the cooling rate to room temperature was changed to 40°C·h^{-1}. All manipulations were carried out inside an argon-field glove box with controlled atmosphere or under vacuum.

7.6 ALUMINUM–POTASSIUM–CESIUM–ANTIMONY

The $K_3Cs_6Sb(AlSb_3)$ quaternary compound, which crystallizes in the hexagonal structure with the lattice parameters $a = 1101.7 \pm 0.2$ and $c = 1158.9 \pm 0.3$ pm, is formed in the

Al–K–Cs–Sb system (Somer et al. 1991b, 1991e). It was prepared from the elements or from the mixture of Cs, GaSb, and KSb at a molar ratio K/Cs/Ga/Sb = 3:8:1:4 in a sealed Nb ampoule at 650–680°C.

7.7 ALUMINUM–LITHIUM–OXYGEN–ANTIMONY

The **$Li_8Al_2Sb_2O_{12}$** quaternary compound, which crystallizes in the monoclinic structure with the lattice parameters $a = 506.8 \pm 1.4$, $b = 876.9 \pm 2.4$, and $c = 513.6 \pm 1.4$ pm and $\beta = 109.309 \pm 0.014°$, is formed in the Al–Li–O–Sb system (Kumar et al. 2012). It was synthesized by heating the stoichiometrically homogenized reactants (Li_2CO_3, Sb_2O_3, Al_2O_3) initially at 650°C for 12 h, followed by heating at 850°C for 24 h.

7.8 ALUMINUM–CALCIUM–BARIUM–ANTIMONY

The **$CaBa_6Al_4Sb_9$** quaternary compound, which crystallizes in the orthorhombic structure with the lattice parameters $a = 1076.66 \pm 0.05$, $b = 1793.78 \pm 0.09$, and $c = 709.75 \pm 0.04$ pm and a calculated density of 5.04 g·cm^{-3} (for $Ca_{0.855}Ba_{6.145}Al_4Sb_9$ composition), is formed in the Al–Ca–Ba–Sb system (He et al. 2016). It was prepared by the direct fusion of the respective elements at 960°C. All reagents and products were handled within an Ar-filled glove box with controlled oxygen and moisture level below 1 ppm.

7.9 ALUMINUM–STRONTIUM–YTTERBIUM–ANTIMONY

The **$Sr_{0.85}Yb_{4.15}Al_2Sb_6$** quaternary compound, which crystallizes in the orthorhombic structure with the lattice parameters $a = 733.13 \pm 0.15$, $b = 2311.8 \pm 0.5$, and $c = 444.97 \pm 0.09$ pm and a calculated density of 6.926 g·cm^{-3}, is formed in the Al–Sr–Yb–Sb system (Todorov et al. 2009).

7.10 ALUMINUM–ZINC–VANADIUM–ANTIMONY

Phase relations in the Al–Zn–V–Sb quaternary system with Zn fixed at 93 at% have been experimentally determined in the entire composition range for two temperatures, 450°C and 600°C, using scanning electron microscopy (SEM) coupled with energy-dispersive X-ray spectroscopy (EDX) and X-ray diffraction (XRD) (Dai et al. 2016). From the experimental results obtained, four four-phase regions and two four-phase regions have been confirmed in the isothermal sections at 450°C and 600°C, respectively (Figure 7.1). The L + ZnVSb field is coexistent with all other phase fields in the section at 450 °C. The solubility of Al in ZnVSb phase is up to 2.1 at%. It dissolves up to 1.5 and 2.2 at% in the V_3Sb phase at 450°C and 600°C, respectively. The vanadium solubility in the AlSb phase is very limited. No new ternary and quaternary compounds were detected in this system. The samples were annealed at 450°C and 600°C for 30 days, and then they were quenched in water.

7.11 ALUMINUM–ZINC–IRON–ANTIMONY

The 450°C isothermal section of the Al–Zn–Fe–Sb quaternary system with Zn fixed at 93 at% has been experimentally studied using XRD and SEM coupled with EDX (Figure 7.2) (Zhu et al. 2013). The L + AlSb field is in equilibrium with other phase fields in this section, except those near the 93Zn + 7Fe corner. The solubility of Sb in $FeZn_{13}$ (ζ), $FeZn_{10}$ (δ),

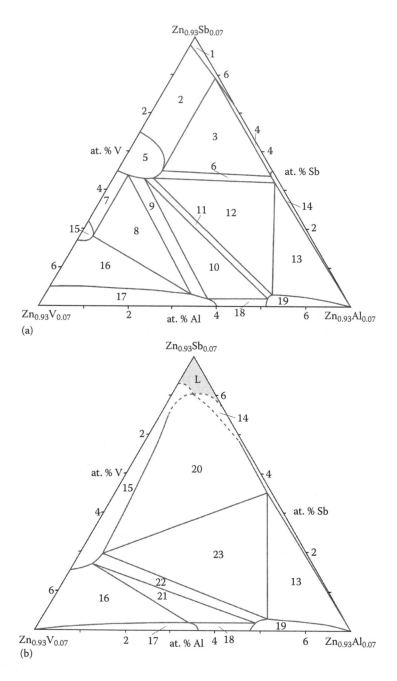

FIGURE 7.1 Isothermal sections of the Al–Zn–V–Sb quaternary system at (a) 450°C and (b), 600°C and at 93 at% Zn: 1, L + Zn_3Sb_2; 2, L + ZnVSb + Zn_3Sb_2; 3, L + ZnVSb + AlSb + Zn_3Sb_2; 4, L + AlSb + Zn_3Sb_2; 5, L + ZnVSb; 6, L + ZnVSb + AlSb; 7, L + ZnVSb + V_3Sb; 8, L + ZnVSb + V_3Sb + Zn_3V; 9, L + ZnVSb + Zn_3V; 10, L + ZnVSb + V_3Sb + Zn_3V; 11, L + ZnVSb + Al_3V; 12, L + ZnVSb + Al_3V + AlSb; 13, L + Al_3V + AlSb; 14, L + AlSb; 15, L + V_3Sb; 16, L + V_3Sb + Zn_3V; 17, L + Zn_3V; 18, L + Zn_3V + Al_3V; 19, L + Al_3V; 20, L + V_3Sb + AlSb, 21, L + V_3Sb + Al_3V + Zn_3V; 22, L + V_3Sb + Al_3V; 23, L + V_3Sb + AlSb + Al_3V. (With kind permission from Springer Science+Business Media: *J. Phase Equilibr. Dif.*, The Zn-rich corner of the Zn–Al–V–Sb quaternary system at 450 and 600°C, 37(5), 2016, 574–580, Dai, Z. et al.)

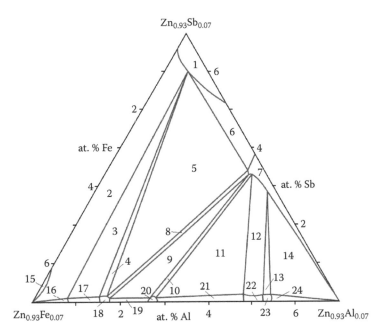

FIGURE 7.2 Isothermal sections of the Al–Zn–Fe–Sb quaternary system at 450°C and at 93 at% Zn: 1, L + Zn$_3$Sb$_2$; 2, L + Zn$_3$Sb$_2$ + ζ; 3, L + δ + ζ + Zn$_3$Sb$_2$; 4, L + δ + Zn$_3$Sb$_2$; 5, L + δ + AlSb + Zn$_3$Sb$_2$; 6, L + Zn$_3$Sb$_2$ + AlSb; 7, L + AlSb; 8, L + δ + AlSb; 9, L+ δ + τ + AlSb; 10, L+ τ + AlSb; 11, L + τ + Fe$_2$Al$_5$ + AlSb; 12, L + Fe$_2$Al$_5$ + AlSb; 13, L + Fe$_2$Al$_5$ + FeAl$_3$ + AlSb; 14, L + FeAl$_3$ + AlSb; 15, ζ + Zn$_3$Sb$_2$; 16, L + ζ; 17, L + ζ + δ; 18, L + δ; 19, L + δ + τ; 20, L + τ; 21, L + τ + Fe$_2$Al$_5$; 22, L + Fe$_2$Al$_5$; 23, L + Fe$_2$Al$_5$ + FeAl$_3$; 24, L + FeAl$_3$. (With kind permission from Springer Science ᛁ Business Media: *J. Phase Equilibr. Dif.*, The Zn-rich corner of the Zn–Fe–Al–Sb quaternary system at 450°C, 34(6), 2013, 474–483, Zhu, Z. et al.)

Al$_6$Fe$_8$Zn$_{86}$ (τ), Fe$_2$Al$_5$, and FeAl$_3$ phases is very limited. The Zn–Fe–Al ternary phase τ was found to be in equilibrium with the L, δ, Fe$_2$Al$_5$, and AlSb phases. The maximum solubilities of Zn in AlSb, Fe$_2$Al$_5$, and FeAl$_3$ are 5.3, 12.3, and 6.2 at%, respectively. Zinc can be dissolved in all compounds existing in the equilibrium alloys. Five four-phase regions and four three-phase regions have been experimentally confirmed. The alloys were prepared using metal powders. The samples were finally annealed at 450°C for 30 days and were water-quenched at the end of the treatment.

7.12 ALUMINUM–CADMIUM–TELLURIUM–ANTIMONY

AlSb–CdTe. The phase diagram is shown in Figure 7.3 (Kuz'mina 1976). Solid solutions over the entire range of concentration are formed in this system. However, there were determined thermal effects on the heating and cooling curves at nearly 450°C, which can correspond to the melting of cadmium antimonide. This assumption is confirmed by the data of XRD. The nonequilibrium state of forming solid solutions exists at low temperatures (Kuz'mina and Khabarov 1969). Solid solutions based on CdTe crystallize in the hexagonal structure at high rates of crystallization. Homogeneous solid solutions were obtained by the high-temperature annealing with their next quenching or by the quenching from the

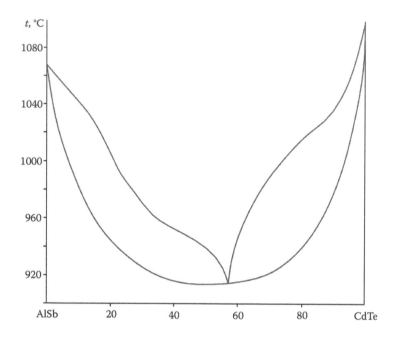

FIGURE 7.3 AlSb–CdTe phase diagram. (From Kuz'mina, G.A., *Izv. AN SSSR. Neorgan. Mater.*, 12 (6), 1121–1122, 1976. With permission.)

temperatures closed to crystallization temperatures. This system was investigated by differential thermal analysis and XRD (Kuz'mina 1976).

7.13 ALUMINUM–MERCURY–TELLURIUM–ANTIMONY

AlSb–HgTe. According to the data of metallography and XRD, the solubility of AlSb in HgTe is not higher than 5 mol% (Burdiyan and Georgitse 1970).

7.14 ALUMINUM–GALLIUM–INDIUM–ANTIMONY

Quasi-subregular solution model with additional ternary parameters has been extended to predict the thermodynamic parameters of the liquid phase in the Al–Ga–In–Sb quaternary system (Sharma and Srivastava 1994). In this system, various two-phase equilibria can be calculated by using the obtained equations. Three-, four-, or invariant five-phase equilibria can also be calculated and represented in the form of isothermal sections. The choice of coordinate of two-dimensional plots used for representing different phase equilibria would depend on the stability range of different phases involved.

AlSb–GaSb–InSb. The liquidus and solidus projections of this quasiternary system are presented in Figure 7.4 (Sharma and Srivastava 1994). The liquidus and solidus of this system was also calculated by Ishida et al. (1989), and the liquidus surface was constructed by Köster and Thoma (1955), but in the last case, the authors believed that the phase diagrams of the AlSb–GaSb and AlSb–InSb systems are the eutectic types. The isothermal sections of the AlSb–GaSb–InSb system at 850°C and 600°C are given in Figure 7.5 (Sharma and Srivastava 1994). The behavior of unstable region for the $Al_yGa_xIn_{1-x-y}Sb$ solid

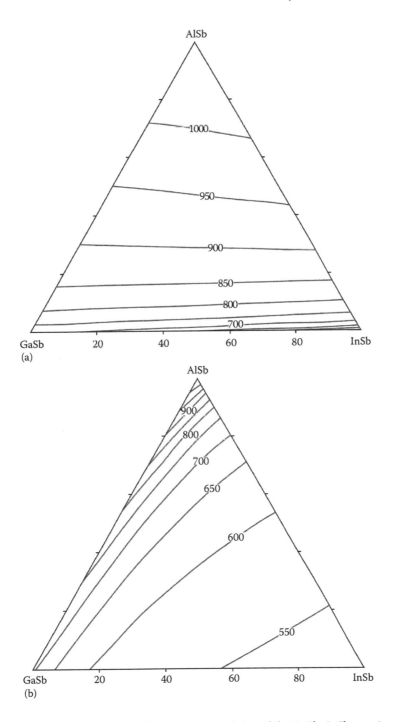

FIGURE 7.4 Liquidus (a) and solidus (b) projections of the AlSb–GaSb–InSb quasiternary system. (With kind permission from Springer Science+Business Media: *J. Phase Equilibr.*, Thermodynamic modeling and phase equilibria calculations of the quaternary Al–Ga–In–Sb system, 15(2), 1994, 178–187, Sharma, R.C., and Srivastava, M.)

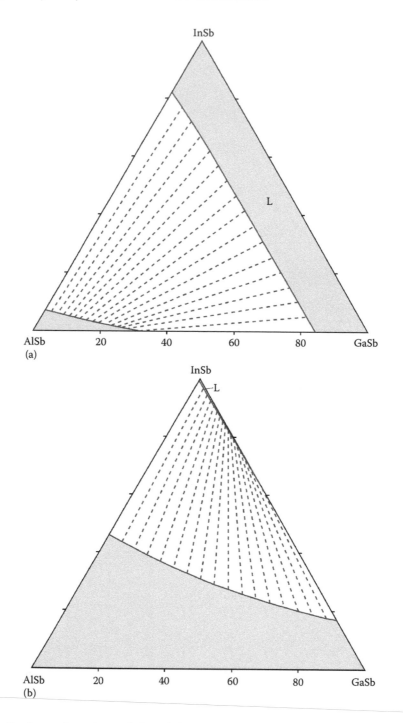

FIGURE 7.5 Isothermal sections of the AlSb–GaSb–InSb quasiternary system at (a) 850°C and (b) 600°C. (With kind permission from Springer Science+Business Media: *J. Phase Equilibr.*, Thermodynamic modeling and phase equilibria calculations of the quaternary Al–Ga–In–Sb system, 15(2), 1994, 178–187, Sharma, R.C., and Srivastava, M.)

solutions was analyzed based on regular approximation for solid solutions (Onabe 1982b). The quaternary critical point for these solid solutions exists at 205°C and has a composition $x = 0.50$ and $y = 0$.

Ingots of the $Al_yGa_xIn_{1-x-y}Sb$ solid solutions ($x < 0.15$) were grown using a directional freeze technique (Zbitnew and Woolley 1981). The results indicated that single-phase solid solution occurs at all compositions in the AlSb–GaSb–InSb quasiternary system, and no miscibility gap was observed. The equations have been found, which fitted well the lattice parameter and energy gap data for the complete alloy system. The final equations are as follows: lattice parameter: a (pm) $= 647.89 - 34.33x - 38.34y - xyz(25 + 325z)$; direct energy gap: E_g (eV) $= 0.095 + 1.76x + 0.28y + 0.345(x^2 + y^2) + 0.085z^2 + xyz(23 - 28y)$; indirect energy gap: E_g (eV) $= 1.0675 + 0.30x + 0.31y + 0.2625x(x^2 + y^2 + z^2) + xyz(-5.9 - 20x)$, where $z = 1 - x - y$.

7.15 ALUMINUM–GALLIUM–TIN–ANTIMONY

AlSb–GaSb–Sn. The liquidus surface of this quasiternary system in the Sn corner is given in Figure 7.6 (Arbenina et al. 1992). The field of the primary crystallization of the $Al_xGa_{1-x}Sb$ solid solution occupies practically all investigation regions. The full hardening of the alloys occurs at $230 \pm 5°C$.

7.16 ALUMINUM–YTTERBIUM–MANGANESE–ANTIMONY

$Yb_{14}Al_xMn_{1-x}Sb_{11}$ quaternary solid solutions exist in the Al–Yb–Mn–Sb system (Cox et al. 2009). They crystallize in the tetragonal structure with the lattice parameters $a = 1658.8 +$

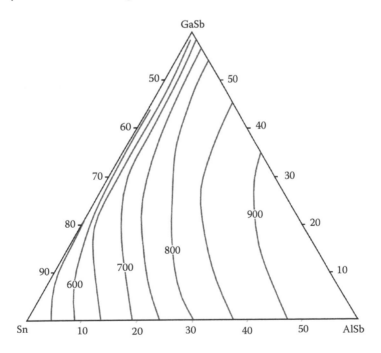

FIGURE 7.6 Liquidus surface of the AlSb–GaSb–Sn quasiternary system in the Sn corner. (From Arbenina, V.V., *Neorgan. Mater.*, 28(3), 513–517, 1992. With permission.)

0.2 and $c = 2191.0 + 0.4$ pm and a calculated density of 8.407 g·cm^{-3} for $x = 0.2$; $a = 1658.3 +$ 0.2, and $c = 2198.5 + 0.4$ pm and a calculated density of 8.366 g·cm^{-3} for $x = 0.4$; $a = 1658.6 + 0.2$, and $c = 2204.1 + 0.4$ pm and a calculated density of 8.329 g·cm^{-3} for $x = 0.6$; $a = 1658.5 + 0.2$, $c = 2204.6 + 0.4$ pm and a calculated density of 8.324 g·cm^{-3} for $x = 0.8$; and $a = 1658.5 + 0.2$, and $c = 2208.2 + 0.4$ pm and a calculated density of 8.297 g·cm^{-3} for $x = 0.95$ at 90 K and $a = 1658.2 + 0.3$ and $c = 2198.9 + 0.6$ pm for $x = 0.4$; $a = 1658.5 + 0.3$ and $c = 2203.7 + 0.9$ pm for $x = 0.6$; $a = 1658.1 + 0.4$ and $c = 2204.5 + 0.8$ pm for $x = 0.8$; and $a = 1658.6 + 0.3$ and $c = 2208.3 + 0.7$ pm for $x = 0.95$ at 300 K. Heat capacity for these materials is independent of composition: $c_p = 663.49 + 0.0898T$ J·mol^{-1}·K^{-1}.

The samples with $x = 0.2, 0.4, 0.6, 0.8$, and 0.95 were prepared by reacting the elements Yb, Sb, Mn, and Al in a Sn flux (Cox et al. 2009). All materials were handled inside a N$_2$-filled drybox. The elements were layered into Al$_2$O$_3$ crucibles in the stoichiometric ratio. Excess (11 times) Mn and Al were used in order to prevent the formation of the binary phase Yb$_{11}$Sb$_{10}$. The samples were sealed in silica glass ampoules under vacuum. The reaction vessels were then heated from room temperature to 500°C in 2 h, held at that temperature for 2 h, and then heated to 1000°C in 2 h. They were held at 1000°C for 6 h and then cooled at 3°C·h^{-1} to 700°C, where the ampoules were inverted and centrifuged at 6500 revolutions per minute for 3–5 min to remove the molten Sn flux from the crystals. After cooling, the reflective silver-colored crystals were removed from the crucible inside a N$_2$-filled dry box.

7.17 ALUMINUM–LEAD–OXYGEN–ANTIMONY

The **Pb$_2$Al$_{0.5}$Sb$_{1.5}$O$_{6.5}$** quaternary compound, which crystallizes in the cubic structure with the lattice parameter $a = 1039.64 \pm 0.01$ pm and a calculated density of 8.45 g·cm^{-3}, is formed in the Al–Pb–O–Sb system (Cascales et al. 1985). It was prepared from a mixture of PbO, Sb$_2$O$_3$, and Al$_2$O$_3$ (molar ratio Pb/Al/Sb = 4:1:3) which, once ground in an agate mortar, was heated in air at 950°C and 1000°C. After each thermal treatment for 24 h, the material was quenched, weighed, and ground. The title compound was obtained as orange-yellow powder.

7.18 ALUMINUM–BISMUTH–OXYGEN–ANTIMONY

The **Bi$_3$AlSb$_2$O$_{11}$** quaternary compound, which crystallizes in the cubic structure with the lattice parameter $a = 943.323 \pm 0.003$ pm (Ismunandar et al. 1996) [$a = 945$ pm (Sleight and Bouchard 1973)], is formed in the Al–Bi–O–Sb system. The title compound was prepared by the solid-state reactions of stoichiometric amounts of Bi$_2$O$_3$, Al$_2$O$_3$, and Sb$_2$O$_3$ at 800°C for 24 h and then at 1100°C for 48 h with regrinding after each heating step (Ismunandar et al. 1996).

7.19 ALUMINUM–VANADIUM–OXYGEN–ANTIMONY

The **AlSbVO$_6$** quaternary compound and **Al$_{1-x}$SbV$_x$O$_4$** $(0 < x < 0.5)$ solid solutions are formed in the Al–V–O–Sb system. AlSbVO$_6$ crystallizes in the tetragonal structure with the lattice parameters $a = 449.45$ and $c = 294.98$ pm and the calculated and experimental densities of 4.96 and 5.21 g·cm^{-3}, respectively (Dąbrowska 2012). This compound was obtained by the conventional solid-state reactions from stoichiometric mixture of the oxides Al_2O_3, V_2O_5, and α-Sb_2O_4 as well as from the reacting mixtures containing $AlSbO_4 + V_2O_5$, $Al_2O_3 + SbVO_5$, $AlVO_4 + \alpha$-Sb_2O_4, or $AlVO_4 + AlSbO_4 + SbVO_5$. Mixtures were carefully homogenized and then shaped into tablets. The tablets were placed in porcelain crucibles and heated in air or Ar atmosphere until a single-phase sample was obtained. All the samples were heated at a temperature range from 550°C to 750°C. On each heating stage, the samples were gradually cooled down to room temperature, weighed, and homogenized by triturating. Brownish $AlSbVO_6$ compound was obtained. It is stable in inert atmosphere up to circa 820°C and decomposes at higher temperature forming $AlSbO_4$ and $VSbO_4$.

$Al_{1-x}SbV_xO_4$ solid solutions also crystallizes in the tetragonal structure with the lattice parameters $a = 452.33 + 0.04$ and $c = 297.46 + 0.05$ pm for $x = 0.2$ and $a = 456.9 + 0.2$ and $c = 301.1 + 0.1$ pm for $x = 0.5$ (Nilsson et al. 1996). Calcination in air at 680°C of mixtures of $Al(OH)_3$, Sb_2O_3, and V_2O_5 leads to the formation of aforementioned solid solutions and α-Sb_2O_4, V_2O_5, $AlVO_4$, and $Sb_{0.9}V_{0.1}O_4$. The formation of these solid solutions required an excess of alumina (>60 at%).

7.20 ALUMINUM–SELENIUM–CHLORINE–ANTIMONY

The ($Sb_{10}Se_{10}$)[$AlCl_4$] quaternary compound, which crystallizes in the triclinic structure with the lattice parameters $a = 947.85 \pm 0.02$, $b = 957.79 \pm 0.02$, $c = 1166.31 \pm 0.03$ pm, $\alpha = 103.622 \pm 0.001°$, $\beta = 110.318 \pm 0.001°$, and $\gamma = 99.868 \pm 0.001$ and a calculated density of 4.20 g·cm^{-3}, is formed in the Al–Se–Cl–Sb system (Ahmed et al. 2011). The title compound was prepared as follows. Stoichiometric amounts of Sb, Se, and $SeCl_4$ (total mass 240 mg) were added to the Lewis acidic ionic liquid 1-n-butyl-3-methylimidazolium chloride/$AlCl_3$ (1:2 molar ratio, 1.5 mL), which immediately turned into a dark brown solution. The reaction mixture was left stirring overnight at room temperature, and then filtered to remove unconverted material. After 4 days, black block-shaped crystals were obtained. The excess ionic liquid was decanted in an Ar-filled glove box, and single crystals were separated. The obtained crystals were washed twice with dry dichloromethane. [$Sb_{10}Se_{10}$][$AlCl_4$] is sensitive toward hydrolysis by moist air.

7.21 ALUMINUM–SELENIUM–BROMINE–ANTIMONY

The ($Sb_7Se_8Br_2$)($Sb_{13}Se_{16}Br_2$)[$AlBr_4$]$_8$ quaternary compound crystallizes in the monoclinic structure with the lattice parameters $a = 2721.4 \pm 0.5$, $b = 938.3 \pm 0.2$, $c = 2291.7 \pm 0.4$ pm, and $\beta = 101.68 \pm 0.01°$ and a calculated density of 4.30 g·cm^{-3} (Ahmed et al. 2014). This compound was obtained by the next procedure. Sb and Se in amounts that correspond to the composition of the target compound (total mass 190 mg) were added to the bromine-based Lewis acidic ionic liquid 1-n-butyl-3-methylimidazolium bromide/$AlBr_3$ (88 mg/ 500 mg) in a glass ampoule. A small amount of $NbCl_5$ (50 mg) was added to the ionic liquid at room temperature, and the solution turned olive green. The glass ampoule was sealed and heated at 160°C for 7 days followed by slow cooling (6°C·h^{-1}) to room temperature. Over the course of the reaction, the solution changed to red. The ampoule was opened inside a glove box, and red platelike crystals were manually separated from the ionic liquid.

7.22 ALUMINUM–TELLURIUM–CHLORINE–ANTIMONY

The **(Sb$_2$Te$_2$)[AlCl$_4$]** quaternary compound crystallizes in the triclinic structure with the lattice parameters $a = 949.30 \pm 0.02$, $b = 1348.24 \pm 0.03$, $c = 1804.97 \pm 0.05$ pm, $\alpha = 78.983 \pm 0.002°$, $\beta = 88.060 \pm 0.001°$, and $\gamma = 88.444 \pm 0.001$ and a calculated density of 3.959 g·cm^{-3} (Beck and Schlüter 2005). To prepare this compound, Sb (0.48 mM), Te (0.72 mM), SbCl$_3$ (0.24 mM), AlCl$_3$ (1.98 mM), and NaCl (0.18 mM) were thoroughly powdered and filled in a glass ampoule, which was evacuated and flame sealed. The ampoule was placed in a horizontally positioned tube furnace at a temperature of 130°C. After 2 weeks, a temperature gradient of about $\Delta T = 20°C$ was applied by placing the ampoule asymmetrically in the oven. The excess of AlCl$_3$ sublimated to the colder tip of the ampoule leaving a dark red viscosous melt in the hotter zone. Dark needle-shaped crystals of the title compound grew from the melt. After cooling, the crystals had to be isolated by mechanical separation.

7.23 ALUMINUM–TELLURIUM–IODINE–ANTIMONY

The **(Sb$_2$Te$_2$)I[AlI$_4$]** quaternary compound crystallizes in the monoclinic structure with the lattice parameters $a = 1024.35 \pm 0.06$, $b = 854.20 \pm 0.05$, and $c = 1746.8 \pm 0.1$ pm and $\beta = 97.905 \pm 0.002°$ (Eich et al. 2015). The title compound was prepared by the next procedure. Sb$_2$Te$_3$ (0.20 mM), SbCl$_3$ (0.16 mM), AlI$_3$ (0.99 mM), and NaI (0.18 mM) were annealed in an evacuated glass ampoule at 100°C for 7 days and at 170°C for additional 5 days. This compound formed black cube-shaped crystals emerging from an orange melt, which solidified at ambient temperature. During the manual separation, the crystals were revealed to be intergrown agglomerates of silvery shiny thin square plate-shaped crystals. An increase yield can be achieved by replacing NaI with CuI or AgI and annealing the ampoules stepwise at 100°C for 7 days, at 170°C for 5 days, and finally at 190°C for 8 days. Cooling down to room temperature with the rate of 6°C·h^{-1} yielded (Sb$_3$Te$_4$)[GaCl$_4$], which is highly moisture sensitive. Tube furnace used for synthesis of this compound was aligned at an angle of about 30° to the horizontal to keep the melt compacted in the hot zone.

CHAPTER 8

Systems Based on GaN

8.1 GALLIUM–HYDROGEN–SODIUM–NITROGEN

The **NaGa(NH$_2$)$_4$** quaternary compound, which crystallizes in the monoclinic structure with the lattice parameters $a = 742.3 \pm 0.6$, $b = 610.7 \pm 0.5$, $c = 1285 \pm 1$ pm, and $\beta = 92°15' \pm 15'$, is formed in the Ga–H–Na–N system (Molinie et al. 1973). This compound was obtained by the action of sodium solution in liquid NH$_3$ on gallium in tubes sealed at ambient temperature.

8.2 GALLIUM–HYDROGEN–PHOSPHORUS–NITROGEN

GaP–NH$_3$. GaP begins to interact with NH$_3$ at 850°C (Zykov and Gaydo 1973). X-ray amorphous phosphorus nitrides are formed in the temperature range from 850°C to 1020°C, and at the higher temperatures GaN is formed. Gallium nitride with stoichiometric composition can be obtained by nitriding of gallium phosphide with ammonia at 1100°C.

8.3 GALLIUM–HYDROGEN–OXYGEN–NITROGEN

[Ga(H$_2$O)$_6$](NO$_3$)$_3$·3H$_2$O and **Ga$_{2.75}$N$_{0.2}$O$_{3.83}$·0.5H$_2$O** quaternary compounds are formed in the Ga–H–O–N system. [Ga(H$_2$O)$_6$](NO$_3$)$_3$·3H$_2$O crystallizes in the monoclinic structure with the lattice parameters $a = 1396.09 \pm 0.06$, $b = 964.98 \pm 0.05$, $c = 1097.43 \pm 0.05$ pm, and $\beta = 95.448 \pm 0.001°$ and a calculated density of 1.886 g·cm^{-3} (Hendsbee et al. 2009). The title compound was prepared by dissolving Ga(NO$_3$)$_3$·H$_2$O (5 g) in a minimum of H$_2$O (approximately 7 mL) and adding three drops of concentrated nitric acid to suppress hydrolysis. The sample was cooled to 248 K, and a seed crystal was introduced to initiate crystallization. A suitable crystal was sealed in a glass capillary to prevent water loss from this hygroscopic material.

Ga$_{2.75}$N$_{0.2}$O$_{3.83}$·0.5H$_2$O crystallizes in the cubic spinel-type structure, which is stable up to temperatures of 500°C in both oxidative and reductive atmospheres (Oberländer et al. 2013). To obtain this compound, Ga metal beads and ethylenediamine were used as raw materials. For the preparation of the gallium–ethylenediamine complex, 25 mM of Ga was mixed with 0.14 M of ethylenediamine and ultrasonically treated at a temperature of 75°C until a black suspension was formed. To this suspension, 97 mM of distilled water was added without stirring. The mixture was then loaded in a 250 mL Teflon-lined stainless steel autoclave and heated for 24 h at a temperature of 200°C. The yellowish products were isolated by centrifugation, washed in ethanol three times, and then dried at a temperature of 60°C in air. Some of the powders were subsequently heat-treated in air, argon, and ammonia atmospheres for 5 h at a temperature of 500°C.

Chemical composition for the as-prepared gallium oxynitrides was $Ga_{2.75}N_{0.2}O_{3.83} \cdot 0.5H_2O$. The gallium oxynitride samples partially released the contained water during the applied heat treatments so that the chemical compositions of the samples annealed at temperatures of 500°C in air, Ar, and NH_3 can be formulated as $Ga_{2.75}N_{0.04}O_{4.04} \cdot 0.2H_2O$, $Ga_{2.75}N_{0.14}O_{4.00} \cdot 0.2H_2O$, and $Ga_{2.75}N_{0.99}O_{2.68} \cdot 0.2H_2O$, respectively. The chemical composition of solvothermally synthesized gallium oxynitride is somewhat hypothetical, since further adsorbates could be present. An energy gap of this gallium oxynitride is 4.88 eV (Oberländer et al. 2013). The small band-gap shift observed for the samples heat-treated in air and Ar to values of 4.67 and 4.66 eV, respectively, is not yet fully understood.

8.4 GALLIUM–LITHIUM–CALCIUM–NITROGEN

The **LiCaGaN₂** quaternary compound, which crystallizes in the monoclinic structure with the lattice parameters $a = 579.0 \pm 0.1$, $b = 696.9 \pm 0.1$, $c = 589.5 \pm 0.1$ pm, and $\beta = 90.037 \pm 0.005°$ and a calculated density of 4.042 g·cm^{-3} at 173 ± 2 K, is formed in the Ga–Li–Ca–N system (Bailey and DiSalvo 2006). The crystals of the title compound were observed in the products of a reaction that attempted to synthesize a quaternary lithium–calcium–chromium–nitride from an excess of sodium and gallium: Na, Ga, Li, Ca, Cr, and NaN_3 were placed into a niobium tube such that the atomic ratios of Na/Ga/Li/Ca/Cr/N₂ were approximately 6:4:0.5:1:1:1.25. The niobium container was sealed in an arc furnace before being sealed under vacuum in a fused silica tube. This silica-encased niobium tube was placed into a muffle furnace, heated to 900°C in 15 h, where it remained for 36 h before the furnace was shut off and allowed to cool at the natural rate (circa 10–20 h to near room temperature). Following the reaction, unreacted Na was removed by evaporation from the products by heating the niobium tube to 350°C under a pressure of ~0.1 Pa for 8 h.

8.5 GALLIUM–LITHIUM–STRONTIUM–NITROGEN

The **LiSrGaN₂** quaternary compound, which crystallizes in the monoclinic structure with the lattice parameters $a = 607.67 \pm 0.06$, $b = 1015.46 \pm 0.09$, $c = 1264.12 \pm 0.12$ pm, and $\beta = 901.366 \pm 0.002°$ and a calculated density of 4.914 g·cm^{-3} at 173 ± 2 K, is formed in the Ga–Li–Sr–N system (Park et al. 2003). This compound was synthesized in a Nb container, made by first welding shut one end of the Nb tube in an Ar atmosphere. The container was loaded with NaN_3, Ga, Sr, and Li in the molar ratio 20:1:1:0.5. The open end of the container was then closed by welding in an Ar atmosphere. The container was finally sealed in fused silica tubing under dynamic vacuum. The reaction container was heated at 50°C·h^{-1} in a muffle furnace to 760°C for 48 h. The temperature was lowered linearly to 200°C, and the furnace was turned off. The solid product was retrieved from the container, after which

Na was separated from products by sublimation at 300°C under a dynamic vacuum. A white powder was obtained after evaporation of Na.

8.6 GALLIUM–LITHIUM–CARBON–NITROGEN

The **LiGa(CN)$_4$** quaternary compound, which crystallizes in the cubic structure with the lattice parameter $a = 587.4 \pm 0.2$ pm, is formed in the Ga–Li–C–N system (Brousseau et al. 1997). The title compound was prepared as follows (Kouvetakis et al. 1996; Brousseau et al. 1997). Ga(CN)$_3$ (8.0 mM) and LiCN (0.5 M in DMF; 20 mL; 10.0 mM) were combined in a Schlenk flask with a magnetic stir bar and refluxed for 24 h. The mixture was then filtered, leaving behind a pale yellow solid. The solvent was removed from the filtrate in vacuo to give the excess LiCN. The pale brown residue was completely dried by heating under high vacuum to provide LiGa(CN)$_4$.

8.7 GALLIUM–LITHIUM–OXYGEN–NITROGEN

The **Li$_2$Ga$_3$NO$_4$** and **Li$_4$GaNO$_2$** quaternary compounds are formed in the Ga–Li–O–N system. Li$_2$Ga$_3$NO$_4$ crystallizes in the hexagonal structure with the lattice parameters $a = 316.74 \pm 0.01$ and $c = 508.54 \pm 0.02$ pm (Kikkawa et al. 2007). It was obtained by ammonia nitridation of a precursor prepared from the addition of citric acid to an aqueous solution of Ga(NO$_3$)$_3$ and LiNO$_3$ at the atomic ratio Li/Ga ≥ 1. The white-colored products formed were contaminated with an impurity of Li$_2$CN$_2$, which was removed by washing with H$_2$O.

The reaction of Li$_3$N with LiGaO$_2$ (molar ratio 1:1) at 700°C gives Li$_4$GaNO$_2$ (Kamler et al. 2000). The same crystalline phase was obtained in the reaction of Li$_3$N with Ga$_2$O$_3$ (molar ratio 2:1) at 600°C. This compound decomposes at 850°C with the formation of Li$_5$GaO$_4$ and Li$_3$GaN$_2$. At 900°C, Li$_3$GaN$_2$ decomposes forming GaN and Li$_3$N.

8.8 GALLIUM–COPPER–CARBON–NITROGEN

The **CuGa(CN)$_4$** quaternary compound, which crystallizes in the cubic structure with the lattice parameter $a = 572.9 \pm 0.5$ pm, is formed in the Ga–Cu–C–N system (Brousseau et al. 1997). To obtain the title compound, Ga(CN)$_3$ (2.0 mM) and CuCN (2.5 mM) were combined in 20 mL of DMF in a Schlenk flask with a magnetic stir bar and heated at 137°C for 24 h. The solvent was then pumped off and the remaining brown solid further annealed at 130°C for 60 h under a dynamic vacuum.

8.9 GALLIUM–MAGNESIUM–STRONTIUM–NITROGEN

The **Sr(Mg$_2$Ga$_2$)N$_4$** quaternary compound, which crystallizes in the tetragonal structure with the lattice parameters $a = 829.25 \pm 0.07$ and $c = 335.85 \pm 0.05$ pm and a calculated density of 4.770 g·cm^{-3} at 173 ± 2 K, is formed in the Ga–Mg–Sr–N system (Park et al. 2008). The synthesis of this compound was carried out in a Nb container. Under argon atmosphere, NaN$_3$ (90.4 mg), Na (202.2 mg), Sr (24.9 mg), Ga (76.9 mg), and Mg (6.9 mg) were loaded in the Nb container. The molar ratio of Na/Sr/Ga/Mg was 40:1:4:1. The container was put into silica tubing and sealed under vacuum. The reaction container was then heated at a rate of 50°C·h^{-1} to 760°C. This temperature was maintained for 48 h and lowered linearly to 200°C over 200 h. Once the temperature reached 200°C, the furnace was turned off, so that it cooled down to room temperature. The sodium was separated from reaction products by evaporation at 300°C under a dynamic vacuum. $Sr(Mg_2Ga_2)N_4$ was obtained as white powder with yellow tint.

8.10 GALLIUM–STRONTIUM–CARBON–NITROGEN

The **$Sr_4GaN_3(CN_2)$** quaternary compound, which crystallizes in the monoclinic structure with the lattice parameters $a = 1347.78 \pm 0.02$, $b = 741.40 \pm 0.01$, $c = 744.40 \pm 0.01$ pm, and $\beta = 98.233 \pm 0.001°$ and a calculated density of 4.532 g·cm^{-3} at 150 ± 1 K, is formed in the Ga–Sr–C–N system (Mallinson et al. 2006). For synthesis of the title compound, the reagents used were Sr pieces, Ga pieces, Na_3N, Na, and graphite. Strontium and gallium, and in some instances graphite, were placed along with Na_3N and Na in a tantalum tube, which was welded closed under 0.1 MPa of purified Ar and then sealed inside an evacuated silica ampoule. The amount of NaN_3 used was limited so as to restrict the maximum possible pressure in the Ta tube to 3.5 MPa. The crystals of $Sr_4GaN_3(CN_2)$ were obtained from a reaction, in which NaN_3 (78 mg), Sr (70 mg), Ga (28 mg), C (5 mg), and Na (200 mg) were placed in the tube. This tube was heated to 900°C at 2°C·min^{-1}, and this temperature was maintained for 48 h, after which the tube was cooled at 0.1°C·min^{-1} to 400°C and then removed from the furnace. The excess sodium present at the end of the reaction was removed by sublimation under a dynamic vacuum of 1 Pa at 50°C. All manipulations of solids were carried out in an Ar-filled glove box.

8.11 GALLIUM–STRONTIUM–OXYGEN–NITROGEN

The **Sr_4GaN_3O** quaternary compound, which crystallizes in the orthorhombic structure with the lattice parameters $a = 740.02 \pm 0.01$, $b = 2433.78 \pm 0.05$, and $c = 740.38 \pm 0.01$ pm and a calculated density of 4.764 g·cm^{-3} at 150 ± 1 K, is formed in the Ga–Sr–O–N system (Mallinson et al. 2006). For synthesis of the title compound, the reagents used were Sr pieces, Ga pieces, Na_3N, and Na. Strontium and gallium were placed along with Na_3N and Na in a tantalum tube, which was welded closed under 0.1 MPa of purified Ar and then sealed inside an evacuated silica ampoule. The amount of NaN_3 used was limited so as to restrict the maximum possible pressure in the Ta tube to 3.5 MPa. The crystals of Sr_4GaN_3O were obtained from a reaction, in which NaN_3 (78 mg), Sr (70 mg), Ga (55 mg), and Na (200 mg) were placed inside the tube. This tube was heated to 900°C at 2°C·min^{-1}, and this temperature was maintained for 48 h, after which the tube was cooled at 0.1°C·min^{-1} to 400°C and then removed from the furnace. The excess sodium present at the end of the reaction was removed by sublimation under a dynamic vacuum of 1 Pa at 50°C. All manipulations of solids were carried out in an Ar-filled glove box.

8.12 GALLIUM–INDIUM–CARBON–NITROGEN

$Ga_xIn_{1-x}(CN)_3$ solid solutions are formed in the Ga–In–C–N quaternary system (Williams et al. 1998). For its obtaining, a suspension of $GaCl_3$ (x mM) and $InCl_3$ ($7.12 - x$ mM) in ether (30 mL) was treated dropwise with a solution of $SiMe_3CN$ (21 mM) in hexane (20 mL). The solution was stirred for 12 h, after which time the colorless precipitate was isolated by filtration, heated in dry hexane for several hours, and then washed with hexane several times to yield the solid solutions. The variation of the lattice parameter with composition obeys Vegard's law ($a = 553.7 \pm 0.2°$, $a = 546.4 \pm 0.3°$, and $a = 542.8 \pm 0.2°$ pm for $x = 0.25$, 0.5, and 0.75, respectively).

8.13 GALLIUM–INDIUM–ARSENIC–NITROGEN

GaN + InAs $\Leftrightarrow$ GaAs + InN. The initial stage of spinodal decomposition of the **$Ga_{1-x}In_xN_yAs_{1-y}$** alloys results in the change of x, y, or both of them in two phases after the phase separation (Asomoza et al. 2002). It was shown that such alloys with $x = 0.28$–0.38 and $y = 0.009$–0.02 are deeply inside the spinodal decomposition range at their growth temperatures and that nitrogen dramatically increases the temperature of the coherent spinodal.

The miscibility gap calculation was carried out for the $Ga_{1-x}In_xN_yAs_{1-y}$ materials grown on a (100)GaAs substrate (Schlenker et al. 1999, 2000). The miscibility gap obtained by neglecting and including strain effects are compared in Figure 8.1. The numbers in the figure show the critical temperatures above which the material is metastable. If the strain effects were neglected, the $Ga_{1-x}In_xN_yAs_{1-y}$ materials are immiscible at room temperature over almost the entire composition range and the maximum spinodal critical temperature can be found at 19 at% of indium and 64 at% of nitrogen. The strain induced during spinodal decomposition reduces the critical temperature. For example, the maximum critical temperature is moved from indium and nitrogen contents of 19 and 64 at%, respectively, neglecting strain effects to 0 and 77 at%, respectively, including strain effects. It is necessary to note that this calculation should be valid for nitrogen contents of a few percent. The calculation predicts that for nitrogen contents below 3 at% and indium contents around 30 at%, the materials are metastable due to stabilization by strain effects. The $Ga_{1-x}In_xN_yAs_{1-y}$ materials on GaN are immiscible at room temperature over almost the entire composition range.

A systematic investigation of nitrogen and indium incorporation in GaAs using Et_3Ga, Me_3In, and di-*tertiary*-butyl-arseno-amine has been presented by Sterzer et al. (2017). Extremely high nitrogen incorporation efficiency in the In containing $Ga_{1-x}In_xN_yAs_{1-y}$ was shown. All samples have been grown in a horizontal gas foil reactor, heated by six halogen lamps. The reactor pressure was held constant at 5 kPa with a flux of Pd-purified H_2 carrier gas. $Ga_{0.92}In_{0.08}N_{0.03}As_{0.97}$ solid solution was obtained.

8.14 GALLIUM–INDIUM–MANGANESE–NITROGEN

GaN–InN–MnN. Phase stability in this quasiternary system was investigated by means of the first-principles full-potential linearized plane-wave method (Miyake et al. 2013). Calculated results predict that the ordered and disordered states are less favorable with respect to the phase separation. The Mn-doped GaN with $x_{Mn} \leq 0.63$ favors the zinc blend structure over the rock salt structure regardless of the In incorporation.

A calculated T-x phase diagram (Figure 8.2) shows a miscibility gap with a critical temperature of about 3230°C for the GaN–InN–MnN quasiternary system (Miyake et al. 2013). It can be seen that the Mn solubility even at high temperatures (circa 1730°C) is very low.

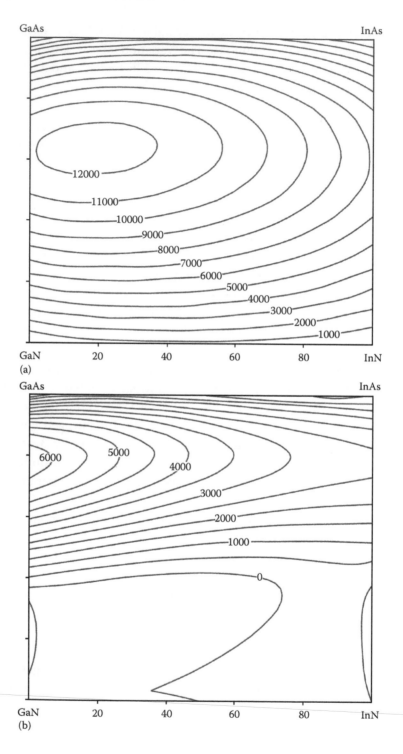

FIGURE 8.1 Miscibility gap calculation for the $Ga_{1-x}In_xN_yAs_{1-y}$ materials: (a) is a contour plot of the miscibility gap calculation neglecting strain effects, and (b) is calculated including strain effects on the GaAs substrate. (Reprinted from *J. Cryst. Growth*, 196(1), Schlenker, D. et al. Miscibility gap calculation for $Ga_{1-x}In_xN_yAs_{1-y}$ including strain effects, 67–70, Copyright (1999), with permission from Elsevier.)

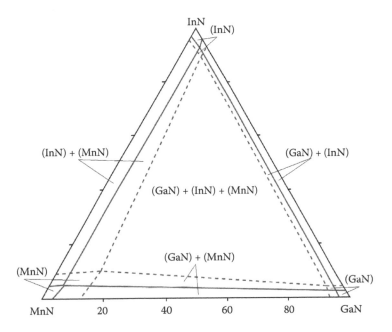

FIGURE 8.2 Calculated phase diagram within zinc blend structure for the GaN–InN–MnN quasiternary system: solid and dashed lines correspond to bi(tri)-nodal lines at 1730°C and 2230°C, respectively. (Reprinted from *J. Cryst. Growth*, 362, Miyake, M. et al. M Structural stability of Mn-doped GaInAs and GaInN alloys, 324–326, Copyright (2013), with permission from Elsevier.)

8.15 GALLIUM–ARSENIC–BISMUTH–NITROGEN

GaN$_y$As$_{1-x-y}$Bi$_x$ alloys were grown on GaAs by molecular beam epitaxy using solid Ga, Bi, and As sources and nitrogen radicals generated from nitrogen gas in radio-frequency (RF) plasma (Huang et al. 2004, 2005; Yoshimoto et al. 2004). Changing the growth temperature is found to be a convenient method for controlling the GaBi molar fraction in the alloy reproducibly. It was shown that GaBi molar fraction into GaN$_y$As$_{1-x-y}$Bi$_x$ was incorporation maintained at a constant value of ~3.0 mol% for various nitrogen flow rates. That is, Bi incorporation was not influenced by changes in nitrogen flow rate while the Ga, Bi, and As fluxes, plasma power, and growth temperature remained constant. When the Ga and As fluxes remained constant, the GaBi molar fraction into these epilayers was found to depend not only on the Bi flux but also on the growth temperature. The GaN molar fraction did not vary with the changes in growth temperature in the range of 35–400°C. The GaBi molar fraction was controlled in a range up to 4.0 mol%. The GaN molar fraction increased up to 8.0 mol% with increasing supply of activated nitrogen generated in RF plasma.

CHAPTER 9

Systems Based on GaP

9.1 GALLIUM–HYDROGEN–OXYGEN–PHOSPHORUS

The **GaPO$_4$·2H$_2$O**, **GaH$_2$P$_3$O$_{10}$**, and **GaPO$_4$·H$_3$PO$_4$·3H$_2$O** quaternary compounds are formed in the Ga–H–O–P system (Tananaev and Chudinova 1962; Mooney-Slater 1966; Pâques-Ledent and Tarte 1969; Chudinova et al. 1978).

GaPO$_4$·2H$_2$O exists in two polymorphic modifications. First modification crystallizes in the monoclinic structure with the lattice parameters $a = 977 \pm 1$, $b = 964 \pm 1$, $c = 968 \pm 1$ pm, and $\beta = 102.7 \pm 0.2°$ and the calculated and experimental densities of 2.998 and 3.00 g·cm^{-3}, respectively (Mooney-Slater 1966). The crystals of this modification were obtained by the very slow dilution or neutralization of an acid solution of the gallium phosphate held at moderately elevated temperatures.

Second modification of this compound crystallizes in the orthorhombic structure with the lattice parameters $a = 989.6$, $b = 859.4$, and $c = 957.2$ pm (Pâques-Ledent and Tarte 1969). Its synthesis was carried out by slow precipitation in an acid medium of the Ga salt with addition of a Na$_2$HPO$_4$ solution. The 0.3 M solution of Ga(NO$_3$)$_3$ (10 mL) was acidified with three or four drops of moderately concentrated H$_3$PO$_4$. The pH was then adjusted to 2 by the addition of Na$_2$HPO$_4$ dilute solution. The suspension was treated in the sealed tube at 90°C.

GaH$_2$P$_3$O$_{10}$ is formed at the interaction of Ga$_2$O$_3$ and 85% H$_3$PO$_4$ in the glassy carbon crucible at the temperature lower than circa 200°C (Chudinova et al. 1978).

9.2 GALLIUM–LITHIUM–OXYGEN–PHOSPHORUS

The **LiGa(PO$_3$)$_4$** and **Li$_9$Ga$_3$(P$_2$O$_7$)$_3$(PO$_4$)$_2$** quaternary compounds are formed in the Ga–Li–O–P system. LiGa(PO$_3$)$_4$ incongruently melts at 655°C (Grunze et al. 1987) and crystallizes in the orthorhombic structure with the lattice parameters $a = 824.4 \pm 0.2$, $b = 904.4 \pm 0.2$, and $c = 1254.0 \pm 0.2$ pm (Palkina et al. 1981).

$Li_9Ga_3(P_2O_7)_3(PO_4)_2$ incongruently melts at 765°C (Poisson et al. 1998) and crystallizes in the trigonal structure with the lattice parameters $a = 972.879 \pm 0.013$ and $c = 1358.27 \pm 0.03$ pm and a calculated density of 2.933 g·cm^{-3} (Liu et al. 2006) [$a = 964.3 \pm 0.1$ and $c = 1359.2 \pm 0.3$ pm (Poisson et al. 1998)]. It was synthesized under hydrothermal conditions. The reaction was carried out with mixtures of Li_2HPO_4 and $GaCl_3$ (0.23 g of Ga dissolved in 2.5 mL of 37% HCl) in a Li/Ga/P molar ratio 1:20:10. The container was about 50% full of solution. The autoclave was placed in an oven with subsequent heating at 220°C for 7 days. The title compound could also be prepared by crystallization in a flux of Li_3PO_4 (Poisson et al. 1998). Starting materials were Li_3PO_4, Ga_2O_3, and $NH_4H_2PO_4$ mixed in molar ratios Li/Ga/P = 60:4:36. The mixture was gradually heated up to 900°C in a Pt crucible and cooled at a rate of 50°C·h^{-1} down to 850°C, of 2°C·h^{-1} between 850°C and 600°C, and finally of ca. 200°C·h^{-1} down to room temperature. In order to isolate this compound, the mixture was treated during 5 days by an aqueous 1 M solution of CH_3COOH which dissolved $Li_4P_2O_7$. A saturated solution of NaCl was then added to the remaining solid to eliminate $LiPO_3$. Colorless platelike crystals of $Li_9Ga_3(P_2O_7)_3(PO_4)_2$ were obtained.

9.3 GALLIUM–SODIUM–POTASSIUM–PHOSPHORUS

The **NaK_2GaP_2** quaternary compound, which crystallizes in the orthorhombic structure with the lattice parameters $a = 661.3 \pm 0.2$, $b = 1449.0 \pm 0.4$, and $c = 640.1 \pm 0.1$ pm and a calculated density of 2.521 g·cm^{-3}, is formed in the Ga–Na–K–P system (Somer et al. 1990b, 1992b). It was synthesized from the elements or from Na, KP, and GaP at 650–680°C in a sealed Nb ampoule.

9.4 GALLIUM–SODIUM–RUBIDIUM–PHOSPHORUS

The **$Na_8RbGa_3P_6$** quaternary compound, which crystallizes in the orthorhombic structure with the lattice parameters $a = 2227.6 \pm 0.3$, $b = 469.47 \pm 0.06$, and $c = 1635.6 + 0.2$ pm and a calculated density of 2.580 g·cm^{-3} at 200 ± 2 K, is formed in the Ga–Na–Rb–P system (He et al. 2012). The starting materials for the synthesis of this compound were elemental Na, Rb, Ga, and P. The elements in a stoichiometric ratio with a total mass of circa 500 mg were loaded into a Nb ampoule, which was subsequently arc-welded under high-purity Ar and then jacketed in a fused silica tube under vacuum. The reaction mixture was heated up to 550°C and equilibrated for 1 week before it was slowly cooled to room temperature. This experiment resulted in black, small crystals of $Na_8RbGa_3P_6$, which were found to be extremely air sensitive and had to be handled with great care. However, besides the title compound, Na_6GaP_3 was also identified. All the manipulations involving alkali metals were performed either inside an Ar-filled glove box or under vacuum.

9.5 GALLIUM–SODIUM–CESIUM–PHOSPHORUS

The **$NaCs_2GaP_2$** quaternary compound, which crystallizes in the orthorhombic structure with the lattice parameters $a = 676.1 \pm 0.2$, $b = 1577.2 \pm 0.5$, and $c = 648.6 \pm 0.2$ pm and a calculated density of 4.010 g·cm^{-3}, is formed in the Ga–Na–Cs–P system (Somer et al. 1992b). It was synthesized from the elements or from Na, Cs_4P_6, and GaP at 680°C in a sealed Nb ampoule.

9.6 GALLIUM–SODIUM–CALCIUM–PHOSPHORUS

The **Na$_3$Ca$_3$GaP$_4$** quaternary compound, which crystallizes in the hexagonal structure with the lattice parameters $a = 917.4 \pm 0.2$ and $c = 703.5 \pm 0.4$ pm and a calculated density of 2.48 g·cm^{-3} at 200 K, is formed in the Ga–Na–Ca–P system (Wang et al. 2017a). The title compound was obtained from the elements. The starting materials were weighed in the glove box and loaded into a Nb tube, which was then closed by arc-welding under high-purity argon gas. The sealed Nb container was subsequently jacketed within an evacuated fused silica tube. The reaction mixture was heated in a muffle furnace to 900°C at a rate of 100°C·h^{-1} and equilibrated at this temperature for 24 h. After that, the sample was cooled to 500°C at a rate of 5°C·h^{-1}, at which point the cooling rate to room temperature was changed to 40°C·h^{-1}. All manipulations were carried out inside an argon-field glove box with controlled atmosphere or under vacuum.

9.7 GALLIUM–SODIUM–STRONTIUM–PHOSPHORUS

The **Na$_3$Sr$_3$GaP$_4$** quaternary compound, which crystallizes in the hexagonal structure with the lattice parameters $a = 937.3 \pm 0.2$ and $c = 737.1 \pm 0.2$ pm, is formed in the Ga–Na–Sr–P system (Somer et al. 1996a). It was synthesized from stoichiometric mixture of Na$_3$P, SrP, and GaP in sealed Nb ampoule at 1100°C.

9.8 GALLIUM–SODIUM–OXYGEN–PHOSPHORUS

The **NaGa(PO$_3$)$_4$** and **Na$_7$Ga$_3$(P$_2$O$_7$)$_3$** quaternary compounds are formed in the Ga–Na–O–P system. NaGa(PO$_3$)$_4$ incongruently melts at 700°C (Grunze et al. 1987).

Na$_7$Ga$_3$(P$_2$O$_7$)$_3$ undergoes a reversible order–disorder transition at 234–236°C (Masquelier et al. 1994). X-ray diffraction (XRD) shows that the α-Na$_7$Ga$_3$(P$_2$O$_7$)$_3$ is a superstructure of the β form. This compound was prepared by crystallization in a flux of sodium arsenate. Starting materials were GaOOH, NaH$_2$PO$_4$, and Na$_4$P$_2$O$_7$. An intimate mixture containing Na$_2$O, Ga$_2$O$_3$, and P$_2$O$_5$ (molar ratio 53:5:42) was placed in a Pt crucible and progressively heated up to 600°C. Cooling was achieved at a rate of 210°C·h^{-1} down to 500°C and then switching off the furnace. The resulting product was washed with H$_2$O.

9.9 GALLIUM–POTASSIUM–CESIUM–PHOSPHORUS

The **K$_{0.69}$Cs$_{5.31}$(Ga$_2$P$_4$)** quaternary compound, which crystallizes in the orthorhombic structure with the lattice parameters $a = 1485.6 \pm 0.4$, $b = 2546.7 \pm 0.6$, and $c = 949.0 \pm 0.2$ pm, is formed in the Ga–K–Cs–P system (Somer et al. 1991h). This compound may be prepared from KP, GaP, and Cs (1:1:3 molar ratio 1:1:3) at 680°C in a sealed steel ampoule. The excess of alkaline metals (K and Cs) was removed by high vacuum distillation at 250°C.

9.10 GALLIUM–POTASSIUM–OXYGEN–PHOSPHORUS

KGaP$_2$O$_7$ and **KGa(PO$_3$)$_4$** quaternary compounds are formed in the Ga–K–O–P system. KGaP$_2$O$_7$ melts at circa 930°C (Chudinova et al. 1978). This compound is formed at the interaction of Ga$_2$O$_3$, K$_2$CO$_3$ and 85% H$_3$PO$_4$ in the glassy carbon crucible within the temperature interval from 150°C to 500°C.

KGa(PO$_3$)$_4$ melts at circa 730°C (Chudinova et al. 1978) [at 710°C (Grunze et al. 1987)] and crystallizes in the monoclinic structure with the lattice parameters a = 513.8 ± 0.3, b = 1229.0 ± 0.5, c = 1680.2 ± 1.3 pm, and β = 101.04 ± 0.05° and an experimental density of 2.55 g·cm^{-3} (Palkina et al. 1979). This compound is formed at the interaction of Ga$_2$O$_3$, K$_2$CO$_3$ and 85% H$_3$PO$_4$ in the glassy carbon crucible within the temperature interval from 310°C to 350°C for 5–10 days (Chudinova et al. 1977).

9.11 GALLIUM–RUBIDIUM–OXYGEN–PHOSPHORUS

The **RbGaP$_2$O$_7$** and **RbGa(PO$_3$)$_4$** quaternary compounds are formed in the Ga–Rb–O–P system. RbGaP$_2$O$_7$ congruently melts at 760°C and RbGa(PO$_3$)$_4$ incongruently melts at circa 700°C, forming Ga(PO$_3$)$_2$ and melt (Chudinova et al. 1979; Grunze et al. 1987). These compounds were obtained at the heating of the mixture of phosphoric acid and Rb$_2$CO$_3$ (Chudinova et al. 1978).

9.12 GALLIUM–CESIUM–OXYGEN–PHOSPHORUS

Some quaternary compounds are formed in the Ga–Cs–O–P system.

CsGaP$_2$O$_7$ is formed as the result of **Cs$_2$GaH$_3$(P$_2$O$_7$)$_2$** decomposition at 570°C (Grunze et al. 1987).

CsGa(PO$_3$)$_4$ incongruently melts at 780°C and crystallizes in the cubic structure with the lattice parameter a = 1437.4 pm and a calculated density of 3.43 g·cm^{-3} (Grunze et al. 1987). Optimal conditions for preparation of this compound are as follows: the ratio of P$_2$O$_5$/Cs$_2$O/Ga$_2$O$_3$ in the initial mixture is equal to 15:7.5:1 (GaCl$_3$, CsCl and H$_3$PO$_4$ were used as raw materials), the temperature is 340–440°C, the heating time is 7–10 days (Chudinova et al. 1987). The heating of **CsGa(H$_2$PO$_4$)$_4$** leads to the formation a mixture of well-formed dodecahedra of the title compound and crystals of **CsGaHP$_3$O$_{10}$** in the form of prismatic plates (Anisimova et al. 1995).

Cs$_2$GaP$_3$O$_{10}$ melts at 735°C (Lutsko et al. 1986) and crystallizes in the monoclinic structure with the lattice parameters a = 949.4 ± 0.6, b = 901.6 ± 0.9, c = 1229.0 ± 3.2 pm and β = 94.97 ± 0.30° (Devi and Vidyasagar 2000). It was synthesized in polycrystalline form by solid-state reactions from stoichiometric mixture of CsNO$_3$, NH$_4$H$_2$PO$_4$, and Ga$_2$O$_3$ (Devi and Vidyasagar 2000). This mixture was heated in open air initially at 400°C for 12 h, and the temperature was raised in steps of 100°C to the maximum value of 700°C, at which temperature the compound was heated for 12 h. Single crystals of Cs$_2$GaP$_3$O$_{10}$ were grown by the flux method using CsPO$_3$ as the flux. The mixture of CsNO$_3$, NH$_4$H$_2$PO$_4$, and Ga$_2$O$_3$ taken in the 6:1 mass ratio of CsPO$_3$ flux to Ga was heated at 700°C for 1 day and then cooled to 550°C at a rate of 3°C·h^{-1}. Platelike colorless crystals were separated by washing away the flux with H$_2$O.

9.13 GALLIUM–SILVER–BARIUM–PHOSPHORUS

The **AgBa$_4$Ga$_5$P$_8$** quaternary compound, which crystallizes in the orthorhombic structure with the lattice parameters a = 729.4 ± 0.9, b = 1803 ± 2, and c = 655.7 ± 0.8 pm; a calculated density of 4.830 g·cm^{-3}; and energy gap of 1.4 eV is formed in the Ga–Ag–Ba–P system

(Pan et al. 2015). The title compounds were synthesized from Pb flux (molar ratio of Ba/Ga/Ag/Pn/Pb = 3:1:1:5:25). The reactants were loaded in an alumina crucible and then sealed in fused silica tube under vacuum. The container was then moved to a programmable furnace; the mixture was first heated to 900°C and homogenized at this temperature for 20 h and then slowly cooled down to 500°C at a rate of 5°C·h^{-1}. Finally, the excessive flux was quickly decanted by centrifuge. $AgBa_4Ga_5P_8$ is stable to air and moisture. The bulk materials remain unchanged after being exposed to ambient air for more than 1 week. Stoichiometric reactions can greatly improve the yield of the target crystals, but a side product of GaP was hardly avoided despite various attempts. The excess Pb flux can be dissolved into the mixture of glacial acetic acid and hydrogen peroxide. All manipulations were performed in an argon-filled glove box.

9.14 GALLIUM–SILVER–TIN–PHOSPHORUS

The influence of silver on the solubility of GaP in Sn at 800°C was investigated by Saidov et al. (1981), and it was shown that Ag increases this solubility.

9.15 GALLIUM–SILVER–SELENIUM–PHOSPHORUS

The **$AgGaP_2Se_6$** quaternary compound, which congruently melts at 450°C and has two different polymorphic modifications, is formed in the Ga–Ag–Se–P system (Pfeiff and Kniep 1992). α-$AgGaP_2Se_6$ crystallizes in the orthorhombic structure with the lattice parameters $a = 1216.9 + 0.5$, $b = 2248.4 ± 0.6$, and $c = 747.3 ± 0.2$ pm; a calculated density of 2.32 g·cm^{-3}, and energy gap of 2.60 eV. β-$AgGaP_2Se_6$ (metastable modification) crystallizes in the trigonal structure with the lattice parameters $a = 637.5 ± 0.6$ and $c = 1332.0 ± 0.1$ pm, a calculated density of 5.05 g·cm^{-3}, and energy gap of 1.91 eV. α-$AgGaP_2Se_6$ was obtained by solid-state reactions from the stoichiometric mixture of the elements in evacuated silica ampoule by heating up to 750°C (10 h) followed by cooling from the melt (30°C·h^{-1}). After quenching the melt to liquid nitrogen, the yellow–red sample was annealed at 400°C for 21 days. β-$AgGaP_2Se_6$ was prepared by cooling down the melt from 750°C to room temperature at a rate of 30°C·h^{-1}. This modification can be converted to the stable α-$AgGaP_2Se_6$ by annealing at 350°C for 14 days.

9.16 GALLIUM–MAGNESIUM–OXYGEN–PHOSPHORUS

$Mg_xGa_{1-x}PO_4$ (0.4 < x < 0.5), which crystallizes in the hexagonal structure with the lattice parameters $a = 1785.9$ and $c = 2714.9$ pm, is formed in the Ga–Mg–O–P system (Bu et al. 1997). This material was made hydrothermally at 180°C with the reaction times between 3 and 5 days. Starting materials included $Ga(NO_3)_3 \cdot xH_2O$ or $Ga_2(SO_4)_3 \cdot xH_2O$, $MgHPO_4 \cdot 3H_2O$, and 85% H_3PO_4.

9.17 GALLIUM–CALCIUM–OXYGEN–PHOSPHORUS

The **$Ca_9Ga(PO_4)_7$** quaternary compound, which crystallizes in the trigonal structure with the lattice parameters $a = 1033.8 ± 0.2$ and $c = 3718 ± 1$ pm, is formed in the Ga–Ca–O–P system (Golubev et al. 1990). This compound was synthesized from the solid-state reaction of $(NH_4)_2HPO_4$ and Ga_2O_3 at 900–1200°C for 30–50 h.

9.18 GALLIUM–STRONTIUM–OXYGEN–PHOSPHORUS

The $Sr_9Ga(PO_4)_7$ quaternary compound, which crystallizes in the monoclinic structure with the lattice parameters $a = 1823.7 \pm 0.3$, $b = 1058.09 \pm 0.06$, $c = 897.8 \pm 0.1$ pm, and $\beta = 132.752 \pm 0.008°$, is formed in the Ga–Sr–O–P system (Belik et al. 2002). This compound was synthesized as white product from a mixture of $SrCO_3$, Ga_2O_3, and $NH_4H_2PO_4$ (molar ratio 9:0.5:7). The mixture was contained in alumina crucibles, heated under air while very slowly raising the temperature from room temperature to 630°C, and allowed to react at 1000–1150°C for 120 h with three intermediated grindings. The products were then quenched to room temperature.

9.19 GALLIUM–ZINC–TIN–PHOSPHORUS

The influence of zinc on the solubility of GaP in Sn at 800°C was investigated by Saidov et al. (1981), and it was shown that Zn increases this solubility.

9.20 GALLIUM–ZINC–OXYGEN–PHOSPHORUS

$Zn_xGa_{1-x}PO_4$ ($0.4 < x < 0.5$) and $ZnGaP_2O_8$ quaternary phases are formed in the Ga–Zn–O–P system (Bu et al. 1997). The first of them crystallizes in the hexagonal structure with the lattice parameters $a = 1784.3$ and $c = 2719.8$ pm and the second crystallizes in the rhombohedral structure with the lattice parameters $a = 1808.0$ and $c = 4195.1$ pm (in hexagonal setting). These compounds were hydrothermally made at 180°C with the reaction times between 3 and 5 days. Starting materials included $Ga(NO_3)_3 \cdot xH_2O$ or $Ga_2(SO_4)_3 \cdot xH_2O$, $Zn(NO_3)_2 \cdot 6H_2O$, and 85% H_3PO_4.

9.21 GALLIUM–ZINC–SULFUR–PHOSPHORUS

GaP–ZnS. The phase diagram is not constructed. Solid solutions with sphalerite structure over the entire range of concentrations are formed in this system (Yim 1969; Sonomura et al. 1973; Voitsehovskiy and Panchenko 1975, 1977). The lattice parameters of these solid solutions linearly change with composition (Yim 1969).

This system was investigated through XRD. The single crystals of $(ZnS)_x(GaP)_{1-x}$ solid solutions were obtained by the chemical transport reactions and by the crystallization from the solutions in the melts of Zn, Ga, and Sn (Yim 1969; Sonomura et al. 1973; Voitsehovskiy and Panchenko 1975, 1977).

9.22 GALLIUM–ZINC–SELENIUM–PHOSPHORUS

GaP–ZnSe. The phase diagram of this system is shown in Figure 9.1 (Shumilin et al. 1977; Ufimtsev et al. 1980). Solid solutions over the entire range of concentration are formed in this system. Lattice parameters of forming solid solutions linearly change with composition (Yim 1969). Sonomura et al. (1973) supposes the presence of immiscibility gap within the interval of 40–90 mol.% ZnSe.

This system was investigated using high-pressure chamber (Shumilin et al. 1977). Liquidus temperatures were determined by sight, and solidus line was calculated using a

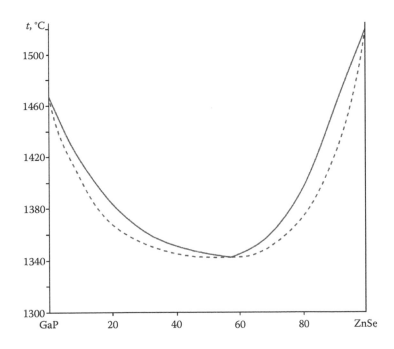

FIGURE 9.1 ZnSe–GaP phase diagram. (From Shumilin, V.P., *AN SSSR. Neorgan. Mater.*, 13(9), 1560–1564, 1977. With permission.)

model of regular solutions. Single crystals of solid solutions were obtained by the chemical transport reactions (Voitsehovskiy and Stetsenko 1976) or by the crystallization from the solutions in the Sn or Zn melts (Sonomura et al. 1973).

ZnSe–Ga–GaP. The liquidus surface of this quasiternary system near Ga corner (Figure 9.2) was constructed using differential thermal analysis, local XRD, measuring of microhardness, and mathematical simulation of experiment (Shumilin et al. 1977).

9.23 GALLIUM–CADMIUM–TIN–PHOSPHORUS

The influence of cadmium on the solubility of GaP in Sn at 800°C was investigated by Saidov et al. (1981), and it was shown that Cd increases this solubility.

9.24 GALLIUM–INDIUM–THALLIUM–PHOSPHORUS

$Ga_xIn_yTl_{1-x-y}P$ solid solutions have been grown by gas-source molecular beam epitaxy on InP substrates (Asahia et al. 1996, 1997). Tl and Ga compositions of these solid solutions were about 0.05–0.2 and 0.1–0.2, respectively. No phase separation was observed in the obtained materials.

9.25 GALLIUM–INDIUM–TIN–PHOSPHORUS

The influence of indium on the solubility of GaP in Sn at 800°C was investigated by Saidov et al. (1981), and it was shown that In increases this solubility.

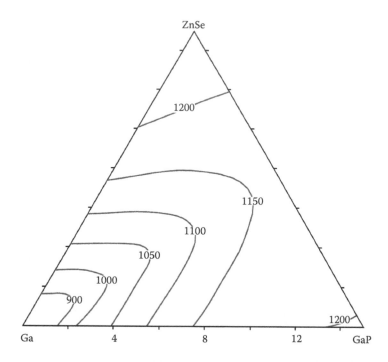

FIGURE 9.2 Liquidus surface of the ZnSe–Ga–GaP quasiternary system near Ga corner. (From Shumilin, V.P., *AN SSSR. Neorgan. Mater.*, 13(9), 1560–1564, 1977. With permission.)

9.26 GALLIUM–INDIUM–ARSENIC–PHOSPHORUS

GaP + InAs ⇔ InP + GaAs. The calculated liquidus and solidus surfaces for this system are given in Figure 9.3, and the miscibility gap is presented in Figure 9.4 (Ishida et al. 1989). According to the data of Stringfellow (1982a, 1982b), the calculated critical point for this system is 824°C and 808°C, respectively [886°C (Onabe 1982c); 826°C (Ishida et al. 1989)]. The behavior of unstable region in this system was analyzed based on regular approximation for solid solutions (Onabe 1982b, 1982c; Stringfellow 1982a, 1982b). The temperature dependence of spinodal isotherms at 200–800°C shows that the unstable regions cover a wide range of compositions. Takahei and Nagai (1981) noted that the immiscibility region in this ternary mutual system below 600°C was suggested according to the results of liquid phase epitaxial growth of $Ga_xIn_{1-x}As_yP_{1-y}$ solid solutions.

Spinodal decomposition was calculated for $Ga_xIn_{1-x}P_yAs_{1-y}$ alloys lattice-matched to GaAs and InP substrates by Asomoza et al. (2001a). Spinodal decomposition in this alloy due to mechanical and chemical origins is presented. It was shown that the spinodal decomposition temperatures are always smaller than their growth temperatures. Earlier data (Müller and Richards 1964) indicated that no miscibility gap is present in this system.

The phase diagram of the Ga–In–As–P quaternary system has been experimentally determined for several As isoconcentration sections at 600°C and 650°C (Nakajima et al. 1978). The liquidus data were obtained by the seed dissolution technique, and the solidus data were determined from liquid phase epitaxial layers grown on InP(111)B substrates by using an

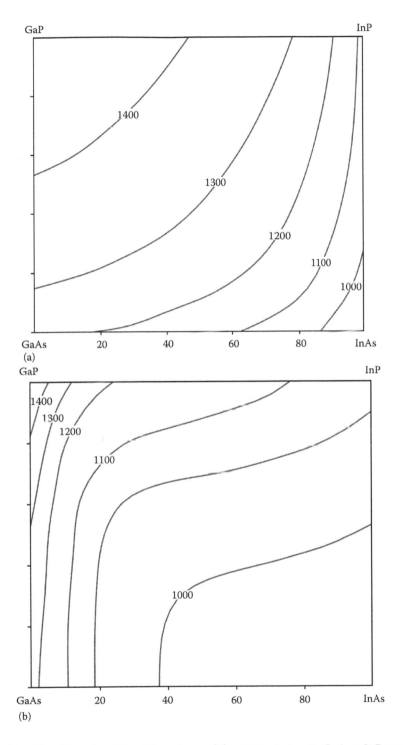

FIGURE 9.3 Liquidus (a) and solidus (b) surfaces of the GaP + InAs ⇔ GaAs + InP ternary mutual system. (Reprinted from *J. Cryst. Growth*, 98(1–2), Ishida, K. et al., Data base for calculating phase diagrams of III-V alloy semiconductors, 140–147, Copyright (1989), with permission from Elsevier.)

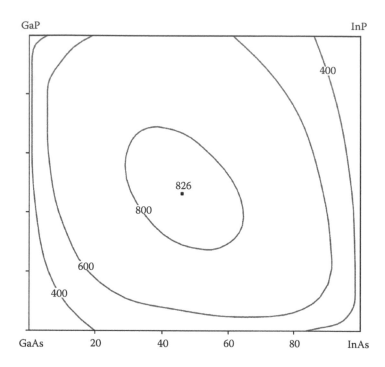

FIGURE 9.4 Miscibility gaps in the GaP + InAs ⇔ GaAs + InP ternary mutual system at 400–800°C. (Reprinted from *J. Cryst. Growth*, 98(1–2), Ishida, K. et al., Data base for calculating phase diagrams of III-V alloy semiconductors, 140–147, Copyright (1989), with permission from Elsevier.)

electron probe microanalysis. The phase diagram was also calculated by using a simple solution model. It was shown that the main effect of the addition of Ga to the melt is an appreciably decreasing of the phosphorus solubility in the liquid, while the presence of arsenic has a less pronounced effect in this range. The solidus isotherms indicate that the distribution coefficient for Ga increases with decreasing growth temperature and/or x^l_{As}, and the distribution coefficient for phosphorus increases with decreasing x^l_{As}. The lattice constants of the $Ga_xIn_{1-x}As_yP_{1-y}$ solid solutions obey Vegard's law.

The experimental results that are on the phase diagram of the Ga–In–As–P system within the temperature interval from 585°C to 636°C were also obtained by Litvak et al. (1999). The liquidus temperatures were determined by visual–polythermal method. Calculations showed that at these temperatures, there are two disjoint closed regions of solid solutions outgoing from the Ga–In–As and Ga–In–P ternary subsystems.

GaAs–InP–InAs. The liquidus and solidus surfaces of this subsystem of the GaP + InAs ⇔ InP +GaAs ternary mutual system (Figure 9.5) were calculated using the theory of subregular solutions (Sirota and Novikov 1985). The obtained results indicated on the formation of continuous solid solutions in this system at high temperatures. The isothermal sections of the GaAs–InP–InAs system at 1180°C, 1130°C, 1030°C, and 980°C are presented in Figure 9.6 (Sirota et al. 1984, Sirota and Novikov 1985).

GaAs–In–InP–InAs. Using the simplex centroid planning, the liquidus surfaces of this system at constant In quantities of 97 and 96 at% were constructed (Figure 9.7)

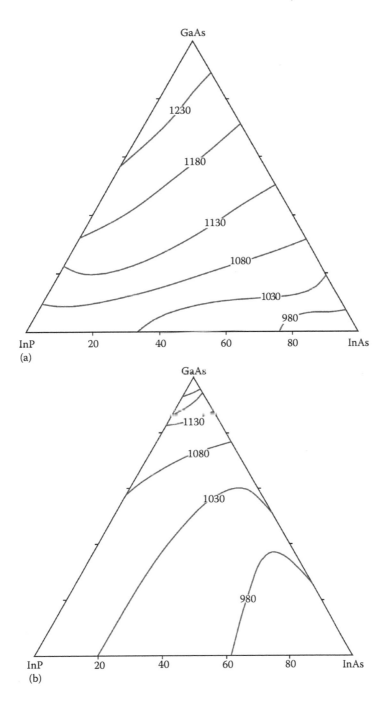

FIGURE 9.5 Liquidus (a) and solidus (b) surfaces of the GaAs–InP–InAs system. (From Sirota, N.N., and Novikov, V.V., *Zhurn. fiz. Khimii*, 59(4), 829–833, 1985. With permission.)

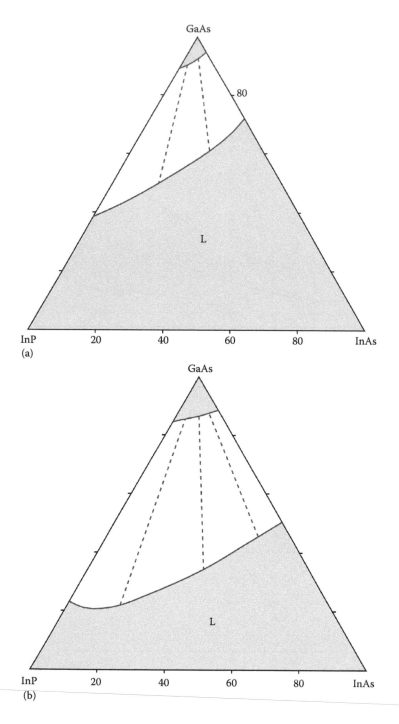

FIGURE 9.6 Isothermal sections of the GaAs–InP–InAs system at 1180°C (a) and 1130°C (b).
(*Continued*)

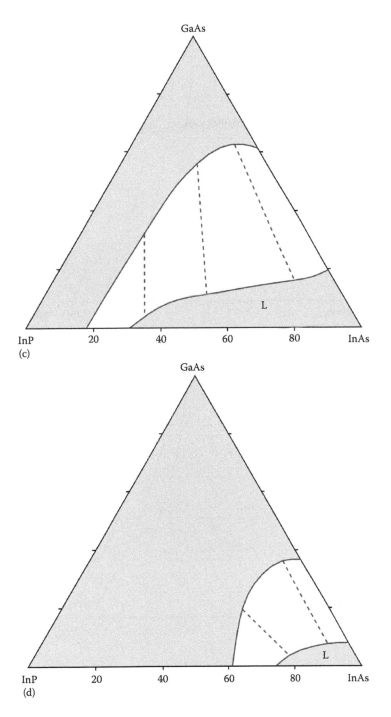

FIGURE 9.6 (CONTINUED) Isothermal sections of the GaAs–InP–InAs system at 1030°C (c) and 980°C (d). (From Sirota, N.N., and Novikov, V.V., *Zhurn. fiz. Khimii*, 59(4), 829–833, 1985. With permission.)

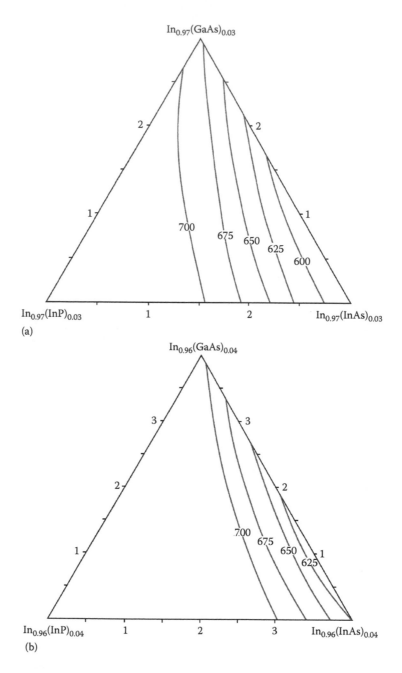

FIGURE 9.7   Isotherms of the liquidus surface of the GaAs–In–InP–InAs system at the constant In quantity: 97 mol.% (a) and 96 mol.% (b). (From Selin, A.A., *Izv. AN SSSR. Neorgan. Mater.*, 18(10), 1693–1696, 1982. With permission.)

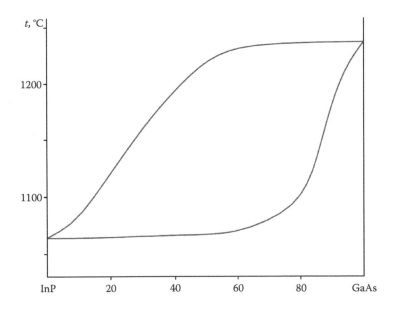

FIGURE 9.8   GaAs–InP phase diagram. (From Sirota, N.N., and Novikov, V.V., *Zhurn. fiz. Khimii*, 59(4), 829–833, 1985. With permission.)

(Selin et al. 1982). Liquidus temperatures were determined by direct visual observation, fixing the temperature, at which the last crystal of the solid solution dissolves in the melt enriched in indium. The analysis of the obtained results shows that for the section with a constant content of indium, an increase in the InP concentration leads to an increase in the liquidus temperature, an increase in the GaAs concentration slightly changes the liquidus temperature, and an increase in the InAs content lowers the liquidus temperature at a constant concentration of the other two compounds.

GaP–InAs. The phase diagram of this system was constructed by Sirota and Bodnar (1974). The continuous solid solutions are formed in this system. Below 227 K, these solid solutions decompose on the solid solution based on GaP and solid solution based on InAs.

GaAs–InP. The phase diagram of this system is given in Figure 9.8 (Sirota and Novikov 1985). The continuous solid solutions are formed in this system (Sirota and Makovetskaya 1963; Sirota and Novikov 1985). The dependence of the lattice parameter on the composition deviates significantly from Vegard's law (negative deviation). The minimum corresponds to a composition of 40 mol.% GaAs ($a = 556.1 \pm 0.2$ pm).

9.27 GALLIUM–INDIUM–ANTIMONY–PHOSPHORUS

GaP + InSb ⇔ InP + GaSb. The solid solution formation in this ternary mutual system is very limited (Müller and Richards 1964). The behavior of unstable region was analyzed based on regular approximation for solid solutions (Onabe 1982b). The temperature

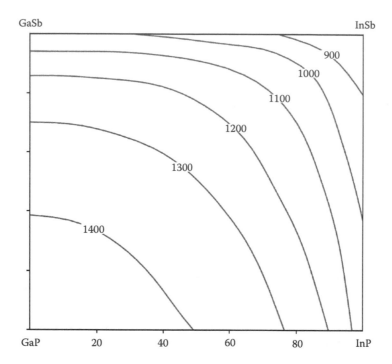

FIGURE 9.9 Liquidus surface of the GaP + InSb ⇔ InP + GaSb ternary mutual system. (Reprinted from *J. Cryst. Growth*, 98(1–2), Ishida, K. et al., Data base for calculating phase diagrams of III-V alloy semiconductors, 140–147, Copyright (1989), with permission from Elsevier.)

dependence of spinodal isotherms shows that the unstable regions cover a wide range of compositions. The calculated liquidus surface for this system is given in Figure 9.9, and the miscibility gap is presented in Figure 9.10 (Ishida et al. 1989). According to the data of Stringfellow (1982b), the calculated critical point for this system is 2197°C [3328°C (Onabe 1982b); 1926°C (Ishida et al. 1989)].

GaP–InSb. A partial solid solubility in this system was suggested by Goryunova and Sokolova (1960).

9.28 GALLIUM–INDIUM–SULFUR–PHOSPHORUS

The Ga_2InPS_3 quaternary compound, which crystallizes in the cubic structure with the lattice parameter $a = 537.1$ pm is formed in the Ga–In–S–P (Robbins and Lambrecht, Jr. 1975). This compound was prepared from a mixture of InP, Ga, and S. The pressed pellets of the mixture were sealed in evacuated quartz tubes and heated at a rate of $10–15°C·h^{-1}$ to 800°C. The samples were held at this temperature for 48 h and cooled to room temperature.

9.29 GALLIUM–INDIUM–SELENIUM–PHOSPHORUS

$GaIn_2PSe_3$ and Ga_2InPSe_3 quaternary compounds, which crystallizes in the cubic structure with the lattice parameter $a = 567.9$ pm (the first compound) and $a = 553.0$ pm (the second compound), are formed in the Ga–In–Se–P (Robbins and Lambrecht, Jr. 1975). These compounds were prepared from a mixture of InP or GaP, Ga or In, and Se.

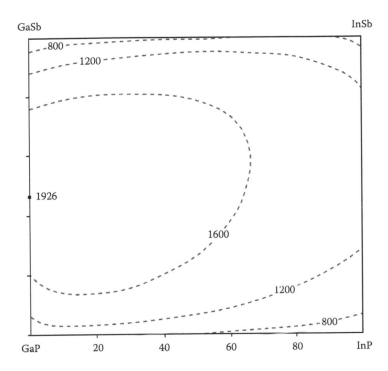

FIGURE 9.10 Miscibility gaps in the GaP + InSb ⇔ InP + GaSb ternary mutual system at 800–1600°C. (Reprinted from *J. Cryst. Growth*, 98(1–2), Ishida, K. et al., Data base for calculating phase diagrams of III-V alloy semiconductors, 140–147, Copyright (1989), with permission from Elsevier.)

The pressed pellets of the mixture were sealed in evacuated quartz tubes and heated at a rate of 10–15°C·h^{-1} to 800°C. The samples were held at this temperature for 48 h and cooled to room temperature.

9.30 GALLIUM–SILICON–TIN–PHOSPHORUS

The influence of silicon on the solubility of GaP in Sn at 800°C was investigated by Saidov et al. (1981), and it was shown that Si increases this solubility.

9.31 GALLIUM–GERMANIUM–TIN–PHOSPHORUS

The influence of germanium on the solubility of GaP in Sn at 800°C was investigated by Saidov et al. (1981), and it was shown that Ge increases this solubility.

9.32 GALLIUM–TIN–LEAD–PHOSPHORUS

The influence of lead on the solubility of GaP in Sn at 800°C was investigated by Saidov et al. (1981), and it was shown that Pb increases this solubility.

9.33 GALLIUM–TIN–CHROMIUM–PHOSPHORUS

The influence of chromium on the solubility of GaP in Sn at 800°C was investigated by Saidov et al. (1981), and it was shown that Cr increases this solubility.

9.34 GALLIUM–TIN–IRON–PHOSPHORUS

The influence of iron on the solubility of GaP in Sn at 800°C was investigated by Saidov et al. (1981), and it was shown that Fe increases this solubility.

9.35 GALLIUM–TIN–NICKEL–PHOSPHORUS

The influence of nickel on the solubility of GaP in Sn at 800°C was investigated by Saidov et al. (1981), and it was shown that Ni increases this solubility.

9.36 GALLIUM–ARSENIC–ANTIMONY–PHOSPHORUS

GaP–GaAs–GaSb. A miscibility gap has been found in this ternary system (Müller and Richards 1964). The behavior of unstable region was analyzed based on regular approximation for solid solutions (Onabe 1982b). The calculated liquidus and solidus surfaces for this system are given in Figure 9.11, and the miscibility gap is presented in Figure 9.12 (Ishida et al. 1989). The temperature dependence of spinodal isotherms shows that the unstable regions cover a wide range of compositions. The thermodynamics of spinodal decomposition in the GaP–GaAs–GaSb quasiternary system have been developed in Stringfellow (1983). Based on the delta-lattice parameter solution model of the free energy of mixing of semiconductor alloys, an analysis has been developed for the calculation of the spinodal surface and the critical temperature for solid alloys. Solid–solid isotherms were presented for this system. Concepts are also developed for the thermodynamic analysis of spinodal decomposition in this quaternary alloys, including the effect of the coherency strain energy. This addition to the free energy of the inhomogeneous solid is shown to completely stabilize the alloys of interest even at temperatures below room temperature. The calculated critical point for this system is 1723°C [1992°C (Onabe 1982b); 1875°C (Ishida et al. 1989)].

9.37 GALLIUM–OXYGEN–COBALT–PHOSPHORUS

CoGa$_{1-x}$PO$_4$ (0.4 < x < 0.5), which crystallizes in the hexagonal structure with the lattice parameters a = 1783.6 and c = 2718.2 pm, is formed in the Ga–O–Co–P system (Bu et al. 1997). This material was hydrothermally made at 180°C with the reaction times between 3 and 5 days. Starting materials included Ga(NO$_3$)$_3$·xH$_2$O or Ga$_2$(SO$_4$)$_3$·xH$_2$O, CoCO$_3$·xH$_2$O, and 85% H$_3$PO$_4$.

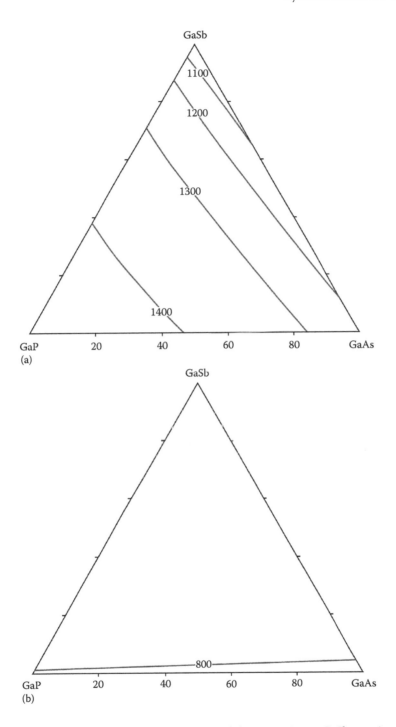

FIGURE 9.11 Liquidus (a) and solidus (b) surfaces of the GaP–GaAs–GaSb quasiternary system. (Reprinted from *J. Cryst. Growth*, 98(1–2), Ishida, K. et al., Data base for calculating phase diagrams of III-V alloy semiconductors, 140–147, Copyright (1989), with permission from Elsevier.)

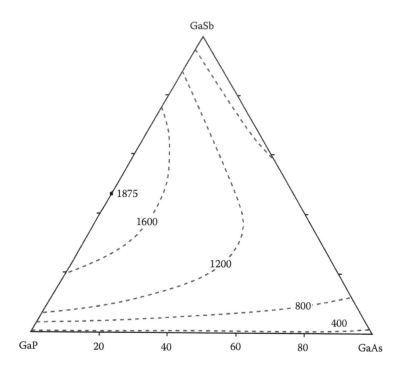

FIGURE 9.12 Miscibility gaps in the GaP–GaAs–GaSb quasiternary system at 400–1600°C. (Reprinted from *J. Cryst. Growth*, 98(1–2), Ishida, K. et al., Data base for calculating phase diagrams of III-V alloy semiconductors, 140–147, Copyright (1989), with permission from Elsevier.)

CHAPTER 10

Systems Based on GaAs

10.1 GALLIUM–HYDROGEN–CHLORINE–ARSENIC

GaAs–HCl–H$_2$. Vapor phase etching of polycrystalline GaAs was applied to the determination of the dominating equilibria in the common GaAs chlorine process using a simple entrainment method (Knobloch et al. 1984). It was established that in the temperature range 550–810°C, the dominating gallium compound in the vapor is GaCl, even at higher HCl inlet concentrations. From the equilibrium constants of the reaction GaAs + HCl = GaCl + 1/4As$_4$ + 1/2H$_2$, the standard enthalpy and the standard entropy of this reaction at 680°C (950 K) have been determined: $\Delta H^0{}_{950}$ = 148.8 ± 1.5 kJ·M^{-1} and $\Delta S^0{}_{950}$ = 131.8 ± 1.3 J·(K·M)$^{-1}$. According to the calculation, the standard enthalpy of this reaction at room temperature is equal $\Delta H^0{}_{298}$ = 150.0 ± 4.0 kJ·M^{-1}.

10.2 GALLIUM–LITHIUM–SODIUM–ARSENIC

The **LiNa$_2$GaAs$_2$** quaternary compound, which crystallizes in the orthorhombic structure with the lattice parameters a = 1205.1 ± 0.2, b = 1403.4 ± 0.2, and c = 590.8 + 0.1 pm, is formed in the Ga–Li–Na–As system (Somer et al. 1995c). This compound was synthesized from a stoichiometric mixture of Na$_3$As, LiAl, and As in evacuated and sealed Nb ampoules at 1000°C.

10.3 GALLIUM–LITHIUM–POTASSIUM–ARSENIC

The **LiK$_2$GaAs$_2$** quaternary compound, which crystallizes in the orthorhombic structure with the lattice parameters a = 1244.5 ± 0.3, b = 1524.3 ± 0.3, and c = 623.2 + 0.1 pm, is formed in the Ga–Li–K–As system (Somer et al. 1995e). This compound was synthesized from a stoichiometric mixture of KAs, Li, and GaAs in evacuated and sealed Nb ampoules at 650°C.

10.4 GALLIUM–LITHIUM–OXYGEN–ARSENIC

The **LiGaAs$_2$O$_7$** quaternary compound, which crystallizes in the monoclinic structure with lattice parameters a = 663.8 ± 0.1, b = 818.1 ± 0.2, c = 469.6 ± 0.1 pm and β = 104.01 ± 0.03° and a calculated density of 4.543 g.cm^{-3}, is formed in the Ga–Li–O–As system (Schwendtner and Kolitsch 2007). This compound was prepared by mild hydrothermal method in

Teflon-lined stainless steel autoclave (220°C, slow furnace cooling). Small, colorless, hemimorphic, pointed crystals of the title compound were prepared (7 days, pH = ~1.5) from a mixture of Li_2CO_3, Ga_2O_3, $H_3AsO_4 \cdot 0.5H_2O$ (molar ratio ~1:1:3) and distilled water. The Teflon container was filled with distilled water to ~70–80% of its inner volume.

10.5 GALLIUM–SODIUM–POTASSIUM–ARSENIC

The **NaK_2GaAs_2** quaternary compound, which crystallizes in the orthorhombic structure with the lattice parameters $a = 673.31 \pm 0.09$, $b = 1480.89 \pm 0.03$, and $c = 657.4 + 0.1$ pm and a calculated density of 3.250 g·cm^{-3}, is formed in the Ga–Na–K–As system (Somer et al. 1990a). This compound may be prepared from KAs, Cs, and GaAs (1:3:1 molar ratio) in evacuated and sealed silica ampoule at 680°C. The excess of alkali metals (K and Cs) was removed by high vacuum distillation at 250°C.

10.6 GALLIUM–SODIUM–RUBIDIUM–ARSENIC

The **$Na_8RbGa_3As_6$** quaternary compound, which crystallizes in the orthorhombic structure with the lattice parameters $a = 2284.3 \pm 0.6$, $b = 478.92 \pm 0.12$, and $c = 1686.1 + 0.4$ pm and a calculated density of 3.342 g·cm^{-3} at 200 ± 2 K, is formed in the Ga–Na–Rb–As system (He et al. 2012). The starting materials were elemental Na, Rb, Ga, and As. The elements in a stoichiometric ratio with a total mass of circa 500 mg were loaded into a Nb ampoule, which was subsequently arc-welded under high-purity Ar and then jacketed in a fused silica tube under vacuum. The reaction mixture was heated up to 550°C and equilibrated for 1 week before it was slowly cooled to room temperature. This experiment resulted in black, small crystals of $Na_8RbGa_3As_6$, which were found to be extremely air sensitive and had to be handled with great care. However, besides the title compound, **$Na_{10}NbGaAs_6$**, which formed from a side reaction with the Nb container, was also identified. All the manipulations involving alkali metals were performed either inside an Ar-filled glove box or under vacuum.

10.7 GALLIUM–SODIUM–CALCIUM–ARSENIC

The **$Na_{3.14\pm0.04}Ca_{2.86}GaAs_4$** quaternary compound, which crystallizes in the hexagonal structure with the lattice parameters $a = 940.69 \pm 0.09$ and $c = 720.39 \pm 0.14$ pm and a calculated density of 3.35 g·cm^{-3} at 200 K, is formed in the Ga–Na–Ca–As system (Wang et al. 2017a). The title compound was synthesized from the elements. The starting materials were weighed in the glove box and loaded into a Nb tube, which was then closed by arc-welding under high-purity argon gas. The sealed Nb container was subsequently jacketed within an evacuated fused silica tube. The reaction mixture was heated in a muffle furnace to 900°C at a rate of 100°C·h^{-1} and equilibrated at this temperature for 24 h. After that, the samples were cooled to 500°C at a rate of 5°C·h^{-1}, at which point the cooling rate to room temperature was changed to 40°C·h^{-1}. All manipulations were carried out inside an argon-field glove box with controlled atmosphere or under vacuum.

10.8 GALLIUM–SODIUM–STRONTIUM–ARSENIC

The **$Na_3Sr_3GaAs_4$** quaternary compound, which crystallizes in the hexagonal structure with the lattice parameters $a = 960.3 \pm 0.1$ and $c = 753.0 \pm 0.1$ pm, is formed in the

Ga–Na–Sr–As system (Somer et al. 1996a). It was synthesized from a stoichiometric mixture of Na_3As, SrAs, and GaAs in sealed Nb ampoule at 1100°C.

10.9 GALLIUM–SODIUM–EUROPIUM–ARSENIC

The $Na_{3.10\pm0.02}Eu_{2.90}GaAs_4$ quaternary compound, which crystallizes in the hexagonal structure with the lattice parameters $a = 955.66 \pm 0.12$ and $c = 743.19 \pm 0.19$ pm and a calculated density of 4.98 g·cm^{-3} at 200 K, is formed in the Ga–Na–Eu–As system (Wang et al. 2017a). The title compound was obtained from the elements. The starting materials were weighed in the glove box and loaded into the Nb tube, which was then closed by arc-welding under high-purity argon gas. The sealed Nb container was subsequently jacketed within evacuated fused silica tube. The reaction mixture was heated in a muffle furnace to 900°C at a rate of 100°C·h^{-1} and equilibrated at this temperature for 24 h. After that, the samples were cooled to 500°C at a rate of 5°C·h^{-1}, at which point the cooling rate to room temperature was changed to 40°C·h^{-1}. All manipulations were carried out inside an argon-field glove box with controlled atmosphere or under vacuum.

10.10 GALLIUM–SODIUM–NIOBIUM–ARSENIC

The $Na_{10}NbGaAs_6$ quaternary compound, which crystallizes in the monoclinic structure with the lattice parameters $a = 832.43 \pm 0.07$, $b = 751.73 \pm 0.06$, and $c = 1354.6 + 0.2$ pm and $\beta = 90.908 \pm 0.001°$ and a calculated density of 3.299 g·cm^{-3} at 200 ± 2 K, is formed in the Ga–Na–Nb–As system (He et al. 2012). The title compound was obtained as by-product at the synthesis of $Na_8RbGa_3As_6$ in a niobium container.

10.11 GALLIUM–SODIUM–OXYGEN–ARSENIC

The $Na_3Ga_2(AsO_4)_3$ and $Na_7Ga_3(As_2O_7)_3$ quaternary compounds are formed in the Ga–Na–O–As system. The first of them crystallizes in the rhombohedral structure with the lattice parameters $a = 1353.1 \pm 0.2$ and $c = 1855.4 \pm 0.2$ pm (in hexagonal setting) (d'Yvoire et al. 1986) and an experimental density of 4.13 g·cm^{-3} (Schwarz and Schmidt 1972). To obtain the title compound, the mixture of GaOOH, $Na_4As_2O_7$, and As_2O_5 (molar ratio 8:3:3) was progressively heated to 790°C and cooled down in several hours (d'Yvoire et al. 1986). Two or three successive heating separated by grinding operations were often necessary to complete the reaction. The product was finally washed with H_2O in order to dissolve the excess sodium arsenates. It was also prepared by the interaction of Na_2CO_3, Ga_2O_3, and $NH_4H_2AsO_4$ (3:2:6 molar ratio) at 750°C for 24 h (Schwarz and Schmidt 1972).

$Na_7Ga_3(As_2O_7)_3$ undergoes a reversible order–disorder transition at 112–114°C (Masquelier et al. 1994). X-ray diffraction (XRD) shows that α-$Na_7Ga_3(As_2O_7)_3$ is a superstructure of the β form. This compound was prepared by crystallization in a flux of sodium arsenate. Starting materials were GaOOH, $Na_4As_2O_7$, and As_2O_5. An intimate mixture containing Na_2O, Ga_2O_3, and As_2O_3 (molar ratio 50:7:43 or 53:7:40) was placed in a Pt crucible and progressively heated up to 650°C or 700°C. Cooling was achieved at a rate of 50°C·h^{-1} or 100°C·h^{-1} down to 500°C, and then the furnace was switched off. The resulting product was washed with H_2O.

10.12 GALLIUM–POTASSIUM–CESIUM–ARSENIC

The $K_{0.77}Cs_{5.23}Ga_2As_4$ quaternary compound, which crystallizes in the orthorhombic structure with the lattice parameters $a = 1503.3 \pm 0.4$, $b = 2616.2 \pm 0.5$, and $c = 969.1 + 0.3$ pm, is formed in the Ga–K–Cs–As system (Somer et al. 1991g). This compound was prepared from KAs, Cs, and GaAs (1:3:1 molar ratio) in sealed Nb ampoule at 680°C. The excess alkaline metals (K and Cs) were removed by high vacuum distillation at 250°C.

10.13 GALLIUM–POTASSIUM–OXYGEN–ARSENIC

The $KGaAs_2O_7$ quaternary compound, which crystallizes in the triclinic structure with the lattice parameters $a = 627.1 \pm 0.1$, $b = 637.6 \pm 0.1$, and $c = 816.9 \pm 0.1$ pm and $\alpha = 96.45 \pm 0.01$, $\beta = 103.86 \pm 0.01°$, and $\gamma = 103.87 \pm 0.01°$ and a calculated density of 4.065 g.cm^{-3}, is formed in the Ga–K–O–As system (Lin and Lii 1996a). To prepare the title compound, KH_2AsO_4 (0.6735 g), $(NH_4)H_2AsO_4$ (0.4433 g), and Ga_2O_3 (0.1847 g) were thoroughly mixed in a 15 mL platinum crucible. The mixture was heated at a rate of 100°C·h^{-1} to 500°C, maintained at this temperature for 6 h, heated to 700°C, maintained at 700°C for 10 h, cooled at 5°C·h^{-1} to 500°C and quenched to room temperature by removing the crucible from the furnace. The flux was dissolved with hot water, and the solid product was obtained by suction filtration. The product contained colorless chunk-shaped crystals of $KGaAs_2O_7$ and a small amount of $GaAsO_4$.

10.14 GALLIUM–RUBIDIUM–CESIUM–ARSENIC

The $Rb_{0.82}Cs_{5.18}Ga_2As_4$ quaternary compound, which crystallizes in the orthorhombic structure with the lattice parameters $a = 1505 \pm 1$, $b = 2638.0 \pm 0.8$, and $c = 974.8 + 0.4$ pm, is formed in the Ga–Rb–Cs–As system (Somer et al. 1992a). This compound was prepared from RbAs, Cs, and GaAs in sealed Nb ampoule at 680°C. The excess alkaline metals (Cs and Rb) were removed by high vacuum distillation at 252°C.

10.15 GALLIUM–CESIUM–OXYGEN–ARSENIC

The $Cs_2Ga_3As_5O_{18}$ quaternary compound, which crystallizes in the monoclinic structure with lattice parameters $a = 2703.60 \pm 0.07$, $b = 510.74 \pm 0.02$, $c = 1251.12 \pm 0.04$ pm and $\beta = 93.86 \pm 0.01°$ and a calculated density of 4.384 g.cm^{-3}, is formed in the Ga–Cs–O–As system (Lin and Lii 1996b). The title compound was synthesized from a flux of CsH_2AsO_4 and $NH_4H_2AsO_4$. A mixture of CsH_2AsO_4 (0.548 g), $NH_4H_2AsO_4$ (0.476 g) and Ga_2O_3 (0.188 g) (Cs/As mole ratio 0.4:1) was placed in a 15 cm^3 Pt crucible and heated at 100°C·h^{-1} to 500°C, maintained at this temperature for 6 h, heated to 750°C, annealed at 750°C for 10 h, then cooled at 5°C.h^{-1} to 500°C, and quenched to room temperature by removing the crucible from the furnace. The flux was dissolved with hot water and the solid product was obtained by suction filtration. The product contained colorless plate-shaped crystals of $Cs_2Ga_3As_5O_{18}$.

10.16 GALLIUM–SILVER–BARIUM–ARSENIC

The $AgBa_4Ga_5As_8$ quaternary compound, which crystallizes in the orthorhombic structure with the lattice parameters $a = 747.69 \pm 0.08$, $b = 1857.66 \pm 0.19$, and $c = 675.90 \pm 0.07$ pm;

a calculated density of 5.679 g·cm^{-3}; and an energy gap of 0.8 eV, is formed in the Ga–Ag–Ba–As system (Pan et al. 2015). The title compounds were synthesized from Pb flux reactions with loading Ba, Ga, Ag, As, and Pb (molar ratio 3:1:1:5:25). The reactants were loaded in an alumina crucible and then sealed in fused silica tube under vacuum. The container was then moved to a programmable furnace; the mixture was first heated to 900°C and homogenized at this temperature for 20 h and then slowly cooled down to 500°C at a rate of 5°C·h^{-1}. Finally, the excessive flux was quickly decanted by centrifuge. AgBa$_4$Ga$_5$As$_8$ is stable to air and moisture. The bulk materials remain unchanged after being exposed to ambient air for more than 1 week. Stoichiometric reactions can greatly improve the yield of the target crystals, but a side product of GaAs was hardly avoided in spite of various attempts. The excess Pb flux can be dissolved into the mixture of glacial acetic acid and hydrogen peroxide. All manipulations were performed in an argon-filled glove box.

10.17 GALLIUM–GOLD–VANADIUM–ARSENIC

All phases that are present in the phase diagrams of binary systems are identified in Au/V/GaAs contacts, but they are formed at lower temperatures down to room temperature (Maksimova et al. 1991). Formation of ternary compounds is also possible. The layered distribution of phases is due to the fact that phases closer to GaAs are enriched with gallium and arsenic.

10.18 GALLIUM–ZINC–SILICON–ARSENIC

2GaAs–ZnSiAs$_2$. The region of solid solutions along the 2GaAs–ZnSiAs$_2$ section extends from the GaAs side to 60-70 mol% of ZnSiAs$_2$, and from the ZnSiAs$_2$ side up to 10–15 mol.% 2 GaAs (Nazarov et al. 1967). The lattice parameters of the forming solid solutions change linearly with composition.

10.19 GALLIUM–ZINC–GERMANIUM–ARSENIC

2GaAs–ZnGeAs$_2$. The region of solid solutions along the 2GaAs–ZnGeAs$_2$ section extends from the GaAs side to 35–40 mol% of ZnGeAs$_2$ (Nazarov et al. 1967). The solubility of GaAs in ZnGeAs$_2$ was not detected. The solid solutions based on GaAs crystallize in the cubic structure of the sphalerite type. In view of the proximity of the lattice parameters of GaAs and ZnGeAs$_2$, there was no significant change in the lattice parameter of this solid solution depending on their composition.

10.20 GALLIUM–ZINC–SELENIUM–ARSENIC

GaAs–ZnSe. The phase diagram of this quasibinary system is shown in Figure 10.1 (Lakeenkov et al. 1975; Vasil'yev and Novikova 1977). Solid solutions over the entire range of concentrations are formed in this system (Goryunova and Fedorova 1959; Kirovskaya and Mulikova 1975). The composition and temperature of azeotropic point are 8 mol% ZnSe and 1230°C, respectively (Lakeenkov et al. 1975) [9 mol% ZnSe and 1233 ± 5°C (Vasil'yev and Novikova 1977)].

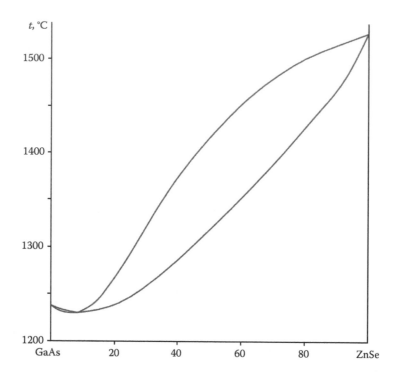

FIGURE 10.1 GaAs–ZnSe phase diagram. (From Lakeenkov, V.M., *Izv. AN SSSR. Neorgan. Mater.*, 11(7), 1311–1312, 1975. With permission.)

The energy gap of the forming solid solutions at 27°C is within the interval of 1.42–1.96 eV (Kirovskaya and Mulikova 1975), increases gradually with increase in ZnSe contents, and has a negative deviation from linearity (Ku and Bodi 1968).

Discrepancies in the determination of concentration dependence of the lattice parameters [corresponding to Vegard's law (Goryunova and Fedorova 1959; Lakeenkov et al. 1975); there is a deviation from Vegard's law with maximum at 50–60 mol% ZnSe (Yim 1969)] can be probably explained by the vicinity of the lattice parameters of ZnSe and GaAs [a_{ZnSe} = 566.87 pm and a_{GaAs} = 565.33 pm (Goryunova and Fedorova 1959)].

Solid solutions in this system were obtained by the melting of ZnSe and GaAs mixtures (Goryunova and Fedorova 1959; Ku and Bodi 1968; Yim 1969) or by the chemical transport reactions and recrystallization from the solutions in the gallium melt and floating-zone refining through the liquid gallium (Ku and Bodi 1968) or by the annealing of fine-dispersed powder of initial binary compounds at 1200°C (Kirovskaya and Mulikova 1975).

This system was investigated through differential thermal analysis (DTA), metallography, XRD, and measuring of microhardness (Goryunova and Fedorova 1959; Kirovskaya and Mulikova 1975; Lakeenkov et al. 1975; Vasil'yev and Novikova 1977).

GaAs–ZnSe–Ga. A part of liquidus surface of this quasiternary system near the Ga corner (Figure 10.2) was constructed according to four vertical sections (Novikova et al. 1974, 1977). Eutectic is degenerated from the Ga-rich side at 26°C in all vertical sections. There is a "valley" on the liquidus surface, which is elongated along the Ga–GaAs subsystem (Lakeenkov et al. 1974; Novikova et al. 1974, 1977). Using XRD, it was determined that solid

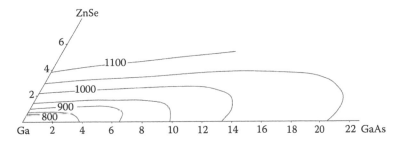

FIGURE 10.2   Part of the liquidus surface of the GaAs–ZnSe–Ga quasiternary system near the Ga corner. (From Lakeenkov, V.M., *Izv. AN SSSR. Neorgan. Mater.*, 11(7), 1311–1312, 1975. With permission.)

materials extracted from the gallium are $(ZnSe)_{1-x}(GaAs)_x$ solid solutions with sphalerite structure. Thus, practically all investigated liquidus surface is a field of $(ZnSe)_{1-x}(GaAs)_x$ solid solution crystallization. This system was investigated through DTA and XRD.

10.21   GALLIUM–ZINC–TELLURIUM–ARSENIC

GaAs–ZnTe. The phase diagram is a eutectic type (Figure 10.3) (Ufimtseva et al. 1973). The eutectic composition and temperature are 53 mol% GaAs and 1142°C, respectively. The solubility of ZnTe in GaAs is equal to 4.0 ± 0.4 mol% at 1180°C and decreases to 1.5 mol% at room temperature [according to the data of Glazov et al. (1975e), the solubility of ZnTe in GaAs reaches 15 mol%]. The solubility of GaAs in ZnTe reaches 12 ± 1.1 mol% at 1180°C and 2.0 ± 0.2 mol% at room temperature (Ufimtseva et al. 1973) [is not higher than 5 mol% (Glazov et al. 1975e)]. The lattice parameters in the homogencity regions change linearly

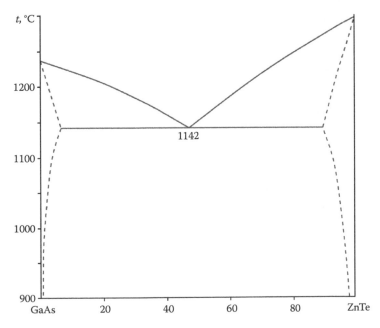

FIGURE 10.3   GaAs–ZnTe phase diagram. (From Ufimtseva, E.V., *Izv. AN SSSR. Neorgan. Mater.*, 9 (4), 587–591, 1973. With permission.)

TABLE 10.1   Liquidus and Solidus Temperatures
of the GaAs–ZnTe System from the GaAs-Rich Side

x_{GaAs}, mol%	t_S, °C	t_L, °C
100	1238	1238
95	1215	1225
90	1210	1220
85	1190	1215

Source: Anishchenko, V.A. et al., *Izv. AN SSSR. Neorgan.
Mater.*, 16(2), 354–355, 1980. With permission.

with composition. The liquidus and solidus temperatures from the GaAs-rich side are given in Table 10.1 (Anishchenko et al. 1980).

This system was investigated through DTA, metallography, local XRD, and measuring of microhardness (Ufimtseva et al. 1973; Glazov et al. 1975e). The ingots were annealed at 1235°C for 100 h. Single crystals of the solid solutions were grown by the oriented crystallization, floating-zone refining, and chemical transport reactions (Anishchenko et al. 1980).

10.22 GALLIUM–CADMIUM–SULFUR–ARSENIC

GaAs–CdS. The phase diagram is not constructed. According to the data of Kirovskaya and Zemtsov (2007) and Kirovskaya et al. (2007c), continuous solid solutions are formed in this quasibinary system. Solid solutions containing up to 50 mol% GaAs crystallize in the hexagonal structure of the wurtzite type and solid solutions based on GaAs (containing >50 mol% GaAs) crystallize in the cubic structure of the sphalerite type. Voitsehovskiy et al. (1970) indicated that the solubility of CdS in GaAs is equal to 10 mol%.

The ingots were obtained by the melting of mixtures from chemical elements at 1250°C without any annealing (Voitsehovskiy et al. 1970).

10.23 GALLIUM–CADMIUM–SELENIUM–ARSENIC

GaAs–CdSe. The phase diagram is a eutectic type (Figure 10.4) (Glazov et al. 1979b, 1983). The eutectic contains 63 mol% CdSe and crystallizes at 1116°C. Maximum solubility of GaAs in CdSe and CdSe in GaAs takes place at the eutectic temperature and is equal to 1.1 and 0.8 mol%, respectively [maximum solubility of CdSe in GaAs is not higher than 1.5 mol% (Glazov et al. 1985)]. According to the data of Voitsehovskiy et al. (1970), the solubility of CdSe in GaAs reaches 10 mol%. The solubility of GaAs in CdSe is also equal to 10 mol% (Kirovskaya et al. 2007c).

This system was investigated through DTA, metallography, and microhardness measuring (Glazov et al. 1979b, 1983, 1985). The ingots were obtained by the melting of mixtures from chemical elements at 1250°C (Voitsehovskiy et al. 1970). The samples were annealed at 900°C, 1000°C, and 1100°C for 700, 600, and 500 h, respectively (Glazov et al. 1985).

10.24 GALLIUM–CADMIUM–TELLURIUM–ARSENIC

GaAs–CdTe. The phase diagram is not constructed. The solubility of CdTe in GaAs is equal to 5 mol% (Voitsehovskiy et al. 1970) and the solubility of GaAs in CdTe reaches

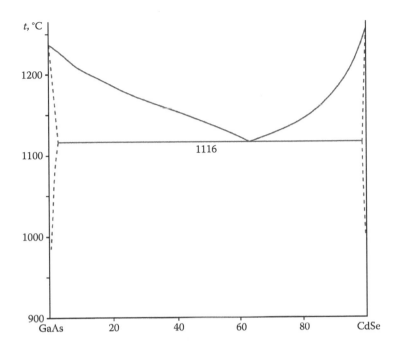

FIGURE 10.4 CdSe–GaAs phase diagram. (From Glazov, V.M., *Izv. AN SSSR. Neorgan. Mater.*, 19 (2), 193–196, 1983. With permission.)

15 mol% (Glazov et al. 1975e). Forming solid solutions crystallize in the cubic structure of sphalerite type.

This system was investigated by metallography (Voitsehovskiy et al. 1970; Glazov et al. 1975e). The ingots were obtained by melting mixtures of chemical elements in stoichiometric ratio at 1250°C (Voitsehovskiy et al. 1970).

10.25 GALLIUM–INDIUM–GERMANIUM–ARSENIC

The liquidus data for the Ga–In–Ge–As system were obtained from solubility experiments in liquid phase epitaxy (LPE) equipment (Müller et al. 1982). In Figure 10.5, some iso-germanium isotherms for the liquid phase in this system are shown. An excellent agreement between experimental and calculated liquidus data is found in the range $x^L_{Ge} \geq 0.1$ and $x^L_{In} > x^L_{Ge}$. The deviations are based on the simplifying assumption of ideal behavior in the binary subsystem Ge–As and the approach for the parameters $\omega_{Ga–Ge}$ and $\omega_{In–Ge}$. For these, the regular solution constants were used, which did not take into consideration the temperature dependence of the interaction parameters.

10.26 GALLIUM–INDIUM–TIN–ARSENIC

The liquidus data for the Ga–In–Sn–As system were obtained from solubility experiments in LPE equipment (Müller et al. 1982). In Figure 10.6, some iso-tin isotherms for the liquid phase in this system are shown. At high Sn concentrations and low temperatures, the deviations between experimental and calculated liquidus data are small. They increase at small Ga concentrations and high temperatures.

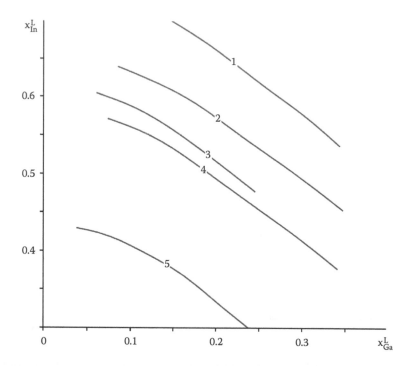

FIGURE 10.5 Liquidus isotherms with constant molecular fraction of Ge in the quaternary Ga–In–Ge–As system: 1, $x_{Ge}^L = 0.085$, 722°C; 2, $x_{Ge}^L = 0.14$, 795°C; 3, $x_{Ge}^L = 0.235$, 745°C; 4, $x_{Ge}^L = 0.235$, 782°C; 5, $x_{Ge}^L = 0.425$, 757°C. (From Müller, A. et al.: Melting diagrams of the quaternary systems Ga–In–As–Ge and Ga–In–As–Sn. *Cryst. Res. Technol.* 1982. 17(10). 1227–1232. Copyright Wiley-VCH Verlag GmbH & Co. KGaA. Reproduced with permission.)

To estimate the equilibrium compositions of the liquid and solid phases, a computational approach was used, followed by verification of its adequacy for some experimental points (Vasil'yev et al. 1985). The isotherms of the liquidus surface of this quaternary system for the Sn-rich melts were determined.

10.27 GALLIUM–INDIUM–LEAD–ARSENIC

GaAs–InAs–Pb. The liquidus isotherms at 600°C and 650°C of this system in the Pb corner are shown in Figure 10.7 (Grebenyuk et al. 1992). The fields of the primary crystallization of $Ga_{1-x}In_xAs$ solid solution and Pb exist in the GaAs–InAs–Pb system. In Figure 10.7, the dashed lines show the isoconcentrates of the components in the solid phase, corresponding to the crystallization of the same composition of the $Ga_{1-x}In_xAs$ solid solution at different temperatures. The experimental results coincide with the thermodynamic simulation.

10.28 GALLIUM–INDIUM–ANTIMONY–ARSENIC

GaAs + InSb ⇔ GaSb + InAs. The liquidus and solidus isotherms in this ternary mutual system were constructed by Nakajima et al. (1977). It was determined that the liquidus temperature becomes higher by an addition of As or Ga in the liquid. The solidus curves crowd near the GaAs–InAs quasibinary section and turn upward at the GaAs-rich corner. This means that the solubility of Sb into the quaternary solid is smaller than that into

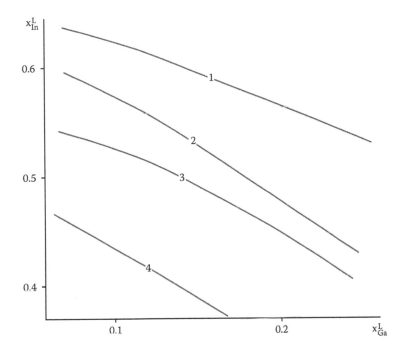

FIGURE 10.6   Liquidus isotherms with constant molecular fraction of Sn in the quaternary Ga–In–Sn–As system: 1, $x_{Sn}^{L} = 0.162$, 800°C; 2, $x_{Sn}^{L} = 0.2625$, 725°C; 3, $x_{Sn}^{L} = 0.2773$, 800°C; 4, $x_{Sn}^{L} = 0.428$, 725°C. (From Müller, A. et al.: Melting diagrams of the quaternary systems Ga–In–As–Ge and Ga–In–As–Sn. *Cryst. Res. Technol.* 1982. 17(10). 1227–1232. Copyright Wiley-VCH Verlag GmbH & Co. KGaA. Reproduced with permission.)

the quasibinary solids and increases as the temperature decreases. The liquidus surface of this system was also constructed by Bagirov et al. (1985) through DTA, XRD, and metallography.

The calculated liquidus and solidus surfaces for the GaAs + InSb ⇔ GaSb + InAs system are given in Figure 10.8, and the miscibility gap is presented in Figure 10.9 (Ishida et al. 1989). According to the data of Stringfellow (1982a) and Stringfellow (1982b), the calculated critical point for this system is 1188°C and 1155°C, respectively [1353°C (Onabe 1982b); 1117°C (Ishida et al. 1989)]. The behavior of unstable regions in this system was analyzed based on regular approximation for solid solutions (Onabe 1982b; Stringfellow 1982a, 1982b). Temperature dependence of spinodal isotherms shows that the unstable regions cover a wide range of compositions.

Spinodal decomposition was calculated for $Ga_xIn_{1-x}As_ySb_{1-y}$ alloys lattice-matched to GaSb and InAs substrates by Asomoza et al. (2001a). The spinodal decomposition in this alloy due to mechanical and chemical origins is presented. It was shown that the spinodal decomposition temperatures are always smaller than their growth temperatures. Earlier data (Müller and Richards 1964) indicated that no miscibility gap is present in this system.

The phase diagram of the Ga–In–As–Sb quaternary system has been determined experimentally and also has been treated on the base of thermodynamic calculations (Dolginov et al. 1978). The liquidus data were obtained by DTA, and solidus data were

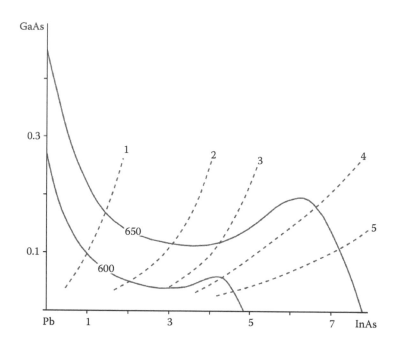

FIGURE 10.7 Liquidus isotherms in the GaAs–InAs–Pb system near the Pb corner: dashed lines show the isoconcentrates of the components in the solid phase, corresponding to the crystallization of the $Ga_{1-x}In_xAs$ solid solution at different temperatures at x = (1) 0.01, (2) 0.05, (3) 0.10, (4) 0.60, and (5) 0.80. (From Grebenyuk, A.M., *Zhurn. Neorgan. Khimii*, 37(1), 201–203, 1992. With permission.)

determined using electron probe microanalysis on LPE layers of $Ga_xIn_{1-x}As_ySb_{1-y}$ on GaSb and InAs substrates. The liquidus and solidus isotherms of the GaAs + InSb ⇔ GaSb + InAs ternary mutual system in the GaSb corner at 500°, 600°C, and 650°C are shown in Figures 10.10 through 10.12. It is clear that a variation in a solid–melt composition, for instance, a variation in the gallium content, results in a greater change of solid solution compositions close to GaSb compared to those close to InAs.

The liquidus and solidus isotherms in this ternary mutual system in the InSb corner at 550°, 675°C, and 600°C were determined by Grebenyuk et al. (1991) and are given in Figure 10.13.

The dependences of the compositions of the coexisting liquid phase and $Ga_xIn_{1-x}Sb_yAs_{1-y}$ solid solutions, isoperiodic to GaSb and InAs at (a) 550°C and (b) 600°C, are shown in Figures 10.14 and 10.15 (Baranov et al. 1994).

Early works of Müller and Richards (1964), Nakajima et al. (1977), and Dolginov et al. (1978) indicated the existence of the miscibility gap in the GaAs + InSb ⇔ GaSb + InAs ternary mutual system. This assumption was confirmed by Karouta et al (1986), Onabe (1982a, 1982b), and Sorokin et al. (2000). A thermodynamic formalism has been developed for the calculation of solid–solid and liquid–solid equilibria (Karouta et al 1986). The predicted limits of the metastable region are in good agreement with experimental data and for a large temperature range (400–800°C), while liquid–solid equilibrium calculations converge with experiments only for the solid composition. The liquidus temperatures of solutions of known compositions were determined by the melt visualization method.

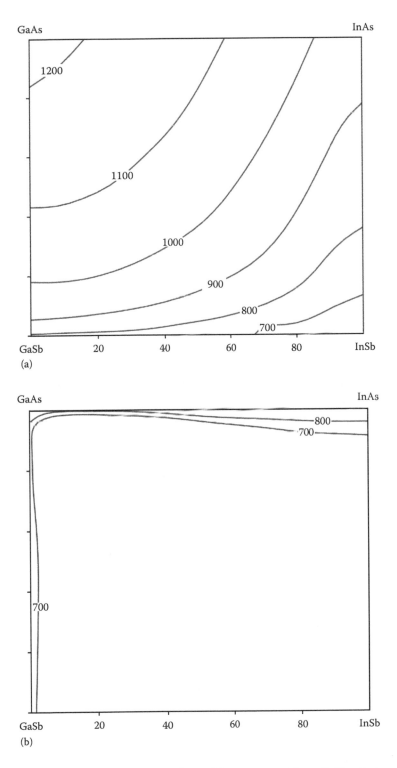

FIGURE 10.8 Liquidus (a) and solidus (b) surfaces of the GaAs + InSb ⇔ GaSb + InAs ternary mutual system. (Reprinted from *J. Cryst. Growth*, 98(1–2), Ishida, K. et al., Data base for calculating phase diagrams of III-V alloy semiconductors, 140–147, Copyright (1989), with permission from Elsevier.)

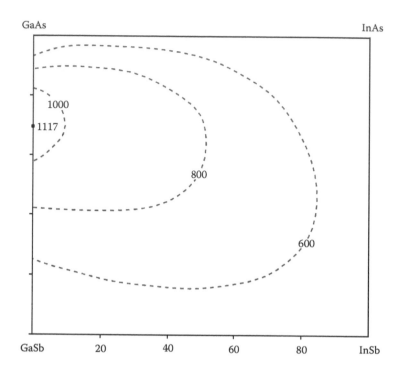

FIGURE 10.9 Miscibility gaps in the GaAs + InSb ⇔ GaSb + InAs ternary mutual system at 600–1000°C. (Reprinted from *J. Cryst. Growth*, 98(1–2), Ishida, K. et al., Data base for calculating phase diagrams of III-V alloy semiconductors, 140–147, Copyright (1989), with permission from Elsevier.)

The spinodal isotherms for the $Ga_xIn_{1-x}As_ySb_{1-y}$ solid solutions were calculated using the strictly regular solution approximation and are presented in Figure 10.16 (Onabe 1982a, 1982b). The dependence of the unstable region extent on the temperature is very clear. A large unstable region, and hence a miscibility gap, is dominant in the composition plane up to a temperature as high as 1000°C. The unstable region is isolated from the GaAs–GaSb quasibinary system above 860°C, which is the critical temperature for this system. The quaternary critical point (summit point of the spinodal surface) is at $x = 0.60$ and $y = 0.58$ with the temperature of 1353°C.

A novel formalism for thermodynamic calculations of a miscibility gap in solid alloys was proposed by Sorokin et al. (2000). The standard thermodynamic functions of binary compounds were employed for the determination of the interaction energy in quaternary alloys. The thermodynamic stability of $Ga_xIn_{1-x}As_ySb_{1-y}$ solid solutions at typical epitaxial temperatures (500–600°C) has been studied. The calculated miscibility and instability gaps are in a good agreement with experimental data on different alloys.

To estimate the regions of $Ga_xIn_{1-x}As_ySb_{1-y}$ solid solutions that cannot be obtained at a temperature below the melting point of InSb, a calculation was made of the component concentrations in the liquid phase, equilibrated at given temperatures with crystals whose composition can be located throughout the concentration square (Selin et al. 1988). The results obtained made it possible to determine the boundaries of the solidus at various

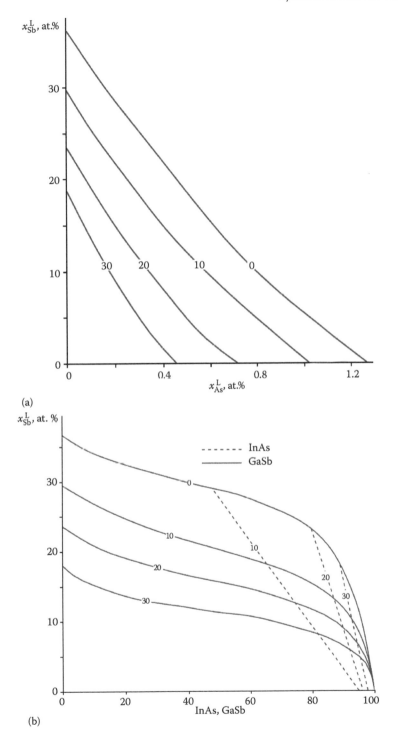

(a)

(b)

FIGURE 10.10 500°C (a) liquidus and (b) solidus isotherms in the GaAs + InSb ⇔ GaSb + InAs ternary mutual system at different Ga concentrations in the melt; the figures on curves refer to the Ga content in the solution–melt system (at%). (Dolginov, L.M. et al.: A study of phase equilibria and heterojunctions in Ga–In–As–Sb quaternary system. *Krist. und Techn.* 1978. 13(6). 631–638. Copyright Wiley-VCH Verlag GmbH & Co. KGaA. Reproduced with permission.)

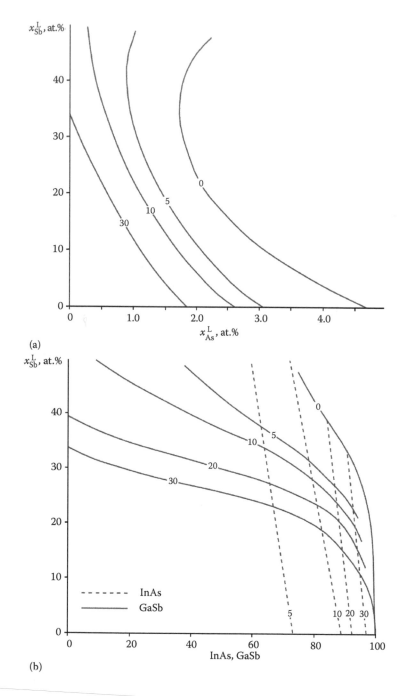

FIGURE 10.11 600°C (a) liquidus and (b) solidus isotherms in the GaAs + InSb ⇔ GaSb + InAs ternary mutual system at different Ga concentrations in the melt; the figures on curves refer to the Ga content in the solution–melt system (at%). (Dolginov, L.M. et al.: A study of phase equilibria and heterojunctions in Ga–In–As–Sb quaternary system. *Krist. und Techn.* 1978. 13(6). 631–638. Copyright Wiley-VCH Verlag GmbH & Co. KGaA. Reproduced with permission.)

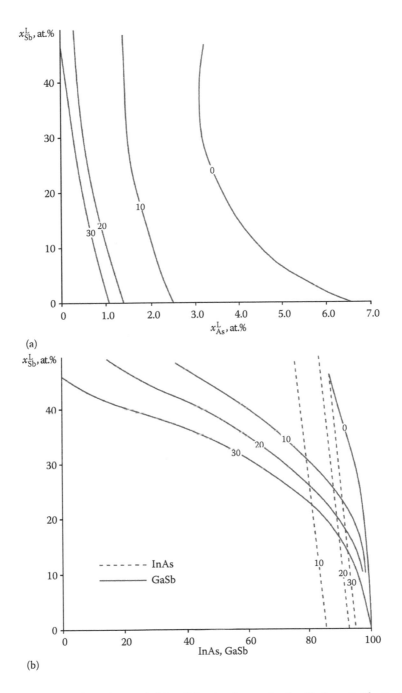

FIGURE 10.12 650°C (a) liquidus and (b) solidus isotherms in the GaAs + InSb ⇔ GaSb + InAs ternary mutual system at different Ga concentrations in the melt; the figures on curves refer to the Ga content in the solution–melt system (at%). (Dolginov, L.M. et al.: A study of phase equilibria and heterojunctions in Ga–In–As–Sb quaternary system. *Krist. und Techn.* 1978. 13(6). 631–638. Copyright Wiley-VCH Verlag GmbH & Co. KGaA. Reproduced with permission.)

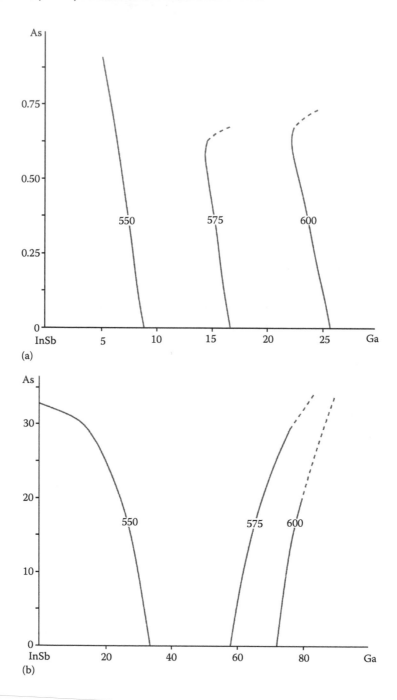

FIGURE 10.13 Liquidus (a) and solidus (b) isotherms in the GaAs + InSb ⇔ GaSb + InAs ternary mutual system in the InSb corner. (From Grebenyuk, A.M., *Zhurn. Neorgan. Khimii*, 36(4), 1067–1071, 1991. With permission.)

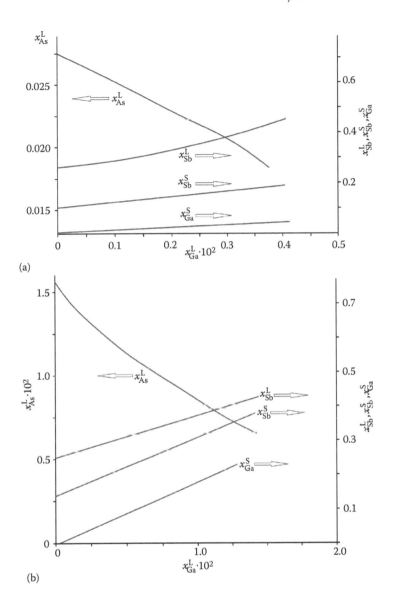

FIGURE 10.14 Dependence of the compositions of the coexisting liquid phase and $Ga_xIn_{1-x}Sb_yAs_{1-y}$ solid solutions, isoperiodic to InAs, at (a) 550°C and (b) 600°C. (From Baranov, A.N., *Zhurn. Prikl. Khimii*, 67(12), 1951–1956, 1994. With permission.)

temperatures. Phase equilibria in the Ga–In–Sb–As system were also obtained both experimentally and using calculation for the $Ga_xIn_{1-x}As_ySb_{1-y}$ solid solutions with the same lattice parameter as for GaSb (Selin and Antipin 1991).

The miscibility gap calculation of Schlenker et al. (2000) predicts that $Ga_xIn_{1-x}As_ySb_{1-y}$ layers, which are lattice-matched to InP and GaSb substrates, are stabilized by strain effects. The calculation also predicts that lattice-matched $Ga_xIn_{1-x}As_ySb_{1-y}$ layers can be grown through the whole composition range on GaSb substrates.

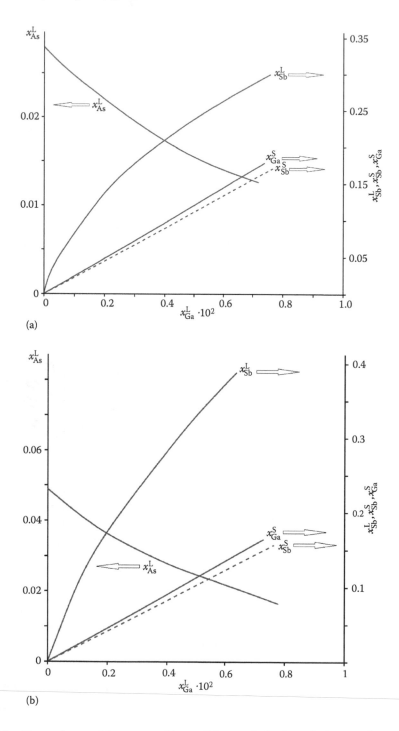

FIGURE 10.15   Dependence of the compositions of the coexisting liquid phase and $Ga_xIn_{1-x}Sb_yAs_{1-y}$ solid solutions, isoperiodic to GaSb, at (a) 550°C and (b) 600°C. (From Baranov, A.N., *Zhurn. Prikl. Khimii*, 67(12), 1951–1956, 1994. With permission.)

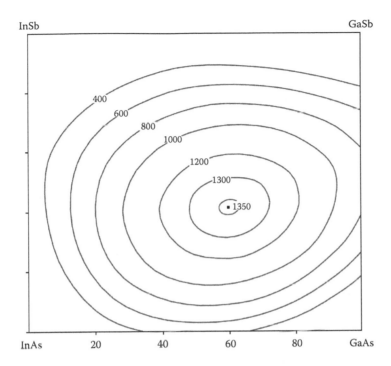

FIGURE 10.16 Spinodal isotherms for the $Ga_xIn_{1-x}Sb_yAs_{1-y}$ solid solutions. (Onabe, K., Unstable region in quaternary $In_{1-x}Ga_xAs_{1-y}Sb_y$ calculated using strictly regular solution approximation, *Jpn. J. Appl. Phys.*, 21(6), Pt. 1, 964, 1982, IOP.)

10.29 GALLIUM–INDIUM–BISMUTH–ARSENIC

$Ga_xIn_{1-x}As_{1-y}Bi_y$ solid solutions were grown on InP substrates by molecular beam epitaxy (MBE) at substrate temperatures between 250°C and 400°C (Bennarndt et al. 2016). Layers with high Bi fractions up to $Ga_yIn_{1-y}As_{0.88}Bi_{0.12}$ were successfully grown. It was determined that the increasing growth temperatures reduce the amount of Bi, which can be incorporated. According to the data of Feng et al. (2007), up to 6 at% GaBi in the $Ga_xIn_{1-x}As$ films have temperatures as low as 260°C. The Bi content decreases rapidly with increasing growth temperature. When the growth temperature is above 300°C, only a small amount of Bi below 1 at% can be incorporated in these films. Up to 2.5–3.1 at% Bi was incorporated in such films by Feng et al. (2005) and Ai et al. (2017). The energy band gap of the films containing up to 7 at% Bi was as low as 0.5 eV (Devenson et al. 2012).

10.30 GALLIUM–INDIUM–SULFUR–ARSENIC

Ga_2InAsS_3 quaternary compound, which crystallizes in the cubic structure with the lattice parameter $a = 538.5$ pm, is formed in the Ga–In–S–As (Robbins and Lambrecht, Jr. 1975). This compound was prepared from a mixture of Ga, In, As, and S. The pressed pellets of the mixture were sealed in evacuated quartz tubes and heated at a rate of 10–15°C·h⁻¹ to 800°C. The samples were held at this temperature for 48 h and cooled to room temperature.

10.31 GALLIUM–INDIUM–SELENIUM–ARSENIC

GaIn$_2$AsSe$_3$ and **Ga$_2$InAsSe$_3$** quaternary compounds, which crystallize in the cubic structure with the lattice parameter $a = 574.4$ pm (the first compound) and $a = 561.1$ pm (the second compound), are formed in the Ga–In–Se–As (Robbins and Lambrecht, Jr. 1975). These compounds were prepared from a mixture of Ga, In, As, and Se. The pressed pellets of the mixture were sealed in evacuated quartz tubes and heated at a rate of 10–15°C·h^{-1} to 800°C. The samples were held at this temperature for 48 h and cooled to room temperature.

10.32 GALLIUM–INDIUM–MANGANESE–ARSENIC

GaAs–InAs–MnAs. Phase stability in this quasiternary system was investigated by means of the first-principles full-potential linearized plane-wave method (Nakamura et al. 2011; Miyake et al. 2013). Calculated results predict that the ordered and disordered states are less favorable with respect to the phase separation. However, the Mn solubility into zinc blend GaAs increases as the In composition increases. The Mn-doped GaAs with $x_{Mn} \leq$ 0.44 favors the zinc blend structure over the rock salt structure regardless of the In incorporation.

The calculated T–x phase diagram (Figure 10.17) shows a miscibility gap with a critical temperature of about 630°C for the GaAs–InAs–MnAs quasiternary system (Miyake et al. 2013). The equilibrium Mn solubility into GaAs is low at low temperatures, namely, 3 at% at 500°C. When the In atoms are incorporated, however, the Mn solubility increases by about twice (when $x_{In} = 0.05$).

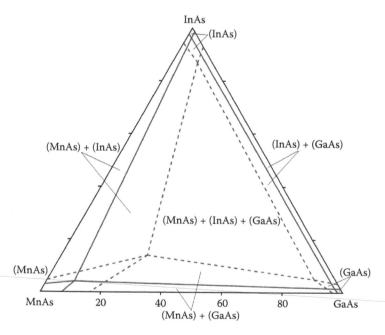

FIGURE 10.17   Calculated phase diagram within the zinc blend structure for the GaAs–InAs–MnAs quasiternary system: solid and dashed lines correspond to bi(tri)-nodal lines at 330°C and 500°C, respectively. (Reprinted from *J. Cryst. Growth*, 362, Miyake, M. et al. M Structural stability of Mn-doped GaInAs and GaInN alloys, 324–326, Copyright (2013), with permission from Elsevier.)

10.33 GALLIUM–INDIUM–IRIDIUM–ARSENIC

Ir_3Ga_5 is in thermodynamic equilibrium with the $Ga_xIn_{1-x}As$ solid solution from $x = 0$ to at least $x = 0.55$ (Swenson et al. 1994).

10.34 GALLIUM–SILICON–TIN–ARSENIC

A theoretical calculation of the $Ga_{0.5}As_{0.5}$–Si–Sn and $Ga_{0.8}As_{0.2}$–Si–Sn of the Ga–Si–Sn–As quaternary system was made in the framework of the model of regular solutions (Brovkin et al. 1981). The calculated liquidus isotherms for these sections are presented in Figure 10.18.

10.35 GALLIUM–SILICON–LEAD–ARSENIC

A theoretical calculation of $Ga_{0.7}As_{0.3}$–Si–Pb and $Ga_{0.8}As_{0.2}$–Si–Pb of the Ga–Si–Pb–As quaternary system was made in the framework of the model of regular solutions (Brovkin et al. 1981). The calculated liquidus isotherms for these sections are presented in Figure 10.19.

10.36 GALLIUM–GERMANIUM–TIN–ARSENIC

A large part of the 800°C liquidus isotherm of the Ga–Ge–Sn–As quaternary system has been determined experimentally, and it has been demonstrated that the compositions of the liquidus surface can be adequately represented by a "simple solution" thermodynamic

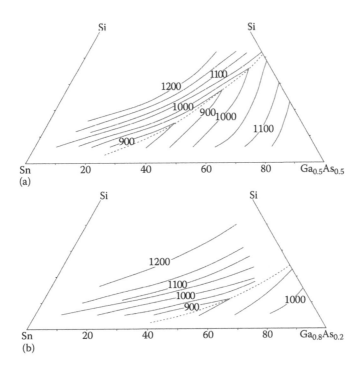

FIGURE 10.18 Liquidus isotherms (°C) for the (a) $Ga_{0.5}As_{0.5}$–Si–Sn and (b) $Ga_{0.8}As_{0.2}$–Si–Sn of the Ga–Si–Sn–As quaternary system. (From Brovkin, V.N., *Izv. AN SSSR. Neorgan. Mater.*, 17(3), 407–411, 1981. With permission.)

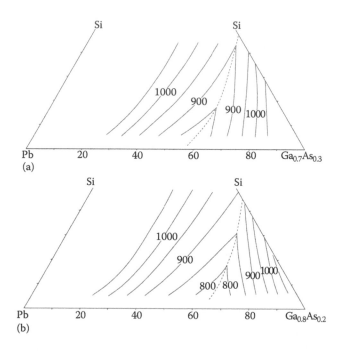

FIGURE 10.19 Liquidus isotherms (°C) for the (a) $Ga_{0.7}As_{0.3}$–Si–Pb and (b) $Ga_{0.8}As_{0.2}$–Si–Pb of the Ga–Si–Pb–As quaternary system. (From Brovkin, V.N., *Izv. AN SSSR. Neorgan. Mater.*, 17(3), 407–411, 1981. With permission.)

model (Panish 1973b). The liquidus compositions along several cuts in the quaternary phase diagram were determined with the seed dissolution technique.

10.37 GALLIUM–LEAD–ANTIMONY–ARSENIC

The liquidus isotherms in the **GaAs–GaSb–Pb** quasiternary system in the Pb corner were determined experimentally and calculated thermodynamically by Grebenyuk et al. (1991) and are presented in Figure 10.20. A good agreement between experimental data and thermodynamic calculation was obtained.

10.38 GALLIUM–TANTALUM–NICKEL–ARSENIC

GaAs–Ta–Ni. The interfacial reactions between Ta–Ni thin films and the (100) GaAs substrate have been analyzed by Auger electron spectroscopy, XRD, and transmission electron microscopy (Lahav and Eizenberg 1984). It was shown that a NiTaAs ternary compound could be formed in this system. At the presence of GaAs, this compound decomposes at temperatures higher than 600°C with formation of NiGa and TaAs.

10.39 GALLIUM–SULFUR–SELENIUM–ARSENIC

The **$Ga_3As_4S_3Se_6$** quaternary compound, which melts congruently at 385–387°C and crystallizes in the monoclinic structure with the lattice parameters $a = 1640$, $b = 846$, and $c = 901$ pm and $\beta = 97°$, is formed in the Ga–S–Se–As system (Il'yasov et al. 1977; Il'yasov and Rustamov 1982, 1983).

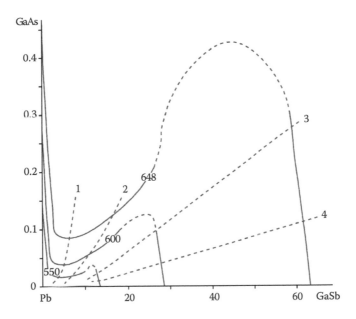

FIGURE 10.20 Liquidus isotherms in the GaAs–GaSb–Pb quasiternary system in the Pb corner: dashed lines show the isoconcentrates of the components in the solid phase, corresponding to the crystallization of the $GaAs_xSb_{1-x}$ solid solution at different temperatures at x = (1) 0.99, (2) 0.94, (3) 0.06, and (4) 0.02. (From Grebenyuk, A.M., *Zhurn. Neorgan. Khimii*, 36(4), 1067–1071, 1991. With permission.)

10.40 GALLIUM–SULFUR–TELLURIUM–ARSENIC

The **Ga₃As₂S₃Te₃** quaternary compound, which melts incongruently at 830°C and has an experimental density of 4.59 g·cm^{-3}, is formed in the Ga–S–Te–As system (Il'yasov 1983).

CHAPTER 11

Systems Based on GaSb

11.1 GALLIUM–LITHIUM–BARIUM–ANTIMONY

The **$Li_{13}Ba_8GaSb_{12}$** quaternary compound, which crystallizes in the monoclinic structure with the lattice parameters $a = 1806.50 \pm 0.10$, $b = 494.07 \pm 0.10$, and $c = 1301.23 \pm 0.10$ pm and $\beta = 126.728 \pm 0.010°$, a calculated density of 4.852 g·cm^{-3}, and an energy gap of 0.0489 ± 0.0001 eV, is formed in the Ga–Li–Ba–Sb system (Todorov and Sevov 2006). The title compound was synthesized from the elements. The corresponding stoichiometric mixture was heated at 860°C for 1 week and then cooled down to room temperature at a rate of 6°C·h^{-1}. All manipulations were carried out in an Ar-filled glove box with a moisture level below 1 part per million (ppm).

11.2 GALLIUM–LITHIUM–OXYGEN–ANTIMONY

The **$Li_8Ga_2Sb_2O_{12}$** quaternary compound, which crystallizes in the monoclinic structure with the lattice parameters $a = 513.49 \pm 0.02$, $b = 887.46 \pm 0.06$, and $c = 515.54 \pm 0.04$ pm and $\beta = 109.376 \pm 0.004°$, is formed in the Ga–Li–O–Sb system (Kumar et al. 2012). It was synthesized by heating the stoichiometrically homogenized reactants (Li_2CO_3, Sb_2O_3, Ga_2O_3) initially at 650°C for 12 h, followed by heating at 850°C for 24 h.

11.3 GALLIUM–SODIUM–POTASSIUM–ANTIMONY

The **$Na_3K_6Sb(GaSb_3)$** quaternary compound, which crystallizes in the hexagonal structure with the lattice parameters $a = 1021.80 \pm 0.07$ and $c = 1059.0 \pm 0.1$ pm, is formed in the Ga–Na–K–Sb system (Somer et al. 1995g). It was prepared from a stoichiometric mixture of K, NaSb, and GaSb in a sealed Nb ampoule at 650°C.

11.4 GALLIUM–SODIUM–CALCIUM–ANTIMONY

The **$Na_{3.98\pm0.01}Ca_{1.02}GaSb_3$** or **$Na_4CaGaSb_3$** quaternary compound, which crystallizes in the monoclinic structure with the lattice parameters $a = 915.13 \pm 0.09$, $b = 898.52 \pm 0.09$, and $c = 1223.05 \pm 0.12$ pm and a calculated density of 3.78 g·cm^{-3} at 200 K, is formed in the Ga–Na–Ca–Sb system (Wang et al. 2015). Single crystals of this compound were first identified from exploratory reactions of Na, Ca, Ga, and Sb in the molar ratio of 3:3:1:4. The metals were weighed in a glove box (total weight circa 500 mg) and loaded into a Nb tube, which was then closed by arc-welding under high-purity Ar gas. The sealed Nb container was subsequently jacketed within an evacuated fused silica tube. The reaction mixture was heated to 900°C at a rate of 100°C·h^{-1} and equilibrated at this temperature for 24 h. Following a cooling step to 500°C at a rate of 5°C·h^{-1}, the product was brought to room temperature over 12 h. After the structure and chemical composition were established by single-crystal work, new reaction with the proper stoichiometry was loaded. It yielded this compound as major product. The produced sample contained small amounts of $Ca_5Ga_2Sb_6$, but the bulk consisted of small, shiny crystals with a dark metallic luster that were the targeted phase. The obtained compound is brittle and air sensitive. All manipulations were carried out inside an argon-field glove box with controlled atmosphere or under vacuum.

11.5 GALLIUM–SODIUM–STRONTIUM–ANTIMONY

The **$Na_3Sr_3GaSb_4$** quaternary compound, which crystallizes in the hexagonal structure with the lattice parameters $a = 1016.4 \pm 0.1$ and $c = 791.4 \pm 0.1$ pm is formed in the Ga–Na–Sr–Sb system (Somer et al. 1996a). It was synthesized from stoichiometric mixture of Na_3Sb, $SrSb$, and $GaSb$ in sealed Nb ampoule at 1100°C.

11.6 GALLIUM–SODIUM–BARIUM–ANTIMONY

The **$Na_2Ba_4Ga_2Sb_6$** quaternary compound, which crystallizes in the orthorhombic structure with the lattice parameters $a = 1234.68 \pm 0.10$, $b = 1066.21 \pm 0.10$, and $c = 1383.44 \pm 0.10$ pm, a calculated density of 5.344 g·cm^{-3}, and energy gap of 0.0137 ± 0.0007 eV, is formed in the Ga–Na–Ba–Sb system (Todorov and Sevov 2006). This compound was synthesized from the elements. The corresponding stoichiometric mixture was heated at 860°C for 1 week and then cooled down to room temperature at a rate of 6°C·h^{-1}. All manipulations were carried out in an Ar-filled glove box with a moisture level below 1 ppm.

11.7 GALLIUM–SODIUM–EUROPIUM–ANTIMONY

The **$Na_{3.59\pm0.01}Eu_{2.41}GaSb_4$** quaternary compound, which crystallizes in the hexagonal structure with the lattice parameters $a = 1011.38 \pm 0.04$ and $c = 780.49 \pm 0.06$ pm and a calculated density of 4.84 g·cm^{-3} at 200 K, is formed in the Ga–Na–Eu–Sb system (Wang et al. 2017a). The title compound was obtained from the elements. The starting materials were weighed in the glove box and loaded into Nb tube, which was then closed by arc-welding under high-purity argon gas. The sealed Nb container was subsequently jacketed within evacuated fused silica tube. The reaction mixture was heated in a muffle furnace to 900°C at a rate of 100°C·h^{-1} and equilibrated at this temperature for 24 h. After that, the sample were cooled to 500°C at a rate of 5°C·h^{-1}, at which point the cooling rate to room temperature was changed to 40°C·h^{-1}. All manipulations were carried out inside an argon-field glove box with controlled atmosphere or under vacuum.

11.8 GALLIUM–POTASSIUM–CESIUM–ANTIMONY

The **$K_3Cs_6Sb(GaSb_3)$** quaternary compound, which crystallizes in the hexagonal structure with the lattice parameters $a = 1097.2 \pm 0.2$ and $c = 1150.1 \pm 0.2$ pm and a calculated density of 4.075 g·cm^{-3}, is formed in the Ga–K–Cs–Sb system (Somer et al. 1991b, 1991f). It was prepared from the elements or from the mixture of Cs, GaSb, and KSb at a molar ratio of K/Cs/Ga/Sb = 3:8:1:4 in a sealed Nb ampoule at 650–680°C.

11.9 GALLIUM–CALCIUM–BARIUM–ANTIMONY

The **CaBa$_6$Ga$_4$Sb$_9$** quaternary compound, which crystallizes in the orthorhombic structure with the lattice parameters a = 1078.83 ± 0.06, b = 1791.10 ± 0.10, and c = 706.21 ± 0.04 pm and a calculated density of 5.50 g·cm^{-3} (for Ca$_{0.367}$Ba$_{6.235}$Ga$_4$Sb$_9$ composition), is formed in the Ga–Ca–Ba–Sb system (He et al. 2016). It was prepared by direct fusion of the respective elements at 960°C. All reagents and products were handled within an Ar-filled glove box with controlled oxygen and moisture level below 1 ppm.

11.10 GALLIUM–STRONTIUM–OXYGEN–ANTIMONY

The **Sr$_2$GaSbO$_6$** quaternary compound, which crystallizes in the cubic structure with the lattice parameter a = 789.2 pm (Wittmann et al. 1981) [in the tetragonal structure with the lattice parameters a = 784 and c = 791 pm (Sleight and Ward 1964)], is formed in the Ga–Sr–O–Sb system. To obtain this compound, the starting materials SrCO$_3$ or Sr(NO$_3$)$_2$, Ga$_2$O$_3$, and Sb$_2$O$_3$ were finely ground and heated up to 1000–1100°C at the rate of 100°C·h^{-1} with the next annealing at these temperatures from 2 to 6 h in corundum crucibles in air (Wittmann et al. 1981). Then, the mixture was annealed at 1200°C for 12 h and at 1300–1350°C two times for 1 day. The title compound could also be prepared at the sintering the mixture of SrO, Ga$_2$O$_3$, and Sb$_2$O$_3$ in air at 1100°C (Sleight and Ward 1964).

11.11 GALLIUM–BARIUM–SULFUR–ANTIMONY

The **Ba$_{23}$Ga$_8$Sb$_2$S$_{38}$** quaternary compound, which crystallizes in the orthorhombic structure with the lattice parameters a = 961.1 ± 0.3, b = 3225 ± 2, and c = 1288.3 ± 0.4 pm and a calculated density of 4.307 g·cm^{-3}, is formed in the Ga–Ba–S–Sb system (Chen et al. 2012). Light yellow crystals of the title compound were synthesized via solid-state reactions by heating of a mixture of the elements Ba, Ga, Sb, and S in an evacuated silica tube. The stoichiometric mixture was loaded into a graphite crucible and then sealed in an evacuated silica tube under vacuum. The mixture was heated to 920°C in 35 h and annealed at this temperature for 100 h, before being cooled to 300°C in 120 h. The single crystals were obtained from the mixture of Ba/Ga/Sb/S = 40:16:8:76 at 920°C. This compound is stable in air. All starting reactants were handled inside an Ar-filled glove box with controlled oxygen and moisture levels below 0.1 ppm, and all manipulations were carried out in the glove box or under vacuum.

11.12 GALLIUM–BARIUM–SELENIUM–ANTIMONY

The **Ba$_2$GaSbSe$_5$** and **Ba$_{23}$Ga$_8$Sb$_2$Se$_{38}$** quaternary compounds are formed in the Ga–Ba–Se–Sb system. Ba$_2$GaSbSe$_5$ crystallizes in the orthorhombic structure with the lattice parameters a = 1268.1 ± 0.3, b = 914.4 ± 0.2, and c = 924.2 ± 0.2 pm, a calculated density of 5.34 g·cm^{-3} at 153 ± 2 K, and energy gap of 2.51 ± 0.2 eV (Hao et al. 2013). To prepare this compound, the reaction mixture of BaSe (2 mM), Sb_2Se_3 (0.5 mM), and Ga_2Se_3 (0.5 mM) were ground and loaded into fused silica tube under an Ar atmosphere in a glove box. The tube was flame sealed under a high vacuum of (10^{-3} Pa) and then placed in a computer-controlled furnace. The samples were heated to 1000°C within 15 h, kept for 48 h, then slowly cooled to 320°C at the rate of 4°C·h^{-1}, and finally cooled to room temperature by switching off the furnace. After the structural identification, the polycrystalline samples of $Ba_2GaSbSe_5$ were synthesized by the stoichiometric reaction of the binary materials in the molar ratio of $BaSe/Sb_2Se_3/Ga_2Se_3$ = 4:1:1 at high temperature in sealed silica tubes evacuated to 10^{-3} Pa. The reaction mixtures were placed in a computer-controlled furnace and then heated to 750°C in 15 h and kept at that temperature for 48 h, and then the furnace was turned off.

$Ba_{23}Ga_8Sb_2Se_{38}$ also crystallizes in the orthorhombic structure with the lattice parameters a = 998.25 ± 0.04, b = 3346.58 ± 0.13, and c = 1338.56 ± 0.5 pm, a calculated density of 5.169 g·cm^{-3}, and energy gap of 2.31 ± 0.2 eV (Yin et al. 2017b). The crystals of the title compound were initially obtained from reaction of a mixture of BaSe (200 mg), Ga_2Se_3 (61 mg), and Sb_2Se_3 (20 mg), which was finely ground and loaded into a fused silica tube. The tube was evacuated, sealed, and placed in a computer-controlled furnace where it was heated to 950°C over 24 h, kept at that temperature for 96 h, cooled to 400°C over 96 h, and then cooled to room temperature by shutting off the furnace. In an optimized synthesis to prepare bulk powder samples of $Ba_{23}Ga_8Sb_2Se_{38}$, a stoichiometric mixture of BaSe, Ga_2Se_3, and Sb_2Se_3 was finely ground and loaded in an evacuated and sealed fused silica tube as before. The tube was heated to 800°C over 24 h, kept at that temperature for 96 h, and then cooled to room temperature by shutting off the furnace.

11.13 GALLIUM–BARIUM–TELLURIUM–ANTIMONY

The **$Ba_2GaSbTe_5$** quaternary compound, which crystallizes in the orthorhombic structure with the lattice parameters a = 1355.3 ± 0.3, b = 959.7 ± 0.2, and c = 977.5 ± 0.2 pm, a calculated density of 5.77 g·cm^{-3} at 153 ± 2 K, and energy gap of 1.66 ± 0.7 eV, is formed in the Ga–Ba–Te–Sb system (Hao et al. 2013). The title compound was obtained by the same way as $Ba_2GaSbSe_5$ was synthesized using BaTe, Sb_2Te_3, and Ga_2Te_3 instead of BaSe, Sb_2Se_3, and Ga_2Se_3, respectively.

11.14 GALLIUM–ZINC–TELLURIUM–ANTIMONY

GaSb–ZnTe. The phase diagram is a eutectic type (Figure 11.1) (Glazov et al. 1975c, 1975d). The eutectic crystallizes at 690°C and contains approximately 10 mol% ZnTe, but according to the data by Burdiyan and Korolevski (1966), Burdiyan (1970, 1971), and Kirovskaya et al. (2007a), the solubility of ZnTe in GaSb is equal 10–20 mol%. The energy gap of forming solid solutions changes almost linearly with the composition (Burdiyan 1970, 1971). The solubility of GaSb in ZnTe reaches 15 mol% (Kirovskaya et al. 2007a) [is not higher than 10 mol% (Glazov et al. 1975c, 1975d)]. This system was investigated through differential thermal analysis (DTA) and metallography (Burdiyan and Korolevski 1966; Glazov et al. 1975c, 1975d).

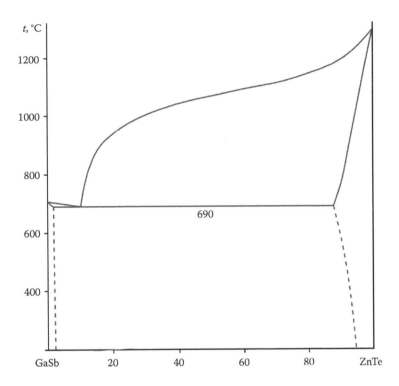

FIGURE 11.1   GaSb–ZnTe phase diagram. (From Glazov, V.M., *AN SSSR. Neorgan. Mater.*, 11(3), 418–423, 1975. With permission.)

11.15 GALLIUM–CADMIUM–TIN–ANTIMONY

GaSb–"CdSnSb₂." Three phases, GaSb, CdSb, and SnSb, are formed at the melting of the alloys of this system (Arbenina et al. 1989). The field of GaSb primary crystallization occupies the concentration region up to more than 90 mol% "CdSnSb₂." The phase relations in this system were also determined, and corresponding vertical section was constructed.

11.16 GALLIUM–CADMIUM–TELLURIUM–ANTIMONY

GaSb–CdTe. The phase diagram is a eutectic type (Figure 11.2) (Glazov et al. 1975d). The eutectic composition and temperature are 95 mol% GaSb and 670°C, respectively. The solubility of GaSb in CdTe is not higher than 10 mol% (Glazov et al. 1975c, 1975d) [solubility reaches 15 mol% (Burdiyan and Makeichik 1966; Kirovskaya et al. 2007a); 20 mol% (Burdiyan 1970, 1971)], and the solubility of CdTe in GaSb is equal to 10 mol% (Kirovskaya et al. 2007a). The energy gap of forming solid solutions changes near linearly with composition (Burdiyan 1970, 1971). This system was investigated by DTA, metallography, and measuring of microhardness (Glazov et al. 1975c, 1975d).

11.17 GALLIUM–MERCURY–TELLURIUM–ANTIMONY

GaSb–HgTe. The phase diagram is not constructed. According to the data of metallography and X-ray diffraction (XRD), the solubility of GaSb in HgTe reaches 10 mol%

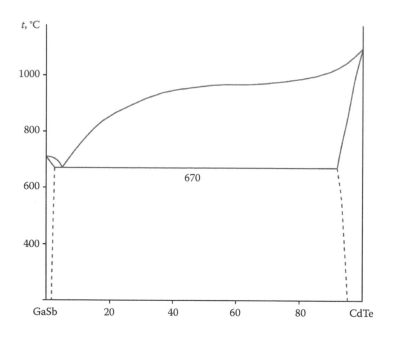

FIGURE 11.2   GaSb–CdTe phase diagram. (From Glazov, V.M., *AN SSSR. Neorgan. Mater.*, 11(3), 418–423, 1975. With permission.)

(Ambros and Burdiyan 1970; Burdiyan 1970, 1971). The energy gap of forming solid solutions changes near linearly with composition (Burdiyan 1970, 1971).

11.18 GALLIUM–INDIUM–LEAD–ANTIMONY

GaSb–InSb–Pb. The liquidus isotherms at 500°C, 550°C, and 600°C of this system in the Pb corner are shown in Figure 11.3 (Grebenyuk et al. 1992). The fields of primary crystallization of $Ga_{1-x}In_xSb$ solid solution and Pb exist in the GaSb–InSb–Pb system. In the Figure 11.3, the dashed lines show the isoconcentrates of the components in the solid phase, corresponding to the crystallization of the same composition of the $Ga_{1-x}In_xSb$ solid solution at different temperatures. The experimental results coincide with the thermodynamic simulation.

11.19 GALLIUM–YTTRIUM–SULFUR–ANTIMONY

The **Y_4GaSbS_9** quaternary compound, which crystallizes in the orthorhombic structure with the lattice parameters $a = 1348.0 \pm 0.4$, $b = 1379.0 \pm 0.4$, and $c = 1399.0 \pm 0.4$ pm, a calculated density of 4.269 g·cm^{-3}, and energy gap of 2.06 eV, is formed in the Ga–Y–S–Sb system (Wang et al. 2017b). To prepare this compound, a mixture (about 300 mg in total) of Y, Ga, Sb, and S in a molar ratio of 4:1:1:10 was loaded in a fused silica tube under vacuum, heating to 950°C over 35 h, and then maintaining there for 6 days followed by cooling to 300°C at 5°C·h^{-1} before switching off the furnace. The raw products were washed first with CS_2 to remove the by-products and dried with ethanol. All the manipulations were performed in an Ar-filled glove box with controlled oxygen and moisture levels below 0.1 ppm.

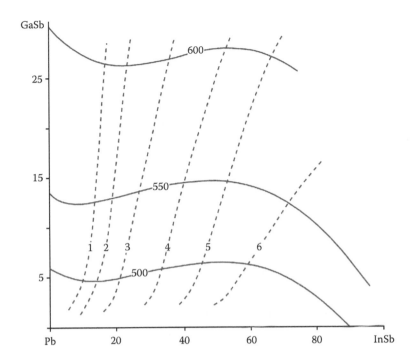

FIGURE 11.3 Liquidus isotherms in the GaSb–InSb–Pb system near the Pb corner: dashed lines show the isoconcentrates of the components in the solid phase, corresponding to the crystallization of the $Ga_{1-x}In_xSb$ solid solution at different temperatures at $x =$ (1) 0.01, (2) 0.02, (3) 0.04, (4) 0.10, (5) 0.20, and (6) 0.40. (From Grebenyuk, A.M., *Zhurn. Neorgan. Khimii*, 37(1), 201–203, 1992. With permission.)

11.20 GALLIUM–LANTHANUM–TIN–ANTIMONY

$LaGa_xSn_ySb_2$ solid solutions are formed in the Ga–La–Sn–Sb quaternary system (Morgan et al. 2002). For low Sn content ($0 \leq y \leq 0.2$), $LaGa_{1-y}Sn_ySb_2$ solid solutions crystallize in the orthorhombic structure of the $SmGaSb_2$-type, as exemplified by $LaGa_{0.92\pm0.03}Sn_{0.08}Sb_2$ ($a = 439.20 \pm 0.04$, $b = 2293.0 \pm 0.2$, $c = 441.38 \pm 0.04$ pm, and a calculated density of 6.820 g·cm^{-3}) and $LaGa_{0.80\pm0.03}Sn_{0.20}Sb_2$ ($a = 438.17 \pm 0.08$, $b = 2290.9 \pm 0.4$, $c = 440.32 \pm 0.04$ pm, and a calculated density of 6.942 g·cm^{-3}).

For higher Sn and lower Ga content, solid solutions crystallize in the orthorhombic structure of the $LaSn_{0.75}Sb_2$-type ($a = 437.83 \pm 0.02$, $b = 2301.73 \pm 0.09$, $c = 442.24 \pm 0.02$ pm, and a calculated density of 6.957 g·cm^{-3} for $LaGa_{0.68\pm0.04}Sn_{0.31\pm0.03}Sb_2$; $a = 437.52 \pm 0.02$, $b = 2300.95 \pm 0.10$, $c = 441.84 \pm 0.02$ pm, and a calculated density of 6.923 g·cm^{-3} for $LaGa_{0.62\pm0.03}Sn_{0.32\pm0.03}Sb_2$; and $a = 437.28 \pm 0.05$, $b = 2302.9 \pm 0.3$, $c = 441.97 \pm 0.05$ pm, and a calculated density of 6.832 g·cm^{-3} for $LaGa_{0.43\pm0.03}Sn_{0.39\pm0.03}Sb_2$).

To prepare these solid solutions, starting materials were La powder, crushed Ga granules, Sn powder, and Sb powder. Reactions were performed on a 0.25 g scale in evacuated fused silica tubes. Generally, successful syntheses involved heating at temperatures between 850°C and 950°C for 2–3 days, cooling to 500°C over 1–4 days, and cooling to 20°C over several hours. $LaGa_xSn_ySb_2$ solid solutions were originally identified as side products in attempts to grow single crystals of $La_{13}Ga_8Sb_{21}$ and $LaGaSb_2$ through the use of a Sn flux.

11.21 GALLIUM–LANTHANUM–SULFUR–ANTIMONY

The $La_2Ga_{0.33}SbS_5$ quaternary compound, which crystallizes in the tetragonal structure with the lattice parameters $a = 751.93 \pm 0.06$ and $c = 1341.26 \pm 0.17$ pm, a calculated density of 5.104 g·cm^{-3}, and energy gap of 1.76 eV, is formed in the Ga–La–S–Sb system (Zhao 2016a). Dark brown crystals of the title compound were prepared as follows. The stoichiometric mixture of La, Ga, Sb, and S was accordingly weighed with an overall loading of about 300 mg and loaded into a silica tube, which was evacuated and sealed. The sample was heated to 950°C in 24 h and kept at that temperature for 96 h, subsequently cooled at 2.5°C·h^{-1} to 200°C before the furnace was turned off. All manipulations were performed inside the glove box.

11.22 GALLIUM–CERIUM–SULFUR–ANTIMONY

The Ce_4GaSbS_9 quaternary compound, which crystallizes in the orthorhombic structure with the lattice parameters $a = 1388.34 \pm 0.09$, $b = 1430.04 \pm 0.11$, and $c = 1441.02 \pm 0.13$ pm and a calculated density of 4.831 g·cm^{-3}, is formed in the Ga–Ce–S–Sb system (Zhao 2016a). Red crystals of the title compound were prepared by the same way as $La_2Ga_{0.33}SbS_5$ was synthesized using Ce instead of La.

11.23 GALLIUM–PRASEODYMIUM–SULFUR–ANTIMONY

The Pr_4GaSbS_9 quaternary compound, which crystallizes in the orthorhombic structure with the lattice parameters $a = 1379.9 \pm 0.3$, $b = 1418.7 \pm 0.3$, and $c = 1432.3 \pm 0.3$ pm and a calculated density of 4.945 g·cm^{-3}, is formed in the Ga–Pr–S–Sb system (Chen et al. 2011). The orange–yellow crystals of the title compound were obtained from the mixture of Pr, Ga, Sb, and S (8:1:1:15 molar ratio) after being heated at 950°C. The optimal conditions to synthesize this compound involved heating of stoichiometry mixtures to 950°C at a rate of 26.5°C·h^{-1}, a dwell period at this temperature for 5 days, and then slowly cooling to 300°C at a rate of 5°C·h^{-1}, at which point, the furnace was switched off. The target quaternary compound together with circa 50% of binary byproducts of Ga_2S_3 and Pr_2S_3 was obtained. Pr_4GaSbS_9 is stable in air at room temperature for long periods (more than 6 months).

11.24 GALLIUM–NEODYMIUM–SULFUR–ANTIMONY

The Nd_4GaSbS_9 quaternary compound, which crystallizes in the orthorhombic structure with the lattice parameters $a = 1375.9 \pm 0.5$, $b = 1411.3 \pm 0.5$, and $c = 1430.0 \pm 0.5$ pm, and a calculated density of 5.057 g·cm^{-3}, is formed in the Ga–Nd–S–Sb system (Chen et al. 2011). The orange–yellow crystals of the title compound were prepared by the same way as Pr_4GaSbS_9 was synthesized using Nd instead of Pr. The target quaternary compound together with circa 50% of binary by-products of Ga_2S_3 and Nd_2S_3 was obtained. Nd_4GaSbS_9 is stable in air at room temperature for long periods (more than 6 months).

11.25 GALLIUM–SAMARIUM–SULFUR–ANTIMONY

The Sm_4GaSbS_9 quaternary compound, which crystallizes in the orthorhombic structure with the lattice parameters $a = 1364.0 \pm 0.3$, $b = 1398.3 \pm 0.3$, and $c = 1416.6 \pm 0.5$ pm, a calculated density of 5.317 g·cm^{-3}, and calculated and experimental energy gaps of 2.20 and 2.23 eV, respectively, is formed in the Ga–Sm–S–Sb system (Chen et al. 2011). The orange–yellow crystals of the title compound were prepared by the same way as Pr$_4$GaSbS$_9$ was synthesized using Sm instead of Pr. The pure Sm$_4$GaSbS$_9$ was obtained. This compound is stable in air at room temperature for long periods (more than 6 months).

11.26 GALLIUM–GADOLINIUM–SULFUR–ANTIMONY

The **Gd$_4$GaSbS$_9$** quaternary compound, which crystallizes in the orthorhombic structure with the lattice parameters $a = 1357.3 \pm 0.2$, $b = 1389.6 \pm 0.2$, and $c = 1409.8 \pm 0.2$ pm, a calculated density of 5.540 g·cm^{-3}, and energy gap of 2.41 eV, is formed in the Ga–Gd–S–Sb system (Chen et al. 2011). The orange–yellow crystals of the title compound were prepared by the same way as Pr$_4$GaSbS$_9$ was synthesized using Gd instead of Pr. The pure Gd$_4$GaSbS$_9$ was obtained. This compound is stable in air at room temperature for long periods (more than 6 months).

11.27 GALLIUM–GADOLINIUM–IRON–ANTIMONY

Quaternary compounds were not found in the Ga–Gd–Fe–Sb system (Gvozdetskyi et al. 2012). According to the data of XRD, at 500°C, GdSb is in thermodynamic equilibrium with GaSb, GdGa$_2$, FeGa$_3$, FeGa, Fe$_{1-x}$Ga$_x$, Fe$_{2.8}$Ga$_{1.2}$, Fe$_3$Ga, Fe, and Gd$_2$Fe$_{17}$, and GaSb is in equilibrium with FeGa$_3$ and Fe$_3$Ga at this temperature. The alloys were annealed in evacuated quartz ampoules at 500°C for 720 h with the next quenching in cold water.

11.28 GALLIUM–TERBIUM–SULFUR–ANTIMONY

The **Tb$_4$GaSbS$_9$** quaternary compound, which crystallizes in the orthorhombic structure with the lattice parameters $a = 1353.0 \pm 0.5$, $b = 1385.0 \pm 0.5$, and $c = 1405.0 \pm 0.5$ pm, a calculated density of 5.629 g·cm^{-3}, and energy gap of 2.44 eV, is formed in the Ga–Tb–S–Sb system (Chen et al. 2011). The orange–yellow crystals of the title compound were prepared by the same way as Pr$_4$GaSbS$_9$ was synthesized using Tb instead of Pr. The pure Tb$_4$GaSbS$_9$ was obtained. This compound is stable in air at room temperature for long periods (more than 6 months).

11.29 GALLIUM–DYSPROSIUM–SULFUR–ANTIMONY

The **Dy$_4$GaSbS$_9$** quaternary compound, which crystallizes in the orthorhombic structure with the lattice parameters $a = 1347.6 \pm 0.2$, $b = 1381.1 \pm 0.2$, and $c = 1401.1 \pm 0.2$ pm, a calculated density of 5.757 g·cm^{-3}, and energy gap of 2.58 eV, is formed in the Ga–Dy–S–Sb system (Chen et al. 2011). The orange–yellow crystals of the title compound were prepared by the same way as Pr$_4$GaSbS$_9$ was synthesized using Dy instead of Pr. The pure Dy$_4$GaSbS$_9$ was obtained. This compound is stable in air at room temperature for long periods (more than 6 months).

11.30 GALLIUM–HOLMIUM–SULFUR–ANTIMONY

The **Ho$_4$GaSbS$_9$** quaternary compound, which crystallizes in the orthorhombic structure with the lattice parameters $a = 1342.7 \pm 0.5$, $b = 1375.6 \pm 0.5$, and $c = 1395.4 \pm 0.5$ pm and a calculated density of 5.874 g·cm^{-3}, is formed in the Ga–Ho–S–Sb system (Chen et al. 2011). The orange–yellow crystals of the title compound were prepared by the same way as Pr$_4$GaSbS$_9$ was synthesized using Ho instead of Pr. The pure Ho$_4$GaSbS$_9$ was obtained. This compound is stable in air at room temperature for long periods (more than 6 months).

11.31 GALLIUM–LEAD–OXYGEN–ANTIMONY

The **Pb$_2$Ga$_{0.5}$Sb$_{1.5}$O$_{6.5}$** quaternary compound, which crystallizes in the cubic structure with the lattice parameter a =1044.33 ± 0.02 pm (Ismunandar et al. 1999) [a =1044.87 ± 0.04 pm and a calculated density of 8.57 g·cm^{-3} (Cascales et al. 1985)], is formed in the Ga–Pb–O–Sb system. It was prepared from a mixture of PbO, Sb$_2$O$_3$, and Ga$_2$O$_3$ (molar ratio Pb/Ga/Sb = 4:1:3) which, once ground in an agate mortar, was heated in air at 680–700°, 800°C, and 900°C (Cascales et al. 1985; Ismunandar et al. 1999). After each thermal treatment for 24 h, the material was quenched, weighed, and ground. The title compound was obtained as orange-yellow powder.

11.32 GALLIUM–BISMUTH–OXYGEN–ANTIMONY

The **Bi$_{1.8}$Ga$_{1.2}$SbO$_7$** or **Bi$_{1.89}$GaSbO$_{6.84}$** and **Bi$_3$GaSb$_2$O$_{11}$** quaternary compounds are formed in the Ga–Bi–O–Sb system. Bi$_3$GaSb$_2$O$_{11}$ crystallizes in the cubic structure with the lattice parameter a = 948.980 ± 0.003 pm (Ismunandar et al. 1996) [a = 949.07 ± 0.02 pm and the calculated and experimental densities of 8.67 and 8.66 g·cm^{-3}, respectively (Sleight and Bouchard 1973)]. This compound was prepared by the solid-state reactions of stoichiometric amounts of Bi$_2$O$_3$, Ga$_2$O$_3$, and Sb$_2$O$_3$ at 800°C for 24 h and then at 1100°C for 48 h with regrinding after each heating step (Ismunandar et al. 1996). Single crystals of the title compound were prepared from a Bi$_2$O$_3$ flux (Sleight and Bouchard 1973). The mixture of Bi$_2$O$_3$, Ga$_2$O$_3$, and Sb$_2$O$_3$ (molar ratio 12:1:1) was heated in air to 1060°C in an open Pt crucible and cooled at 6°C·h^{-1} to 800°C, and the furnace was cooled. After washing the product with nitric acid, only transparent yellow crystals of Bi$_3$GaSb$_2$O$_{11}$ remained.

Bi$_{1.8}$Ga$_{1.2}$SbO$_7$ also crystallizes in the cubic structure with the lattice parameter a = 1038.870 ± 0.005 pm (Ellert et al. 2016) (a = 1038.07 ± 0.04 pm for Bi$_{1.89}$GaSbO$_{6.84}$ [Ismunandar et al. 1999]) and energy gap of 2.90 eV (Egorysheva et al. 2017). Bi$_{1.8}$Ga$_{1.2}$SbO$_7$ was synthesized by a convenient solid-state method (Ellert et al. 2016). High-purity Bi$_2$O$_3$, Ga$_2$O$_3$, and Sb$_2$O$_3$ were used as starting materials. The stoichiometric mixture of oxides was ground in acetone for approximately 30 min. The synthesis was carried out in Pt crucibles inside a muffle furnace. The samples were heated in two steps, first at 650°C for 24 h and subsequently at 800°C for 24 h. Each sample was reground after calcinations. Subsequently, the samples were annealed at 930–980°C for 5–15 days and reground again.

Bi$_{1.89}$GaSbO$_{6.84}$ was also prepared by the solid-state reactions of stoichiometric amounts of Bi$_2$O$_3$, Ga$_2$O$_3$, and Sb$_2$O$_3$ (Ismunandar et al. 1999). The mixture of the starting materials was heated at 800°C, 900°C, and 1000°C for 15 h each, and then finally at 1100°C for 24 h. The samples were reground after all calcinations. The formation of the title compound occurs over a limited temperature range. Heating the constitute oxides at or below 1000°C yielded Bi$_3$GaSb$_2$O$_{11}$ together with unreacted Bi$_2$O$_3$ and Ga$_2$O$_3$. At a temperature above 1200°C, a complicated mixture of products was obtained mainly consisting of Ga$_2$O$_3$.

Heating the mixture at 1100°C resulted in the formation of a highly crystalline yellow sample of $Bi_{1.89}GaSbO_{6.84}$.

11.33 GALLIUM–SULFUR–SELENIUM–ANTIMONY

According to the data of Safarov and Salmanov (1975), the $GaSbS_2Se$, $GaSb_2S_3Se$, $GaSb_4S_6Se$, $Ga_3Sb_4S_6Se_3$, and $Ga_4Sb_2S_3Se_4$ quaternary compounds are formed in the Ga–S–Se–Sb system. $GaSbS_2Se$ incongruently melts and crystallizes in the monoclinic structure with the lattice parameters $a = 1031$, $b = 807$ and, $c = 663$ pm, and $\beta = 92°30'$ and an experimental density of 5.6 $g·cm^{-3}$.

$GaSb_2S_3Se$ also incongruently melts and crystallizes in the monoclinic structure with the lattice parameters $a = 1962$, $b = 789$, and $c = 419$ pm, and $\beta = 90°$ and an experimental density of 5.2 $g·cm^{-3}$.

$GaSb_4S_6Se$ congruently melts and crystallizes in the orthorhombic structure with the lattice parameters $a = 1144$, $b = 1417$, and $c = 370$ pm and an experimental density of 4.6 $g·cm^{-3}$.

$Ga_3Sb_4S_6Se_3$ also incongruently melts and crystallizes in the orthorhombic structure with the lattice parameters $a = 1909$, $b = 2384$, and $c = 403$ pm and an experimental density of 6.9 $g·cm^{-3}$.

$Ga_4Sb_2S_3Se_4$ congruently melts and crystallizes in the monoclinic structure with the lattice parameters $a = 1330$, $b = 404$, and $c = 1630$ pm, and $\beta = 94°30'$ and an experimental density of 6.7 $g·cm^{-3}$.

11.34 GALLIUM–SULFUR–TELLURIUM–ANTIMONY

The $GaSb_4Te_6S$ and $Ga_4Sb_2Te_3S_4$ quaternary compounds, which incongruently melt at 563°C and 577°C, respectively, are formed in the Ga–S–Te–Sb system (Safarov et al. 1988).

11.35 GALLIUM–SULFUR–CHLORINE–ANTIMONY

The $(Sb_7S_8Cl_2)[GaCl_4]$ quaternary compound, which crystallizes in the monoclinic structure with the lattice parameters $a = 1700.25 \pm 0.12$, $b = 1199.18 \pm 0.11$, $c = 861.05 \pm 0.12$ pm, and $\beta = 93.19 \pm 0.01°$ and a calculated density of 3.44 $g·cm^{-3}$ at 123 K, is formed in the Ga–S–Cl–Sb system (Eich et al. 2014). To prepare this compound, Sb_2S_3 (0.36 mM), $SbCl_3$ (0.16 mM), $GaCl_3$ (0.99 mM), and CuCl (1.18 mM) were added to a glass ampoule under an Ar atmosphere. The ampoule was evacuated and flame sealed. After 10 days at 100°C, orange crystals with the appearance of needles of hexagonal cross section formed in a mushy yellow–green melt. The title compound is air sensitive.

11.36 GALLIUM–SELENIUM–CHLORINE–ANTIMONY

The $(Sb_7Se_8)[GaCl_4]_2[Ga_2Cl_7]_3$ and $(Sb_7Se_8Cl_2)[GaCl_4]$ quaternary compounds are formed in the Ga–Se–Cl–Sb system (Eich et al. 2014). The first of them crystallizes in the triclinic structure with the lattice parameters $a = 1085.08 \pm 0.05$, $b = 1550.27 \pm 0.08$, and $c = 1893.24 \pm 0.11$ pm; $\alpha = 79.85 \pm 0.01°$, $\beta = 84.04 \pm 0.01°$, and $\gamma = 70.70 \pm 0.01$; and a calculated density of 3.45 $g·cm^{-3}$ at 100 K. To synthesize this compound, Se (0.36 mM), Sb (0.24 mM), $SbCl_3$ (0.12 mM), $GaCl_3$ (0.99 mM), and CuCl (0.18 mM) were added to a glass ampoule under an Ar atmosphere. The ampoule was evacuated and flame sealed. After 7 days at 100°C, orange, rod-shaped crystals appeared in a solidified orange melt. The title compound is air sensitive.

$(Sb_7Se_8Cl_2)[GaCl_4]$ crystallizes in the orthorhombic structure with the lattice parameters $a = 1220.27 \pm 0.07$, $b = 1700.89 \pm 0.09$, and $c = 1773.5 \pm 0.1$ pm and a calculated density of 3.95 g·cm^{-3} at 100 K. To prepare this compound, Se (0.36 mM), Sb (0.24 mM), $SbCl_3$ (0.12 mM), and $GaCl_3$ (0.99 mM) were added to a glass ampoule under an Ar atmosphere and heated to 100°C, whereupon a melt formed. After 16 days, orange rod- and cube-shaped crystals appeared in a green mushy melt. The addition of tetraphenylphosphonium chloride (0.18 mM) improved the yield, crystal size, and crystal quality. The melt was dark red and of relative low viscosity, which allowed it to be decanted from the crystals. With this additive compound, the reaction time is shortened to 7 days, and the temperature interval for a successful synthesis is broader (90–140°C). The obtained compound is air sensitive.

11.37 GALLIUM–TELLURIUM–CHLORINE–ANTIMONY

The **$Ga_7Sb_7Te_8Cl_{26}$, $Ga_8Sb_7Te_8Cl_{29}$, $(SbTe_4)[Ga_2Cl_7]$, $(Sb_2Te_2)[GaCl_4]$**, and **$(Sb_3Te_4)[GaCl_4]$** quaternary compounds are formed in the Ga–Te–Cl–Sb system. $Ga_7Sb_7Te_8Cl_{26}$ crystallizes in the triclinic structure with the lattice parameters $a = 1075.93 \pm 0.01$, $b = 1560.11 \pm 0.02$, and $c = 1886.12 \pm 0.02$ pm; $\alpha = 68.071 \pm 0.001°$, $\beta = 83.777 \pm 0.001°$, and $\gamma = 88.556 \pm 0.001$; and a calculated density of 3.74 g·cm^{-3} at 123 ± 2 K (Eich et al. 2013). To synthesize this compound, in the glove box, weighed amounts of starting materials were filled in glass ampoules, which were evacuated and flame sealed under vacuum. Tube ovens used for syntheses were aligned in an angle of about 30° to the horizontal to keep the melt compacted in the hot zone. After some days, the ampoules were visually inspected. If the formation of crystals was observed, the hot melt was decanted from the solids. After cooling to ambient temperature, the ampoules were opened in the glove box. Since this compound is very sensitive toward air, samples were stored under Ar atmosphere. Te (0.36 mM), Sb (0.24 mM), $SbCl_3$ (0.12 mM), and $GaCl_3$ (0.99 mM) were added to a glass ampoule under an Ar atmosphere. Then, NaCl (0.18 mM) was added. The ampoules were evacuated and flame sealed. After 15 days at 100°C, orange rod-shaped crystals appeared in the black reaction melt. After opening the ampoule in the glove box, the orange crystals had to be mechanically separated from adherent solidified melt.

$Ga_8Sb_7Te_8Cl_{29}$ crystallizes in the monoclinic structure with the lattice parameters $a = 1813.57 \pm 0.02$, $b = 1058.4 \pm 0.1$, $c = 3685.24 \pm 0.04$ pm, and $\beta = 115.41 \pm 0.07°$ and a calculated density of 3.60 g·cm^{-3} at 123 ± 2 K (Eich et al. 2013). This compound was prepared by the same way as $Ga_7Sb_7Te_8Cl_{26}$ was synthesized using CuCl instead of NaCl and annealing the mixture at 100°C for 60 days instead of 15 days. With the respective amount of InCl (0.18 mM), the title compound was already obtained after 5 days at 100°C. After opening the ampoule in the glove box, the orange crystals had to be mechanically separated from adherent solidified melt.

$(SbTe_4)[Ga_2Cl_7]$ also crystallizes in the monoclinic structure with the lattice parameters $a = 714.76 \pm 0.05$, $b = 1885.4 \pm 0.2$, $c = 1286.99 \pm 0.08$ pm, and $\beta = 94.240 \pm 0.004°$ (Eich et al. 2015). To prepare this compound, Te (0.36 mM), Sb (0.24 mM), $SbCl_3$ (0.12 mM),

GaCl$_3$ (0.99 mM), tetraphenylphosphonium chloride (0.16 mM), and TeO$_2$ (0.09 mM) were placed in a glass ampoule, which was flame sealed under dynamic vacuum. After 60 days at 50°C, the title compound was obtained as black needles. (SbTe$_4$)[Ga$_2$Cl$_7$] is highly moisture sensitive. Tube furnace used for the synthesis of this compound was aligned at an angle of about 30° to the horizontal to keep the melt compacted in the hot zone.

(Sb$_2$Te$_2$)[GaCl$_4$] crystallizes in the triclinic structure with the lattice parameters a = 944.44 ± 0.02, b = 1341.86 ± 0.02, and c = 1791.79 ± 0.03 pm and α = 75.372 ± 0.001°, β = 87.831 ± 0.001°, and γ = 88.556 ± 0.001 (Eich et al. 2015). The title compound was synthesized as follows. Under Ar atmosphere, Te (0.72 mM), Sb (0.48 mM), SbCl$_3$ (0.24 mM), GaCl$_3$ (1.98 mM), and NaCl (0.18 mM) were filled in a glass ampoule, which was evacuated, flame sealed, and placed in a horizontal tube furnace. After 13 days at 125°C, silvery shiny rod-shaped crystals of this compound were obtained. (Sb$_2$Te$_2$)[GaCl$_4$] is highly moisture sensitive. Tube furnace used for synthesis of this compound was aligned at an angle of about 30° to the horizontal to keep the melt compacted in the hot zone.

(Sb$_3$Te$_4$)[GaCl$_4$] crystallizes in the monoclinic structure with the lattice parameters a = 1445.1 ± 0.2, b = 419.58 ± 0.04, c = 1289.9 ± 0.1 pm, and β = 111.797 ± 0.004° (Eich et al. 2015). The title compound was prepared by the next procedure. Te (0.36 mM), Sb (0.24 mM), SbCl$_3$ (0.12 mM), GaCl$_3$ (0.99 mM), and tetraphenylphosphonium chloride (0.16 mM) were filled in a glass ampoule, which was evacuated and sealed. On annealing at temperatures between 50°C and 140°C, crystals appeared within 13 weeks in the form of silvery, metallic shiny rods. Tetraphenylphosphonium chloride may be replaced by NaCl. At slightly higher reaction temperatures between 130°C and 160°C, crystallization starts after 3 days. (Sb$_3$Te$_4$)[GaCl$_4$] is highly moisture sensitive. Tube furnace used for synthesis of this compound was aligned at an angle of about 30° to the horizontal to keep the melt compacted in the hot zone.

11.38 GALLIUM–COBALT–TELLURIUM–ANTIMONY

The Ga$_x$Co$_4$Sb$_{11.7}$Te$_3$ (x = 0.1 and 0.2) quaternary compounds, which crystallize in the cubic structure with the lattice parameter a = 905.1 pm, for x = 0.1, and a = 905.4 pm, for x = 0.2, are formed in the Ga–Co–Te–Sb system (Zhao et al. 2012a). The samples of this compound were prepared as follows. Stoichiometric quantities of the constituent pure elements Te (shot), Co (powder), Sb (powder), and Ga (shot) were loaded in a quartz tube with carbon depositing on the inner wall, and sealed under a pressure of 0.1 Pa. The samples were melted at 1080°C for 18 h and were then quenched in salt water. The ingots were annealed at 700°C for 144 h to give sufficient time for Ga to fill the void in the skutterudite structure. The obtained solid product was ground into fine powder and then sintered into a dense solid by hot-press sintering at 590°C for 12 min in a graphite die.

CHAPTER 12

Systems Based on InN

12.1 INDIUM–LITHIUM–STRONTIUM–NITROGEN

The **$Sr_6In_4(In_{0.32}Li_{0.92})N_{2.49}$** quaternary compound, which crystallizes in the cubic structure with the lattice parameter $a = 1437.52 \pm 0.04$ at 173 ± 2 K, is formed in the In–Li–Sr–N system (Bailey et al. 2005). To obtain the title compound, two initial reactions were performed. In both cases, Na (200 mg), Sr (88 mg), In (70 mg), NaN_3 (70 mg) (molar ratios of $Na/Sr/In/N_2 = 16.0:1.6:1:2.6$), and Li (3.5 mg) were loaded into a Nb tube. The Nb containers were sealed under 0.1 MPa of Ar in an arc furnace and then were themselves sealed under vacuum in fused silica tubes. These starting materials were heated to 800°C in 15 h; they remained at that temperature for 24 h and then were linearly cooled to 200°C over 200 h, at which point the furnace was shut off and allowed to cool naturally. Upon completion of the reactions, unreacted sodium was removed by sublimation from the products by heating the Nb tubes to 350°C under a pressure of approximately 0.1 Pa for 8 h.

12.2 INDIUM–LITHIUM–BARIUM–NITROGEN

$LiIn_2Ba_3N_{0.83}$ and **$Li_{35}In_{45}Ba_{39}N_9$** quaternary compounds are formed in the In–Li–Ba–N system. $LiIn_2Ba_3N_{0.83}$ crystallizes in the cubic structure with the lattice parameter $a = 1491.3 \pm 0.2$ and a calculated density of 5.289 g·cm^{-3} (Smetana et al. 2010). A sample with the stoichiometric composition $LiIn_2Ba_3N$ (472.2 mg Ba, 278.6 mg In, 8.5 mg Li, and 44.8 mg BaN_6) resulted in high yield of the title compound. In order to obtain single crystals of better quality, another sample with a 50% lithium excess was prepared. All samples were heated to 650°C and annealed at this temperature for 50 h, followed by cooling to 350°C at a rate of 1°C·h^{-1}, further annealing for 1 month, and cooling to room temperature at a rate of 3°C·h^{-1}.

$Li_{35}In_{45}Ba_{39}N_9$ crystallizes in the tetragonal structure with the lattice parameters $a = 1529.9 \pm 0.2$ and $c = 3068.2 \pm 0.6$ pm and a calculated density of 5.045 g·cm^{-3} (Smetana et al. 2010). A sample (300 mg Ba, 257.2 mg In, 23.5 mg Li, and 10.9 mg NaN_3) with the overall composition $3Na + 20Li + Li_{40}In_{40}Ba_{39}N_9$ was prepared in order to obtain the compound isostructural with $Li_{80}Ba_{39}N_9$, yielding single crystals of this subnitride. All samples were treated as in the case of $LiIn_2Ba_3N_{0.83}$ preparation.

Due to the extreme sensitivity of some educts and the products to air, all handling was performed under purified argon using the Schlenk technique or a glove box. Reactions were performed in closed Ta containers, which were arc-welded and sealed inside Pyrex glass ampoules.

12.3 INDIUM–LITHIUM–OXYGEN–NITROGEN

The **Li_4InNO_2** quaternary compound is formed in the In–Li–O–N system (Tokarzewski and Podsiadło 1998). The formation of this compound was observed during the reactions of Li_2O with InN, Li_3N with In_2O_3, and Li_3N with $LiInO_2$. In the reaction of Li_2O and InN (molar ratio 2:1), Li_4InNO_2 is formed at 480°C. The same compound is formed in the reaction of Li_3N with In_2O_3 (molar ratio 2:1) at 460°C. In the reaction of Li_3N with $LiInO_2$ (molar ratio 1:1), the title compound was obtained at 250°C.

12.4 INDIUM–SODIUM–CARBON–NITROGEN

The **$NaIn(NCN)_2$** quaternary compound, which crystallizes in the orthorhombic structure with the lattice parameters $a = 961.30 \pm 0.06$, $b = 716.84 \pm 0.05$, and $c = 603.65 \pm 0.04$ pm and a calculated density of 3.479 g·cm^{-3}, is formed in the In–Na–C–N system (Dronskowski 1995). Single crystals of this compound can be synthesized from a reaction between InBr and NaCN at 400°C, followed by chemical transport at 400–500°C. These crystals seem to exhibit a moderate sensitivity with respect to air or humidity. Therefore, all subsequent operations were performed under protective gas (argon).

12.5 INDIUM–COPPER–STRONTIUM–NITROGEN

The **$Cu_3Sr_8In_4N_5$** quaternary compound, which crystallizes in the orthorhombic structure with the lattice parameters $a = 381.61 \pm 0.05$, $b = 1243.7 \pm 0.2$, and $c = 1890.2 \pm 0.2$ pm and a calculated density of 5.260 g·cm^{-3}, is formed in the In–Cu–Sr–N system (Yamane et al. 2003). Single crystals of this compound were prepared from Sr, Cu, In, and Na (molar ratio 2:1:1:6) in a BN crucible. The crucible was placed into a stainless steel container and sealed in the glove box. After heating to 750°C in a furnace, a N_2 or Ar + N_2 mixture was introduced into the container. The total pressure in the container was 7 MPa. The sample was heated at this temperature for 1 h and then cooled to 550°C at a rate of 2°C·h^{-1}. Below 550°C, the sample was cooled to room temperature in the furnace by shutting off the power. The products in the crucible were washed in liquid NH_3 to dissolve away the Na flux. All materials were handled in an Ar-filled glove box.

12.6 INDIUM–COPPER–BARIUM–NITROGEN

The **$Cu_3Ba_8In_4N_5$** and **$Cu_2Ba_{14}In_4N_7$** quaternary compounds are formed in the In–Cu–Ba–N system. $Cu_3Ba_8In_4N_5$ crystallizes in the orthorhombic structure with the lattice parameters $a = 407.81 \pm 0.06$, $b = 1258.8 \pm 0.2$, and $c = 1980.4 \pm 0.3$ pm and a calculated density of 5.941 g·cm^{-3} (Yamane et al. 2002a). Single crystals of this compound were prepared by heating Ba, Cu, and In in a Na flux at 750°C under 7 MPa of N_2 and by slow cooling from this temperature. All manipulations were carried out in an Ar-filled glove box.

$Cu_2Ba_{14}In_4N_7$ crystallizes in the monoclinic structure with the lattice parameters $a = 932.25 \pm 0.19$, $b = 800.89 \pm 0.17$, $c = 2165.1 \pm 0.5$ pm, and $\beta = 102.263 \pm 0.005°$ and a calculated density of 5.481 g·cm^{-3} (Yamane et al. 2002b). To obtain this compound, Ba, Cu, In, and Na (molar ratio 6:1:1:6) were weighed and placed in a BN crucible. The crucible was placed in a stainless steel container and sealed in the glove box. The container was connected to a nitrogen gas feed line. After heating to 750°C, N_2 was introduced into the container and the pressure was maintained at 7 MPa of N_2 with a pressure regulator. The sample was heated at this temperature for 1 h and then cooled from 750°C to 550°C at a rate of 2°C·h^{-1} under 7 MPa of N_2. Below 550°C, the sample was cooled to room temperature by shutting off the electric power to the furnace. The products in the crucible were washed with liquid NH_3 to dissolve the Na flux. Part of the melt surface was covered with a crust of fine granular crystals. Behind the crust and at the side of the melt, single platelet crystals of $Ba_{14}Cu_3In_4N_7$ with a black metallic luster were obtained. Fine grains of the intermetallics were deposited at the bottom of the crucible. Single crystals of the title compound were grown by slow cooling from 750°C at 7 MPa of N_2 using a sodium flux. The obtained single crystals are not stable in air. All manipulations were carried out in an Ar-filled glove box.

12.7 INDIUM–CARBON–SULFUR–NITROGEN

The **In(SCN)$_3$** quaternary compound, which crystallizes in the cubic structure with the lattice parameter $a = 1200 \pm 5$ pm and decomposes at about 150°C, is formed in the In–C–S–N system (Goggin et al. 1966). The title compound could be prepared by the next three methods:

1. InCl$_3$ (2.2 g) was dissolved in dry EtOH (100 mL), and a dry ethanolic solution (300 mL) of 0.1 N NaSCN was added. NaCl immediately precipitated. The solution was filtered and evaporated under a vacuum. The residue was extracted with ether and on removal of the solvent gave In(SCN)$_3$ as a cream solid.

2. A suspension of Hg(SCN)$_2$ (1 g) in H$_2$O (20 mL) was shaken with an excess of indium shavings (1 g) for 9 days, during the course of which Hg precipitated. The solution was filtered and the filtrate shaken with an equal volume of ether. The ether layer was separated under a vacuum, leaving a white solid residue. The residue was extracted with dry EtOH (5 mL) and the title compound recovered by evaporation of the solvent.

3. In(CN)$_3$ (0.5 g) and sulfur (0.25 g) were mixed in a dry box. The mixture was sealed into a tube under a vacuum and heated at 130°C for 2 h. The tube was opened in a dry box, and the sulfur excess was extracted with CS$_2$.

CHAPTER 13

Systems Based on InP

13.1 INDIUM–HYDROGEN–OXYGEN–PHOSPHORUS

Some quaternary compounds are formed in the In–H–O–P system.

InPO$_4$·H$_2$O crystallizes in the triclinic structure with the lattice parameters a = 543.42 ± 0.06, b = 555.08 ± 0.04, c = 654.46 ± 0.05 pm and α = 97.593 ± 0.006°, β = 94.558 ± 0.006°, and γ = 107.565 ± 0.006° and a calculated density of 4.088 g·cm^{-3} (Tang and Lachgar 1998). To synthesize this compound, KNO$_3$ (0.5 mM), InCl$_3$ (0.5 mM), 85% H$_3$PO$_4$ (3 mM), and CuCO$_3$·Cu(OH)$_2$ (6.25 × 10^{-2} mM) were sealed under vacuum in a thick-walled Pyrex tube (40% filling). The reaction was carried out at 200°C for 60 h, and then the furnace was cooled to room temperature. The product was isolated as colorless needle-like crystals, which were filtered under vacuum, washed using copious amounts of cold water and acetone, and dried in air at room temperature. The title compound loses one H$_2$O molecule in the temperature range between 370°C and 480°C to yield InPO$_4$.

InPO$_4$·2H$_2$O is characterized by two polymorphic modifications. α-InPO$_4$·2H$_2$O crystallizes in the orthorhombic structure with the lattice parameters a = 1036, b = 884, and c = 1019 pm and a calculated density of 3.45 g·cm^{-3} (Mooney-Slater 1961). The addition of phosphate ions to an aqueous solution of In(NO$_3$)$_3$ results in the immediate precipitation of a voluminous white amorphous product. These precipitates were dissolved in HNO$_3$ and crystallized out by very slow dilution or neutralization at temperatures of about 75°C. The crystals appeared as shot, stout bipyramidal prisms. Several hours of heating at circa 750°C [900°C (Deichman et al. 1970)] were required to entirely convert a sample to the anhydrous form.

β-InPO$_4$·2H$_2$O crystallizes in the monoclinic structure with the lattice parameters a = 545.51 ± 0.03, b = 1022.93 ± 0.04, c = 888.61 ± 0.03 pm, and β = 91.489 ± 0.004° and a calculated density of 3.294 g·cm^{-3} (Tang et al. 2001) [a = 545.08 ± 0.12, b = 1022.29 ± 0.12, c = 888.30 ± 0.17 pm, and β = 91.50 ± 0.02° and a calculated density of 3.300 g·cm^{-3} (Sugiyama et al. 1999)]. This modification was obtained using hydrothermal synthesis techniques. A mixture of InCl$_3$, Me$_4$NOH·5H$_2$O, 85% H$_3$PO$_4$, and H$_2$O (molar ratio

1:2:4:280) was sealed under vacuum in a thick-walled borosilicate tube (36% filling). The reaction mixture was heated at 190°C for 2 days in a tubular furnace, followed by slow cooling to room temperature. The products were filtered, washed with H_2O and acetone, and dried at room temperature. The major phase was a white powder, which was found to be β-$InPO_4 \cdot 2H_2O$. The title compound was found as colorless rod-like crystals.

β-$InPO_4 \cdot 2H_2O$ could be also prepared as follows (Sugiyama et al. 1999). A mixture of H_3PO_4, $In(OH)_3$, HF, H_2O, and quinuclidine ($C_7H_{13}N$) (molar ratio 2:1:2:80:2) was sealed in a Teflon-lined autoclave. The autoclave was heated to 170°C, kept at this temperature for 10 days, and then quenched in water. The crystals obtained were filtered off, washed with distilled water, and dried at 50°C. The starting chemical composition appears to be important for the synthesis of the title compound; for example, the formation of the orthorhombic phase (α-$InPO_4 \cdot 2H_2O$) was favored in the products synthesized from a mixture with a molar ratio of 1.2:1:2:80:2.

$InPO_4 \cdot H_3PO_4 \cdot 4H_2O$ was prepared at 0°C and is stable in the air (Deichman et al. 1969). The final products of the title compound dehydration are $In_4(P_2O_7)_3$ and $In(PO_3)_3$ (Deichman et al. 1970).

$InPO_4 \cdot 2H_3PO_4 \cdot 4H_2O$ [$In(H_2PO_4)_3 \cdot H_2O$] was prepared at 25°C and decomposes readily in the air with formation of $InPO_4 \cdot H_3PO_4 \cdot 4H_2O$ (Deichman et al. 1969). The final product of this compound dehydration is $In(PO_3)_3$ (Deichman et al. 1970).

$In(H_2PO_2)_3$ was prepared by the interaction of $In_2(SO_4)_3 \cdot 5H_2O$ and $NaH_2PO_2 \cdot H_2O$ in aqueous solution and is stable up to 105°C (Ensslin et al. 1947).

$In_2O_3 \cdot P_2O_5 \cdot 4H_2O$ exists as a stable solid phase at 70°C in the form of the white crystalline powder (Brownlow et al. 1960).

$In_2O_3 \cdot 2P_2O_5 \cdot 11H_2O$ exists as a stable solid phase at 25°C in the form of white irregular microcrystals (Brownlow et al. 1960).

$In_2O_3 \cdot 3P_2O_5 \cdot 7H_2O$ exists as a stable solid phase at 25°C in the form of rhombohedral plates with beveled edges (Brownlow et al. 1960).

$InHP_2O_7 \cdot 5H_2O$ was prepared at the interaction of the aqueous solutions of $In_2(SO_4)_3$ and $Na_4P_2O_7$ at the $In^{3+}/P_2O_7^{4-}$ ratio of 1:1 with the next washing of the precipitate by water and ethanol (Guzairov et al. 1964). The heating of this compound up to 600–700°C leads to the formation of $InHP_2O_7$.

$In_4(P_2O_7)_3 \cdot H_2O$ was prepared by the heating of $In_4(P_2O_7)_3 \cdot 8H_2O$ at 250°C (Ensslin et al. 1947).

$In_4(P_2O_7)_3 \cdot 4H_2O$ was prepared by the heating of $In_4(P_2O_7)_3 \cdot 8H_2O$ at 175°C (Ensslin et al. 1947).

$In_4(P_2O_7)_3 \cdot 8H_2O$ was obtained by precipitation of an aqueous solution of $In_2(SO_4)_3 \cdot 5H_2O$ with $Na_4P_2O_7 \cdot 10H_2O$ (Ensslin et al. 1947). Its experimental density is 4.510 g·cm^{-3}.

$In_4(P_2O_7)_3 \cdot 14H_2O$ was synthesized at the interaction of $InCl_3$ and $Li_4P_2O_7$ or $K_4P_2O_7$ in the aqueous solution (Deichman et al. 1967a, 1967b).

$In_4(P_2O_7)_3 \cdot 21H_2O$ was prepared at the interaction of the aqueous solutions of $In_2(SO_4)_3$ and $Na_4P_2O_7$ at the $In^{3+}/P_2O_7^{4-}$ ratio of 4:3 with the next washing of the precipitate by water and ethanol (Guzairov et al. 1964). The heating of this compound up to 600–700°C leads to the formation of $In_4(P_2O_7)_3$.

13.2 INDIUM–LITHIUM–OXYGEN–PHOSPHORUS

The **LiInP$_2$O$_7$** and **Li$_3$In$_2$(PO$_4$)$_3$** quaternary compounds are formed in the In–Li–O–P system. LiInP$_2$O$_7$ crystallizes in the monoclinic structure with the lattice parameters $a =$ 708.4 ± 0.2, $b =$ 843.6 ± 0.2, $c =$ 490.8 ± 0.3 pm, and β = 110.75 ± 0.02° and a calculated density of 3.58 g·cm^{-3} (Tranqui et al. 1987).

Li$_3$In$_2$(PO$_4$)$_3$ has a polymorphic transition at circa 100°C, and low-temperature modification crystallizes in the hexagonal structure with the lattice parameters $a =$ 843.2 ± 0.2 and $c =$ 2326.8 ± 0.6 pm and a calculated density of 3.72 g·cm^{-3} (Genkina et al. 1987). Its single crystals were grown from the solution in the melt.

High-temperature modification of the title compound crystallizes in the monoclinic structure with the lattice parameters $a =$ 859.2 ± 0.2, $b =$ 890.8 ± 0.2, $c =$ 1229.0 ± 0.3 pm, and β = 90.0 ± 0.2° and a calculated density of 3.78 g·cm^{-3} (Tran Qui and Hamdoune 1987) [$a =$ 857.5 ± 0.2, $b =$ 888.8 ± 0.2, $c =$ 1227.1 ± 0.3 pm, and β = 89.90 ± 0.04° (Hamdoune et al. 1986)]. To obtain β-Li$_3$In$_2$(PO$_4$)$_3$, stoichiometric mixtures of Li$_2$CO$_3$, In$_2$O$_3$, and NH$_4$H$_2$PO$_4$, finely ground, were progressively heated up to 1000°C in Pt crucibles for 2–5 h. In order to complete the reaction and to improve the crystallinity of the prepared compounds, several heating cycles followed by grinding operations were systematically repeated. Single crystals were grown by the flux method.

13.3 INDIUM–SODIUM–POTASSIUM–PHOSPHORUS

The **NaK$_2$InP$_2$** quaternary compound, which crystallizes in the orthorhombic structure with the lattice parameters $a =$ 664.9 ± 0.2, $b =$ 1489.5 ± 0.3, and $c =$ 662.7 ± 0.1 pm and a calculated density of 2.813 g·cm^{-3}, is formed in the In–Na–K–P system (Somer et al. 1991c, 1992b). It may be prepared from a stoichiometric mixture of the elements or InP, Na, and KP in an evacuated and sealed silica tube at 680°C. A crucible was used to avoid reaction with the silica ampoule.

13.4 INDIUM–SODIUM–CALCIUM–PHOSPHORUS

The **Na$_3$Ca$_3$InP$_4$** quaternary compound, which crystallizes in the hexagonal structure with the lattice parameters $a =$ 935.80 ± 0.03 and $c =$ 714.77 ± 0.05 pm and a calculated density of 2.62 g·cm^{-3} at 200 K, is formed in the In–Na–Ca–P system (Wang et al. 2017a). The title compound was prepared from the elements. The starting materials were weighed in the glove box and loaded into a Nb tube, which was then closed by arc-welding under high-purity argon gas. The sealed Nb container was subsequently jacketed within an evacuated fused silica tube. The reaction mixture was heated in a muffle furnace to 900°C at a rate of 100°C·h^{-1} and equilibrated at this temperature for 24 h. After that, the samples were cooled to 500°C at a rate of 5°C·h^{-1}, at which point the cooling rate to room temperature was changed to 40°C·h^{-1}. All manipulations were carried out inside an argon-field glove box with controlled atmosphere or under vacuum.

13.5 INDIUM–SODIUM–STRONTIUM–PHOSPHORUS

The **Na$_{3.26±0.02}$Sr$_{2.74}$InP$_4$** quaternary compound, which crystallizes in the hexagonal structure with the lattice parameters $a =$ 951.14 ± 0.08 and $c =$ 743.65 ± 0.12 pm and a calculated density of 3.16 g·cm^{-3} at 200 K, is formed in the In–Na–Sr–P system (Wang et al. 2017a). The title compound was obtained from the elements. The starting materials were weighed in the glove box and loaded into an Nb tube, which was then closed by arc-welding under high-purity argon gas. The sealed Nb container was subsequently jacketed within an evacuated fused silica tube. The reaction mixture was heated in a muffle furnace to 900°C at a rate of 100°C·h^{-1} and equilibrated at this temperature for 24 h. After that, the samples were cooled to 500°C at a rate of 5°C·h^{-1}, at which point the cooling rate to room temperature was changed to 40°C·h^{-1}. All manipulations were carried out inside an argon-field glove box with controlled atmosphere or under vacuum.

13.6 INDIUM–SODIUM–EUROPIUM–PHOSPHORUS

The **Na$_{3.40±0.04}$Eu$_{2.60}$InP$_4$** quaternary compound, which crystallizes in the hexagonal structure with the lattice parameters a = 949.59 ± 0.12 and c = 735.05 ± 0.18 pm and a calculated density of 4.12 g·cm^{-3} at 200 K, is formed in the In–Na–Eu–P system (Wang et al. 2017a). The title compound was synthesized from the elements. The starting materials were weighed in the glove box and loaded into an Nb tube, which was then closed by arc-welding under high-purity argon gas. The sealed Nb container was subsequently jacketed within an evacuated fused silica tube. The reaction mixture was heated in a muffle furnace to 900°C at a rate of 100°C·h^{-1} and equilibrated at this temperature for 24 h. After that, the samples were cooled to 500°C at a rate of 5°C·h^{-1}, at which point the cooling rate to room temperature was changed to 40°C·h^{-1}. All manipulations were carried out inside an argon-field glove box with controlled atmosphere or under vacuum.

13.7 INDIUM–SODIUM–OXYGEN–PHOSPHORUS

The **Na$_3$In(PO$_4$)$_2$**, **Na$_3$In$_2$(PO$_4$)$_3$**, and **Na$_8$In$_4$(P$_2$O$_7$)$_5$** quaternary compounds are formed in the In–Na–O–P system. Na$_3$In(PO$_4$)$_2$ exists in two polymorphic modifications, and the phase transition takes place at 700°C (Zhizhin et al. 2000). Low-temperature modification crystallizes in the monoclinic structure with the lattice parameters a = 712.7 ± 0.1, b = 1822.0 ± 0.1, c = 861.6 ± 0.1 pm, and β = 143.36 ± 0.01° [a = 514.1 ± 0.1, b = 1819.7 ± 0.2, c = 713.6 ± 0.1 pm, and β = 92.355 ± 0.002° (Lii 1996)]. It was obtained by hydrothermal synthesis from In$_2$O$_3$, Na$_2$HPO$_4$·2H$_2$O, NaH$_2$PO$_4$·2H$_2$O (Na/P and Na/In starting molar ratios 1.667–1.778 and 1.75, respectively), and H$_2$O at 200°C and 2 MPa for 12–24 h [at 600°C and 2.4 GPa for 40 h (Lii 1996)] (Zhizhin et al. 2000). Formation of α-Na$_3$In(PO$_4$)$_2$ as white sediment was observed. The product was filtered off, washed with H$_2$O, rinsed with EtOH, and dried at 60°C.

High-temperature modification is stable up to 910°C and also crystallizes in the monoclinic structure with the lattice parameters a = 863.35± 0.01, b = 545.50 ± 0.01, c = 704.82 ± 0.01 pm, and β = 90.295 ± 0.001° (Zhizhin et al. 2000). It was prepared by sintering a stoichiometric ratio of In$_2$O$_3$, (NH$_4$)$_2$HPO$_4$, and Na$_2$CO$_3$ at 850°C for 120 h in air.

Na$_3$In$_2$(PO$_4$)$_3$ crystallizes in the monoclinic structure with the lattice parameters a = 1245.0 ± 0.1, b = 1278.6 ± 0.1, c = 659.20 ± 0.07 pm, and β = 114.174 ± 0.002° and a calculated density of 4.048 g·cm^{-3} (Lii and Ye 1997). To obtain this compound, high-temperature, high-pressure hydrothermal synthesis was performed in gold ampoules contained in an autoclave where pressure was provided by water pumped by a compressed air-driven intensifier. Crystals were obtained by heating a mixture of $Na_2HPO_4 \cdot 2H_2O$ (0.5607 g), In_2O_3 (0.1388 g), and 0.7 mL of 3 M H_3PO_4 (Na/P molar ratio is equal 1.2) to 600°C. The autoclave was soaked at this temperature for 36 h, achieving a pressure of 0.24 GPa. It was then cooled to 275°C at 5°C·h^{-1} and quenched to ambient temperature by removing the autoclave from the furnace. The product was filtered off, washed with water, rinsed with methanol, and dried in a desiccator at ambient temperature. The product contained colorless rod crystals of the title compound. $Na_3In_2(PO_4)_3$ could be also prepared by melting a mixture of $InPO_4$ and $Na_3PO_4 \cdot 10H_2O$ at 1130°C (Ensslin et al. 1947).

$Na_8In_4(P_2O_7)_5$ was prepared from the melt (Ensslin et al. 1947).

13.8 INDIUM–POTASSIUM–OXYGEN–PHOSPHORUS

The $KInP_2O_7$ and $K_3In(PO_4)_2$ quaternary compounds are formed in the In–K–O–P system. $KInP_2O_7$ was synthesized from a mixture of 85% H_3PO_4, In_2O_3, and K_2CO_3 [molar ratio 15:(7.5–10):1], which was heated in vitreous glassy carbon crucibles at 150–400°C and with occasional stirring for 12 days (150°C) and 2 days (400°C) (Avaliani et al. 1979). The resulting crystalline product was washed from the melt with water, acetone, and ether.

$K_3In(PO_4)_2$ crystallizes in the monoclinic structure with the lattice parameters $a = 1565.37 \pm 0.05$, $b = 973.57 \pm 0.04$, $c = 1119.75 \pm 0.04$ pm, and $\beta = 90.124 \pm 0.005°$ and a calculated density of 3.284 g·cm^{-3} (Arakcheeva et al. 2003) [$a = 1564.11 \pm 0.01$, $b = 1119.09 \pm 0.01$, $c = 969.81 \pm 0.01$ pm, and $\beta = 90.119 \pm 0.001°$ (Zhizhin et al. 2002)]. The title compound was prepared by solid-phase reaction from stoichiometric amounts of In_2O_3, K_2HPO_4, and K_2CO_3 at 900°C for 120 h (Zhizhin et al. 2002). The synthesis was carried out in alundum crucibles in air. Every 24 h, samples were ground in order to ensure homogenization of the mixture. The samples were annealed at higher temperatures (up to 1250°C) with the aim of preparing single crystals and elucidating the character of melting of the relevant compounds. A rapid cooling of the melt to temperatures below 1200°C led to the formation of transparent, light-yellow crystals in the form of prismatic needles. The crystals prepared were washed from the melt with a large amount of warm distilled water and dried in air at 50°C. Single crystals of $K_3In(PO_4)_2$ were also prepared by sintering a stoichiometric ratio of In_2O_3, KH_2PO_4, and K_2CO_3 at 1200°C for 4 h in air (Arakcheeva et al. 2003).

13.9 INDIUM–POTASSIUM–SULFUR–PHOSPHORUS

The $KInP_2S_7$ quaternary compound, which crystallizes in the monoclinic structure with the lattice parameters $a = 875.06 \pm 0.09$, $b = 996.85 \pm 0.09$, $c = 631.51 \pm 0.08$ pm, and $\beta = 98.906 \pm 0.013°$ and a calculated density of 2.686 g·cm^{-3}, is formed in the In–K–S–P system (Kopnin et al. 2000). The synthesis of this compound was carried out in a dry box under N_2 atmosphere. K_2S_3 (0.57 mM), P_2S_5 (1.14 mM), In (1.14 mM), and S (0.57 mM) were mixed, sealed under vacuum, and heated to 500°C at 4°C·h^{-1}. The temperature was kept at 500°C for 1 week and then cooled to 300°C at 2°C·h^{-1} and down to 100°C at 4°C·h^{-1}. The yellow plate-like crystals obtained were found to be neither air nor water sensitive. The samples was not homogeneous and contained $InPS_4$ and $In_4(P_2S_6)_3$ as impurities.

13.10 INDIUM–POTASSIUM–SELENIUM–PHOSPHORUS

The $K_4In_2(PSe_5)_2(P_2Se_6)$ quaternary compound, which crystallizes in the monoclinic structure with the lattice parameters $a = 1115.64 \pm 0.01$, $b = 2287.71 \pm 0.01$, $c = 1265.25 \pm 0.02$ pm, and $\beta = 109.039 \pm 0.001°$ and a calculated density of 3.858 g·cm^{-3}, is formed in the In–K–Se–P system (Chondroudis and Kanatzidis 1998). The reaction of In with a molten mixture of $K_2Se+P_2Se_5+Se$ yielded brick-red, needle-like crystals of the title compound. A mixture of In (0.25 mM), P_2Se_5 (0.5 mM), K_2Se (0.5 mM), and Se (2.5 mM) was sealed under vacuum in a Pyrex tube and heated to 480°C for 3 days, followed by cooling to 150°C at 4°C·h^{-1}. The excess $K_xP_ySe_z$ flux was removed by washing with dimethylformamide (DMF) under N_2 atmosphere. The product was then washed with tri-n-butylphosphine to remove residual elemental Se and then ether to reveal analytically pure, brick-red needles. The obtained crystals are air and water stable.

13.11 INDIUM–RUBIDIUM–OXYGEN–PHOSPHORUS

The $RbInP_2O_7$ and $Rb_3In(PO_4)_2$ quaternary compounds are formed in the In–Rb–O–P system. $RbInP_2O_7$ was synthesized from a mixture of 85% H_3PO_4, In_2O_3, and Rb_2CO_3 [molar ratio 15:(7.5–10):1], which was heated in vitreous glassy carbon crucibles at 250–400°C and with occasional stirring for 12 days (250°C) and 2 days (400°C) (Avaliani et al. 1979). The resulting crystalline product was washed from the melt with water, acetone, and ether.

Rb$_3$In(PO$_4$)$_2$ crystallizes in the monoclinic structure with the lattice parameters $a = 996.5 \pm 0.2$, $b = 1161.2 \pm 0.2$, $c = 1590.2 \pm 0.2$ pm, and $\beta = 90.30 \pm 0.03°$ (Zhizhin et al. 2002). This compound was prepared by solid-phase reaction from stoichiometric amounts of In_2O_3, Rb_2HPO_4, and Rb_2CO_3 at 900°C for 120 h. The synthesis was carried out in alundum crucibles in air. Every 24 h, samples were ground in order to ensure homogenization of the mixture. The samples were annealed at higher temperatures (up to 1250°C) with the aim of preparing single crystals and elucidating the character of melting of the relevant compounds. A slow cooling of the melt to temperatures from 950°C to 900°C (5°C·h^{-1}) led to the formation of transparent, light-yellow crystals in the form of prismatic needles. The crystals prepared were washed from the melt with a large amount of warm distilled water and dried in air at 50°C.

13.12 INDIUM–CESIUM–OXYGEN–PHOSPHORUS

The $Cs_3In_3P_{12}O_{36}$ (cesium indium cyclododecaphosphate) quaternary compound, which crystallizes in the cubic structure with the lattice parameter $a = 1488.5 \pm 0.2$ pm and a calculated density of 3.405 g·cm^{-3}, is formed in the In–Cs–O–P system (Murashova and Chudinova 2001). To prepare this compound, Cs_2CO_3, In_2O_3, and 85% H_3PO_4 (Cs/In/P molar ratio is 5:1:15) were used as starting chemicals. Syntheses were performed at 350–400°C in glassy carbon crucibles. Small colorless crystals of $CsInHP_3O_{10}$ grew in just 5 days. After 3 weeks, they began to convert into well-faced crystals of $Cs_3In_3P_{12}O_{36}$. The reaction almost reached completion in 30 days. The resultant crystals were washed with cold water. To isolate the title compound from the samples containing both compounds, these were treated with hot HCl, which dissolves $CsInHP_3O_{10}$ and does not react with $Cs_3In_3P_{12}O_{36}$.

13.13 INDIUM–CESIUM–SELENIUM–PHOSPHORUS

The **$Cs_5In(P_2Se_6)_2$** quaternary compound, which crystallizes in the tetragonal structure with the lattice parameters $a = 1388.6 \pm 0.1$ and $c = 759.7 \pm 0.2$ pm and a calculated density of 4.196 g·cm^{-3}, is formed in the In–Cs–Se–P system (Chondroudis et al. 1998). It was synthesized from a mixture of In (0.4 mM), Cs_2Se (1.00 mM), P (1.60 mM), and Sc (3.80 mM), which was sealed under vacuum in a quartz tube and heated to 750°C for 2 days, followed by cooling to 150°C at 25°C·h^{-1}. The product was washed with degassed DMF and anhydrous ether, which revealed irregular, dark orange crystals of the title compound. The latter are air and water stable.

13.14 INDIUM–COPPER–OXYGEN–PHOSPHORUS

The **$Cu_3In_2(PO_4)_4$** quaternary compound, which crystallizes in the monoclinic structure with the lattice parameters $a = 890.67 \pm 0.06$, $b = 882.71 \pm 0.05$, $c = 788.15 \pm 0.05$ pm, and $\beta = 108.393 \pm 0.005°$ and a calculated density of 4.519 g·cm^{-3}, is formed in the In–Cu–O–P system (Gruß and Glaum 2001). The synthesis of the title compound was made of Cu powder (4.86 mM) and In ingot (3.24 mM), which were dissolved in hot 65% HNO_3 (20 mL). After combining with a solution of $(NH_4)_2HPO_4$ (6.49 mM) in circa 20 mL of H_2O, the obtained solution was evaporated to dryness and the residue was then annealed in Pt crucible at 800°C for 4 days. As a result, $Cu_3In_2(PO_4)_4$ was obtained. Single crystals of this compound were grown by chemical vapor transport using Cl_2 as a transport agent.

13.15 INDIUM–COPPER–SELENIUM–PHOSPHORUS

The **$CuInP_2Se_6$** quaternary compound, which melts congruently at $650 \pm 5°C$ [incongruently at 642°C (Pfeiff and Kniep 1992)] and crystallizes in the trigonal structure with the lattice parameters $a = 639.5 \pm 0.2$ and $c = 1334.0 \pm 0.1$ pm [$a = 639.2 \pm 0.0$, $c = 1333.8 \pm 0.1$ pm, a calculated density of 10.0 g·cm^{-3}, and energy gap 1.78 eV (Pfeiff and Kniep 1992)], is formed in the In–Cu–Se–P system (Motrya et al. 2009; Potoriy et al. 2010). The homogeneity region of this compound at 610°C is circa 30 mol% in the direction of $In_4(P_2Se_6)_3$ and approximately 18 mol% $CuInSe_2$ along the $CuInSe_2$–"P_2Se_4" section. At 400°C, this region is 18–22 mol% in the direction of $In_4(P_2Se_6)_3$ and approximately 20 mol% in the direction of $CuInSe_2$. It was obtained by solid-state reactions from the stoichiometric mixture of the elements in evacuated silica ampoule by heating up to 750°C (10 h) followed by quenching from the melt (30°C·h^{-1}) (Pfeiff and Kniep 1992). After quenching the melt to liquid nitrogen, the dark-red sample was annealed at 500°C for 14 days. Single crystals of the title compound were grown by chemical vapor transport using I_2 or CuI as a transport agent or by directional crystallization from the melt (Motrya et al. 2009; Potoriy et al. 2010).

13.16 INDIUM–SILVER–OXYGEN–PHOSPHORUS

The **$AgInP_2O_7$** quaternary compound, which crystallizes in the monoclinic structure with the lattice parameters $a = 748.67 \pm 0.03$, $b = 826.20 \pm 0.03$, $c = 983.83 \pm 0.05$ pm, and $\beta = 112.038 \pm 0.002°$ and a calculated density of 4.670 g·cm^{-3}, is formed in the In–Ag–O–P system (Zouihri et al. 2011). The title compound in the form of single crystals was prepared by stoichiometric reaction of $AgNO_3$, $(NH_4)_2HPO_4$, and In_2O_3 in B_2O_3 flux. The mixture was heated at 500°C under ambient atmosphere for 6 h and at 700°C for 2 h with intermediate grindings to ensure complete reaction. Subsequent melting at 1050°C followed by slow cooling to room temperature at a rate of $12°C·h^{-1}$ resulted in colorless crystals of $AgInP_2O_7$.

13.17 INDIUM–SILVER–SULFUR–PHOSPHORUS

The **$AgInP_2S_6$** quaternary compound, which crystallizes in the trigonal structure with the lattice parameters $a = 618.2 ± 0.2$ and $c = 1295.7 ± 0.2$ pm, is formed in the In–Ag–S–P system (Ouili et al. 1987). It was synthesized from the elements, except for indium, which was introduced through In_2S_3, and heated in evacuated silica tubes for 1 week at 700°C.

13.18 INDIUM–SILVER–SELENIUM–PHOSPHORUS

The **$AgInP_2Se_6$** quaternary compound, which melts congruently at 673°C and crystallizes in the trigonal structure with the lattice parameters $a = 648.3 ± 0.1$, $c = 1333.0 ± 0.4$ pm, a calculated density of 10.4 $g·cm^{-3}$, and energy gap 1.79 eV, is formed in the In–Ag–Se–P system (Pfeiff and Kniep 1992). It was obtained by solid-state reactions from the stoichiometric mixture of the elements in evacuated silica ampoule by heating up to 750°C (10 h) followed by cooling from the melt ($30°C·h^{-1}$). After quenching the melt to liquid nitrogen, the dark-red sample was annealed at 500°C for 14 days.

13.19 INDIUM–CALCIUM–OXYGEN–PHOSPHORUS

The **$Ca_9In(PO_4)_7$** quaternary compound, which formed in the In–Ca–O–P system, is characterized by a first-order $β⇔β′$ phase transition of the ferroelectric type at 629°C (Morozov et al. 2002). The phase transition is accompanied by a jump in the temperature dependence of the a and c lattice parameters. Both modifications crystallize in the trigonal structure with the lattice parameters $a = 1040.08 ± 0.01$ and $c = 3727.2 ± 0.1$ pm for low-temperature modification [$a = 1040.2 ± 0.2$ and $c = 3726 ± 1$ pm (Golubev et al. 1990)] and $a = 1046.11 ± 0.02$ and $c = 3787.4 ± 0.1$ pm at 700°C for high-temperature modification. The title compound was prepared from a stoichiometric mixture of $Ca_2P_2O_7$, $CaCO_3$, and In_2O_3 by a ceramic technique in an Al_2O_3 crucible at 1000°C for 90 h in air (Morozov et al. 2002). It could be also synthesized from the solid-state reaction of $(NH_4)_2HPO_4$ and In_2O_3 at 900–1200°C for 30–50 h (Golubev et al. 1990).

13.20 INDIUM–STRONTIUM–OXYGEN–PHOSPHORUS

The **$Sr_3In(PO_4)_3$** and **$Sr_9In(PO_4)_7$** quaternary compounds are formed in the In–Sr–O–P system. $Sr_3In(PO_4)_3$ crystallizes in the cubic structure with the lattice parameter $a = 998$ pm (Blasse 1970). It was prepared by usual ceramic methods using a mixture of $SrHPO_4$ and In_2O_3. Firing temperatures varied from 950°C to 1150°C.

$Sr_9In(PO_4)_7$ exhibits an antiferroelectric phase transition at 500°C (Stefanovich et al. 2004; Deyneko et al. 2016). Low-temperature modification crystallizes in the monoclinic structure with the lattice parameters $a = 1454.493 ± 0.008$, $b = 1066.321 ± 0.005$,

$c = 1837.147 \pm 0.011$ pm, and $\beta = 114.726 \pm 0.001°$ at 27°C (Deyneko et al. 2016) [$a = 1804.25 \pm 0.02$, $b = 1066.307 \pm 0.004$, $c = 1837.14 \pm 0.02$ pm, and $\beta = 132.9263 \pm 0.0005°$ (Belik et al. 2002); $a = 1454.6$, $b = 1066.3$, $c = 1837.2$ pm, and $\beta = 114.7°$ (Stefanovich et al. 2004)]. High-temperature modification crystallizes in the rhombohedral structure with the lattice parameters $a = 906.5$ and $\alpha = 72.23°$ or $a = 1068.561 \pm 0.010$ and $c = 1992.605 \pm 0.014$ (in hexagonal setting) at 550°C (Stefanovich et al. 2004; Deyneko et al. 2016). The title compound was synthesized as white product from a mixture of $SrCO_3$, In_2O_3, and $NH_4H_2PO_4$ (molar ratio 9:0.5:7) (Belik et al. 2002; Stefanovich et al. 2004; Deyneko et al. 2016). The mixture was contained in alumina crucibles, heated under air while raising the temperature very slowly from room temperature to 630°C, and allowed to react at 1000–1150°C for 120 h with three intermediated grindings. The products were then quenched to room temperature.

13.21 INDIUM–BARIUM–OXYGEN–PHOSPHORUS

The $Ba_3In(PO_4)_3$ quaternary compound, which crystallizes in the cubic structure with the lattice parameter $a = 1036$ pm, is formed in the In–Ba–O–P system (Blasse 1970). It was prepared by usual ceramic methods using a mixture of $BaHPO_4$ and In_2O_3. Firing temperatures varied from 950°C to 1150°C.

13.22 INDIUM–ZINC–TELLURIUM–PHOSPHORUS

The solid solution regions based on InP in the InP–Zn–Te subsystem at 750–1000°C are shown in Figure 13.1 (Glazov et al. 1974). The ingots were homogenized at 750°C, 800°C, 850°C, 900°C, and 1000°C for 1400, 1050, 700, 350, and 200 h, respectively. During annealing, the alloys were periodically quenched in water. The obtained samples were investigated through metallography and measurements of microhardness.

InP–ZnTe. The phase diagram of this system is a eutectic type (Figure 13.2) (Glazov et al. 1972, 1973). The eutectic composition and temperature are 48 mol% ZnTe and 420 ± 10°C, respectively. The solubility of ZnTe in InP at the eutectic temperature is equal to 3 mol%, and the solubility of InP in ZnTe at the same temperature reaches 25 mol%. This system was investigated through differential thermal analysis (DTA) and metallography.

13.23 INDIUM–CADMIUM–SULFUR–PHOSPHORUS

InP–CdS. Solubility of CdS in InP reaches 7 mol% (Kirovskaya and Timoshenko 2007).

13.24 INDIUM–CADMIUM–TELLURIUM–PHOSPHORUS

The solid solution regions based on InP in the InP–Cd–Te subsystem at 750–1000°C are shown in Figure 13.3 (Glazov et al. 1974). The ingots were homogenized at 750°C, 800°C, 850°C, and 900°C for 1400, 1050, 700, and 350, respectively. During annealing, the alloys were periodically quenched in water. The obtained samples were investigated through metallography and measurements of microhardness.

InP–CdTe. The phase diagram is a eutectic type (Figure 13.4) (Glazov et al. 1972, 1973, 1975a). The eutectic composition and temperature are 50 mol% InP and 410 ± 10°C,

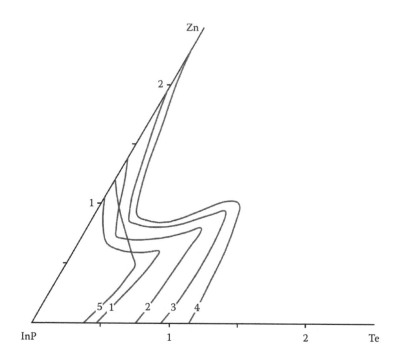

FIGURE 13.1 Solid solution regions based on InP in the InP–Zn–Te subsystem at (1) 750°C, (2) 800°C, (3) 850°C, (4) 900°C, and (5) 1000°C. (From Glazov, V.M., *Izv. AN SSSR. Neorgan. Mater.*, 10(1), 144–145, 1974. With permission.)

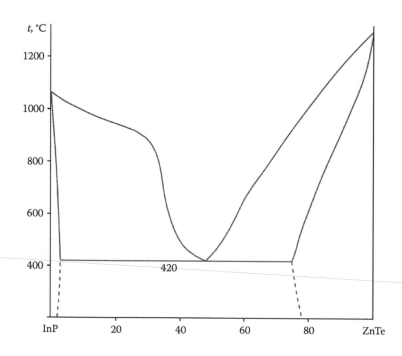

FIGURE 13.2 Phase diagram of the InP–ZnTe system. (From Glazov, V.M., *Izv. AN SSSR. Neorgan. Mater.*, 9(11), 1883–1889, 1973. With permission.)

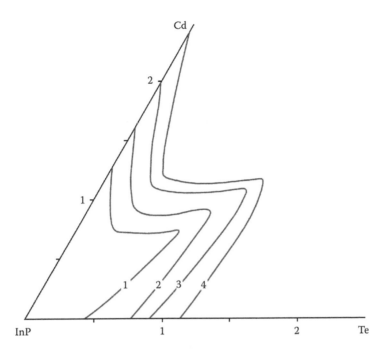

FIGURE 13.3   Solid solution regions based on InP in the InP–Cd–Te subsystem at (1) 750°C, (2) 800°C, (3) 850°C, and (4) 900°C. (From Glazov, V.M., *Izv. AN SSSR. Neorgan. Mater.*, 10(1), 144–145, 1974. With permission.)

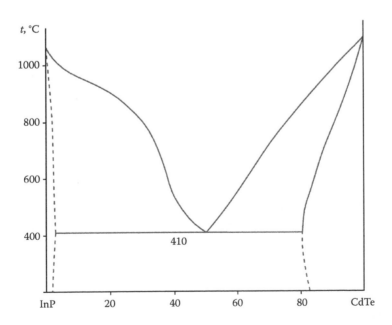

FIGURE 13.4   Phase diagram of the InP–CdTe system. (From Glazov, V.M., *Izv. AN SSSR. Neorgan. Mater.*, 9(11), 1883–1889, 1973. With permission.)

respectively. The solubility of CdTe in InP is equal to 3 mol%, and the solubility of InP in CdTe reaches 20–25 mol%. This system was investigated by DTA and metallography.

13.25 INDIUM–TIN–IODINE–PHOSPHORUS

The $Sn_{10}In_{14}P_{22}I_8$ and $Sn_{14}In_{10}P_{21.2}I_8$ quaternary compounds are formed in the In–Sn–I–P system (Shatruk et al. 2001). $Sn_{10}In_{14}P_{22}I_8$ crystallizes in the cubic structure with the lattice parameter $a = 1004.50 \pm 0.07$ pm and a calculated density of 5.535 g·cm^{-3}, and $Sn_{14}In_{10}P_{21.2}I_8$ crystallizes in the tetragonal structure with the lattice parameters $a = 2474.5 \pm 0.3$ and $c = 1106.7 \pm 0.1$ pm and a calculated density of 5.521 g·cm^{-3}. Both compounds were prepared by heating the respective mixtures of Sn, In, red phosphorus, and SnI_4 in sealed silica tubes under vacuum at 500°C for 5 days. The samples were then reground and heated in sealed silica tubes under vacuum at 300°C for another 14 days, followed by slow cooling (about 20°C·h^{-1}) to room temperature. After the second annealing, the samples appeared as black air-stable homogeneous powders.

13.26 INDIUM–ARSENIC–ANTIMONY–PHOSPHORUS

InP–InAs–InSb. A miscibility gap has been found in this ternary system (Müller and Richards 1964). The behavior of an unstable region was analyzed based on regular approximation for solid solutions (Onabe 1982b). The calculated liquidus and solidus surfaces for this system are given in Figure 13.5, and the miscibility gap is presented in Figure 13.6 (Ishida et al. 1989). Temperature dependence of spinodal isotherms shows that the unstable regions cover a wide range of compositions (Onabe 1982b; Ishida et al. 1989). The thermodynamics of spinodal decomposition in the InP–InAs–InSb quasiternary system have been developed in Stringfellow (1983). Based on the delta-lattice parameter solution model of the free energy of mixing of semiconductor alloys, an analysis has been developed for the calculation of the spinodal surface and the critical temperature for solid alloys. Solid–solid isotherms were presented for this system. Concepts are also developed for the thermodynamic analysis of spinodal decomposition in this quaternary alloys, including the effect of the coherency strain energy. This addition to the free energy of the inhomogeneous solid is shown to completely stabilize the alloys of interest even at temperatures below room temperature. The calculated critical point for this system is 1046°C [1238°C (Onabe 1982b); 1335°C (Ishida et al. 1989)].

The liquidus surface of this quasiternary system was also constructed by Semenova et al. (2006, 2009). It was shown that the liquidus surface descends with a gentle slope (but not monotonically) from the most refractory component (InP) to the InAs–InSb side of the concentration triangle. The existence of a more gently sloping segment with InP molar fractions within 0.5–0.6 implies a possibility of liquid–liquid phase separation. The liquidus surface rather strongly descends to lower temperatures near the InP–InSb and InP–InAs sides of the concentration triangle in agreement with the phase diagrams of these quasi-binary systems. The eutectic immiscibility is field looped inside the diagram; because the eutectic in the InAs–InP–InSb quasiternary is degenerate; however, the quaternary $InP_xAs_ySb_{1-x-y}$ solid solution field occupies the greatest part of the liquidus surface.

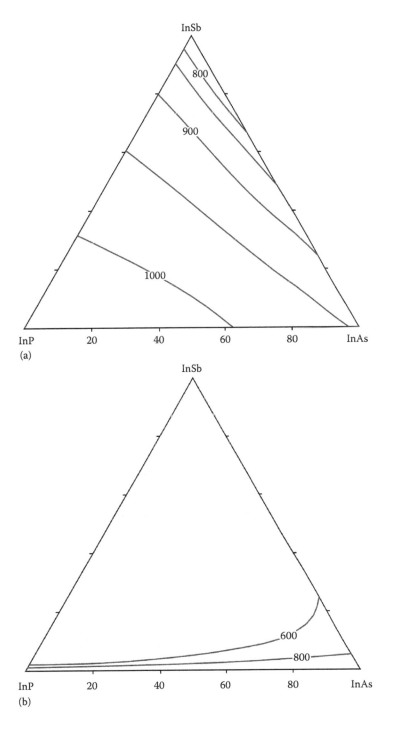

FIGURE 13.5 Liquidus (a) and solidus (b) surfaces of the InP–InAs–InSb quasiternary system. (Reprinted from *J. Cryst. Growth*, 98(1–2), Ishida et al., Data base for calculating phase diagrams of III-V alloy semiconductors, 140–147, Copyright (1989), with permission from Elsevier.)

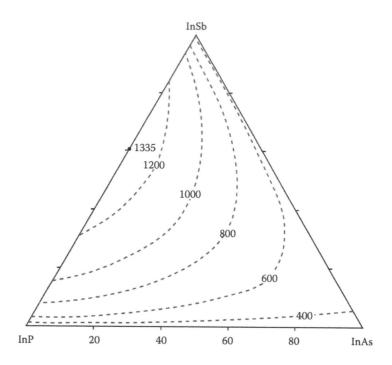

FIGURE 13.6 Miscibility gaps in the InP–InAs–InSb quasiternary system at 600–1000°C. (Reprinted from *J. Cryst. Growth*, 98(1–2), Ishida et al., Data base for calculating phase diagrams of III-V alloy semiconductors, 140–147, Copyright (1989), with permission from Elsevier.)

The development of the InP–InAs–InSb quasiternary system with projections of isothermal sections at 30°C, 130°C, 230°C, 330°C, 430°C, and 530°C was also investigated by Semenova et al. (2008, 2009). It was shown that the two-phase field (a mixture of $InAs_{1-x}Sb_x$ and InP_xAs_{1-x} solid solutions) decreases with increasing temperature, although occupying a considerable part of the concentration triangle even at moderate subsolidus temperatures.

Phase equilibria in the In–As–Sb–P system were also determined by liquid phase epitaxy and are shown in Figure 13.7 (Litvak et al. 1994). It can be seen that with an increase in temperature of the two-phase equilibrium, the maximum available concentration of InP in the $InP_xAs_ySb_{1-x-y}$ solid solution decreases. For example, at 550°C according to the calculations, the maximum achievable concentration of InP is 34–37 mol%, at 600°C it is equal to 25–26 mol%, and at 650°C this concentration is only 17–19 mol%. The liquidus isotherms for the InP–InAs–InSb quasiternary system in the phosphorus corner determined through DTA are given in Figure 13.8 (Vigdorovich et al. 1980). The composition of the $InP_xAs_ySb_{1-x-y}$ solid solution, containing 60, 70, and 80 mol% InP, is also presented.

The solubility of InAs in the In–As–Sb–P melt at 710°C determined by the visual thermal analysis and through the dissolution of a substrate source is presented in Figure 13.9 (Kuznetsov et al. 1992). The Sb concentration in the liquid phase was equal to 37 mol%, and the phosphorus concentration was within the interval from 0 to 0.6 mol%.

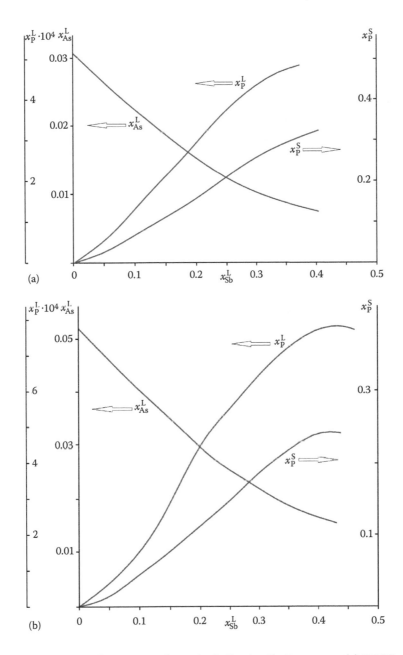

FIGURE 13.7 Composition of coexisting phases in the In–As–Sb–P system at (a) 550°C and (b) 600°C: the solid phase has the same lattice constant as InAs has. (*Continued*)

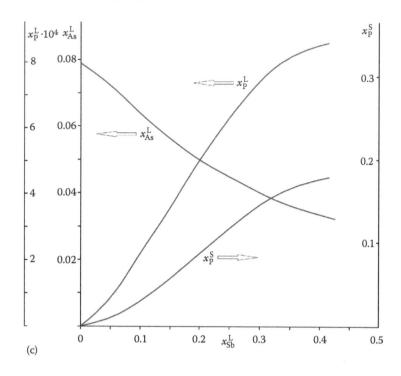

(c)

FIGURE 13.7 (CONTINUED)   Composition of coexisting phases in the In–As–Sb–P system at (c) 650°C: the solid phase has the same lattice constant as InAs has. (From Litvak, A.M., *Zhurn. Prikl. Khimii*, 67(12), 1957–1960, 1994. With permission.)

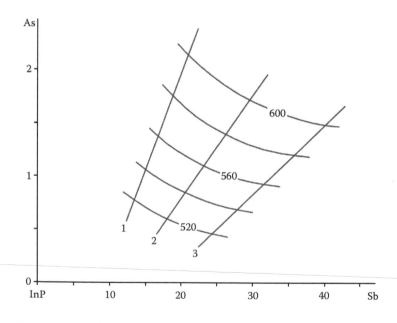

FIGURE 13.8   Composition of the $InP_xAs_ySb_{1-x-y}$ solid solutions [(1) $x = 0.8$ and $y = 0.062$; (2) $x = 0.7$ and $y = 0.093$; (3) $x = 0.6$ and $y = 0.124$] and liquidus isotherms for the InP–InAs–InSb quasiternary system in the InP corner. (From Vigdorovich, V.N., *Dokl. AN SSSR*, 252(6), 1423–1426, 1980. With permission.)

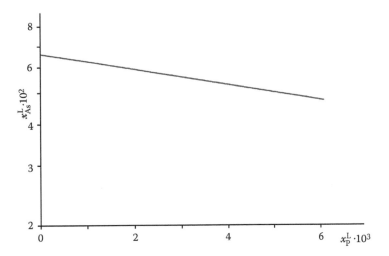

FIGURE 13.9 Solubility of InAs in the In–As–Sb–P melt at 710°C. (From Kuznetsov, V.V., *Kristallografiya*, 37(4), 998–1002, 1992. With permission.)

CHAPTER 14

Systems Based on InAs

14.1 INDIUM–HYDROGEN–OXYGEN–ARSENIC

Some compounds are formed in this quaternary system.

InAsO$_4$·H$_2$O crystallizes in the triclinic structure with the lattice parameters $a = 550.0 \pm 0.1$, $b = 572.0 \pm 0.1$, and $c = 668.5 \pm 0.1$ pm and $\alpha = 98.90 \pm 0.03$, $\beta = 94.60 \pm 0.03°$, and $\gamma = 107.55 \pm 0.03°$ and a calculated density of 4.598 g·cm^{-3} (Kolitsch and Schwendtner 2005). The title compound was prepared by a hydrothermal method (Teflon-lined stainless steel bomb, 220°C, 7 days, and slow furnace cooling) from mixtures of distilled water, arsenic acid, In$_2$O$_3$, and Li$_2$CO$_3$. The final pH value of the reacted solutions was about 1. InAsO$_4$·H$_2$O forms tiny colorless pseudorhombohedral crystals.

InAsO$_4$·2H$_2$O crystallizes in the orthorhombic structure with the lattice parameters $a = 909.0 \pm 0.4$, $b = 1034.4 \pm 0.4$, and $c = 1046.8 \pm 0.4$ pm (Chen et al. 2002) [$a = 1047.1 \pm 0.3$, $b = 909.2 \pm 0.2$, and $c = 1034.1 \pm 0.2$ pm (Botelho et al. 1994); $a = 1047.8 \pm 0.1$, $b = 909.98 \pm 0.08$, and $c = 1034.5 \pm 0.1$ pm and a calculated density of 3.903 g·cm^{-3} (Tang et al. 2001)]. The title compound was obtained using hydrothermal techniques. A mixture of InCl$_3$, As$_2$O$_5$·xH$_2$O ($x \approx 3$), Me$_4$NOH, CsNO$_3$, Cd(C$_2$H$_3$O$_2$)·2H$_2$O, and water in a molar ratio 4:1:2:4:8:776 was sealed under vacuum in a thick-walled Pyrex tube (~20% filled) (Tang et al. 2001). The reaction was carried out at 160°C for 12 days followed by slow cooling to room temperature. The products were filtered, washed with water and acetone, and dried at room temperature. Two phases were found to be present: InAsO$_4$·2H$_2$O in about 80% yield, and **CsIn(HAsO$_4$)$_2$**. In the absence of Cd(C$_2$H$_3$O$_2$)·2H$_2$O, only CsIn(HAsO$_4$)$_2$ forms and only an unidentified microcrystalline material forms in the absence of CsNO$_3$ and Cd(C$_2$H$_3$O$_2$)·2H$_2$O. InAsO$_4$·2H$_2$O has been also synthesized from the hydrothermal reactions of H$_2$O, InCl$_3$·4H$_2$O, H$_3$AsO$_4$, and (C$_2$H$_4$)$_3$N$_2$ (1,4-diazabicyclo[2.2.2]octane) (molar ratio 1:1:1:133) at 200°C for 2 days (Chen et al. 2002).

InAsO$_4$·2H$_2$O could be also prepared if to the solution containing InCl$_3$ and H$_3$AsO$_4$ (molar ratio 1:1.2) NaOH was added to obtain the solution with pH = 2.7 (Deichman et al. 1974). The solution with the precipitate was heated on a water bath with stirring for 1–2 h. The precipitate was washed by decantation with hot solution of 0.05 M H$_3$AsO$_4$ and then wrings out under pressure and dried in air. This compound was also obtained by the interaction of $In(OH)_3$ and H_3AsO_4. $In(OH)_3$ was dissolved at heating in small excess (circa 6%) of H_3AsO_4. The precipitate was quickly washed with cold 50% EtOH, wrings out under pressure, and air-dried.

$InAsO_4 \cdot 2H_2O$ is known in nature as the mineral yanomatite (orthorhombic structure, $a = 1044.6 \pm 0.6$, $b = 908.5 \pm 0.4$, and $c = 1034.5 \pm 0.6$ pm and a calculated density of 3.876 g·cm^{-3}) (Botelho et al. 1994; Jambor and Roberts 1995).

$InAsO_4 \cdot 2H_3AsO_4 \cdot 2H_2O$ or $In(H_2AsO_4)_3 \cdot 2H_2O$ and $InAsO_4 \cdot 2H_3AsO_4 \cdot 5H_2O$ or $In(H_2AsO_4)_3 \cdot 5H_2O$ were obtained by Ezhova et al. (1976). The second compound is strongly hygroscopic and decomposes at 830°C to form $InAsO_4$.

14.2 INDIUM–LITHIUM–POTASSIUM–ARSENIC

The LiK_2InAs_2 quaternary compound, which crystallizes in the orthorhombic structure with the lattice parameters $a = 1285.3 \pm 0.3$, $b = 1551.4 \pm 0.3$, and $c = 639.2 \pm 0.2$ pm, is formed in the In–Li–K–As system (Somer et al. 1995b). It was synthesized from a stoichiometric mixture of KAs, LiIn, and As in sealed niobium ampoules at 650°C.

14.3 INDIUM–SODIUM–POTASSIUM–ARSENIC

The NaK_2InAs_2 quaternary compound, which crystallizes in the orthorhombic structure with the lattice parameters $a = 677.1 \pm 0.8$, $b = 1529 \pm 3$, and $c = 678.3 \pm 0.9$ pm and a calculated density of 3.461 g·cm^{-3}, is formed in the In–Na–K–As system (Somer et al. 1991a, 1992b; Carrillo-Cabrera et al. 1993). This compound was prepared from the stoichiometric mixture of Na, KAs, and InAs in an evacuated and sealed tube. A crucible of Al_2O_3 was used to avoid reaction with the silica ampoule.

14.4 INDIUM–SODIUM–STRONTIUM–ARSENIC

The $Na_{3.34\pm0.02}Sr_{2.66}InAs_4$ quaternary compound, which crystallizes in the hexagonal structure with the lattice parameters $a = 974.26 \pm 0.12$ and $c = 760.5 \pm 0.2$ pm and a calculated density of 3.85 g·cm^{-3} at 200 K, is formed in the In–Na–Eu–As system (Wang et al. 2017a). To prepare the title compound, the elements were weighed in the glove box and loaded into an Nb tube, which was then closed by arc-welding under high-purity argon gas. The sealed Nb container was subsequently jacketed within an evacuated fused silica tube. The reaction mixture was heated in a muffle furnace to 900°C at a rate of 100°C·h^{-1} and equilibrated at this temperature for 24 h. After that, the samples were cooled to 500°C at a rate of 5°C·h^{-1}, at which point the cooling rate to room temperature was changed to 40°C·h^{-1}. All manipulations were carried out inside an argon-field glove box with controlled atmosphere or under vacuum.

14.5 INDIUM–SODIUM–OXYGEN–ARSENIC

The $Na_3In_2(AsO_4)_3$ quaternary compound, which crystallizes in the monoclinic structure with the lattice parameters $a = 1260.25 \pm 0.01$, $b = 1316.99 \pm 0.01$, $c = 683.35 \pm 0.01$ pm, and $\beta = 113.7422 \pm 0.0005°$, is formed in the In–Na–O–As system (Khorari et al. 1997) [$a = 1260.37 \pm 0.02$, $b = 1317.11 \pm 0.02$, $c = 683.42 \pm 0.01$ pm, and $\beta = 113.743 \pm 0.001°$ and a calculated density of 4.576 g·cm^{-3} (Lii and Ye 1997)]. The title compound can be synthesized, either by a conventional solid-state reaction (starting from $NaHCO_3$, In_2O_3, and $NH_4H_2AsO_4 \cdot 7H_2O \cdot$) or by a chemical attack of the reagents ($NaHCO_3$, In_2O_3, As_2O_3) by nitric acid, followed by evaporation to dryness (Khorari et al. 1997). The dry mixture was slowly heated up to 650°C (1 day), and then to 700°C (1 day), and 800°C (1 day) with intervening mixing and grinding.

$Na_3In_2(AsO_4)_3$ could be also prepared through high-temperature, high-pressure hydrothermal synthesis in gold ampoules contained in an autoclave where pressure was provided by water pumped by a compressed air-driven intensifier (Lii and Ye 1997). Crystals were obtained by heating a mixture of $NH_4H_2AsO_4 \cdot 7H_2O \cdot$ (0.5616 g), In_2O_3 (0.0833 g), and 0.4 mL of 3 M H_3AsO_4 (Na/As molar ratio is equal 1.2) to 600°C. The autoclave was soaked at this temperature for 36 h, achieving a pressure of 0.24 GPa. It was then cooled to 275°C at 5°C·h^{-1} and quenched to ambient temperature by removing the autoclave from the furnace. The product was filtered off, washed with water, rinsed with methanol, and dried in a desiccator at ambient temperature. The product was a colorless microcrystalline powder of the title compound.

14.6 INDIUM–SODIUM–CHLORINE–ARSENIC

InAs–NaCl. After standing for 24 h a melt containing 30 mass% InAs and 70 mass% NaCl, a delamination was observed (Koppel et al. 1967). The salt layer contains about 1 mass% In and less than 0.1 mass% As. The solubility of InAs in NaCl is practically absent.

14.7 INDIUM–SODIUM–BROMINE–ARSENIC

InAs–NaBr. After standing for 24 h a melt containing 30 mass% InAs and 70 mass% NaBr, a delamination was observed (Koppel et al. 1967). The salt layer contains about 1 mass% In and less than 0.1 mass% As. The solubility of InAs in NaBr is practically absent.

14.8 INDIUM–POTASSIUM–BARIUM–ARSENIC

The KBa_2InAs_3 quaternary compound, which crystallizes in the orthorhombic structure with the lattice parameters $a = 1012.9 \pm 0.2$, $b = 2520.8 \pm 0.4$, and $c = 1388.4 \pm 0.3$ pm and a calculated density of 4.897 g·cm^{-3}, is formed in the In–K–Ba–As system (Gascoin and Sevov 2002c). It was made from a mixture of corresponding elements (molar ratio K/Ba/In/As was 1:2:1:3), which was heated at a milder temperature of 500°C and slowly cooled. The mixture was loaded in a Nb container that was then sealed by arc-welding under Ar. This vessel was placed in a fused-silica ampoule that was then flame-sealed under vacuum. This compound is extremely air and moisture sensitive and crystallizes as dark-gray thin plates with a coal-like luster. All manipulations were performed inside a nitrogen-filled glove box with moisture level below 1 ppm.

14.9 INDIUM–POTASSIUM–GERMANIUM–ARSENIC

The $K_5In_5Ge_5As_4$ and $K_8In_8Ge_5As_{17}$ quaternary compounds are formed in the In–K–Ge–As system (Shreeve-Keyer et al. 1997). $K_5In_5Ge_5As_4$ crystallizes in the monoclinic structure with the lattice parameters $a = 4000 \pm 1$, $b = 392.5 \pm 0.2$, $c = 1029.9 \pm 0.3$ pm, and $\beta = 99.97 \pm 0.02°$ and a calculated density of 4.549 g·cm^{-3}. Silver needle-like crystals of this compound were obtained as follows. The solids K_3As_7 (0.779 mM), Ba (0.175 mM), In (2.73 mM), and

Ge (1.71 mM) were combined in a mole ratio of 5:1:16:10 and then sealed in an evacuated quartz ampoule. This mixture was heated to 750°C over 12 h. The temperature was held at 750°C for 6 h, then slowly cooled to 550°C at 4°C·h⁻¹, followed by cooling to ambient temperature over 6 h. The title compound is extremely air and moisture sensitive. All manipulations were performed under a high-purity inert gas atmosphere.

$K_8In_8Ge_5As_{17}$ also crystallizes in the monoclinic structure with the lattice parameters $a = 1839.6 \pm 0.7$, $b = 1908.7 \pm 0.7$, $c = 2335.9 \pm 0.4$ pm, and $\beta = 105.71 \pm 0.02°$ and a calculated density of 4.444 g·cm⁻³. To obtain black cubic-shaped crystals of this compound, K (2.99 mM), In (1.79 mM), Ge (1.20 mM), and As (4.79 mM) were combined in a mole ratio of 5:3:2:8 and then sealed into an evacuated quartz ampoule. This mixture was heated to 871°C over 8 h, where the temperature was held for another 6 h. X-ray quality crystals were obtained by slowly cooling the tube to 482°C over 20 h and then to ambient temperature over 15 h. $K_8In_8Ge_5As_{17}$ is also extremely air and moisture sensitive; therefore, all manipulations were performed under a high-purity inert gas atmosphere.

14.10 INDIUM–POTASSIUM–NIOBIUM–ARSENIC

The **$K_{10}InNbAs_6$** quaternary compound, which crystallizes in the monoclinic structure with the lattice parameters $a = 910.7 \pm 0.1$, $b = 828.78 \pm 0.08$, $c = 1513.9 \pm 0.1$ pm, and $\beta = 91.112 \pm 0.009°$, is formed in the In–K–Nb–As system (Gascoin and Sevov 2003b). It was prepared from direct reactions of the pure elements. All manipulations were carried out in a glove box with typical moisture and oxygen levels below 1 ppm. The mixture was loaded in a Nb container that was enclosed and sealed in an evacuated fused-silica ampoule and heated at 600°C for a week and cooled slowly. $K_{10}InNbAs_6$ crystallizes as very thin and shiny plates with dark gray to black color and very smooth surfaces.

14.11 INDIUM–POTASSIUM–CHLORINE–ARSENIC

InAs–KCl. After standing for 24 h a melt containing 30 mass% InAs and 70 mass% KCl, a delamination was observed (Koppel et al. 1967). The salt layer contains about 1 mass% In and less than 0.1 mass% As. The solubility of InAs in KCl is practically absent.

14.12 INDIUM–POTASSIUM–BROMINE–ARSENIC

InAs–KBr. After standing for 24 h a melt containing 30 mass% InAs and 70 mass% KBr, a delamination was observed (Koppel et al. 1967). The salt layer contains about 1 mass% In and less than 0.1 mass% As. The solubility of InAs in KBr is practically absent.

14.13 INDIUM–POTASSIUM–IODINE–ARSENIC

InAs–KI. After standing for 24 h a melt containing 30 mass% InAs and 70 mass% KI, a delamination was observed (Koppel et al. 1967). The salt layer contains about 1 mass% In and less than 0.1 mass% As. The solubility of InAs in KI is practically absent.

14.14 INDIUM–RUBIDIUM–OXYGEN–ARSENIC

The **$RbInAs_2O_7$** quaternary compound, which crystallizes in the triclinic structure with lattice parameters $a = 784.5 \pm 0.2$, $b = 867.8 \pm 0.2$, and $c = 1049.2 \pm 0.2$ pm and $\alpha = 88.85 \pm$

0.03, $\beta = 89.93 \pm 0.03°$, and $\gamma = 74.31 \pm 0.03°$ and a calculated density of 4.47 g·cm^{-3}, is formed in the In–Rb–O–As system (Schwendtner 2006). The title compound was synthesized under hydrothermal conditions in a Teflon-lined stainless steel autoclave at 220°C under autogeneous pressure. Small, colorless, lenticular crystals were grown from a mixture of Rb_2CO_3, In_2O_3, H_3AsO_4, and distilled water. The autoclave was filled to about 60–80% of its inner volume and was then heated to 220°C, kept at this temperature for 7 days, and slowly cooled to room temperature overnight. Initial and final pH values were about 1 and 1, respectively. The reaction products were washed thoroughly with distilled water, filtered, and slowly dried at room temperature. The crystals of $RbInAs_2O_7$ were associated with small hexagonal plates of **$RbIn(HAsO_4)_2$**.

14.15 INDIUM–CESIUM–NIOBIUM–ARSENIC

The **$Cs_7In_3NbAs_5$**, **$Cs_{13}In_6Nb_2As_{10}$**, and **$Cs_{24}In_{12}Nb_2As_{18}$** quaternary compounds are formed in the In–Cs–Nb–As system. $Cs_7In_3NbAs_5$ crystallizes in the triclinic structure with the lattice parameters $a = 881.01 \pm 0.09$, $b = 966.4 \pm 0.1$, and $c = 1474.8 \pm 0.2$ pm and $\alpha = 85.421 \pm 0.002°$, $\beta = 85.961 \pm 0.002°$, and $\gamma = 86.421 \pm 0.002°$ and a calculated density of 4.642 g·cm^{-3} (Gascoin and Sevov 2002a, 2002b). To prepare this compound, a stoichiometric mixture of the elements was loaded in a tubular Nb container sealed at one end by arc-welding and then the other end was also arc-welded. The container was enclosed in a fused-silica ampoule, which was flame-sealed under vacuum. The assembly was heated at 800°C and 500°C for 3 days at each temperature and was then slowly cooled to room temperature at a rate of 5°C·h^{-1}. The obtained crystals are black, irregular, and with smooth, coal-like surfaces. All manipulations were performed inside a nitrogen-filled glove box with moisture level below 1 ppm.

$Cs_{13}In_6Nb_2As_{10}$ also crystallizes in the triclinic structure with the lattice parameters $a = 955.64 \pm 0.05$, $b = 962.88 \pm 0.05$, and $c = 1390.71 \pm 0.07$ pm and $\alpha = 83.7911 \pm 0.0008°$, $\beta = 80.2973 \pm 0.0008°$, and $\gamma = 64.9796 \pm 0.0008°$ and a calculated density of 4.874 g·cm^{-3} (Gascoin and Sevov 2003a). This compound was obtained as follows. Mixtures of Cs, In, As, and Nb with the nominal composition "$Cs_{20}Nb_3In_9As_{15}$" were loaded in Nb containers that were then sealed by arc-welding under Ar. The containers were enclosed in fused-silica ampoules, which were then flame-sealed under vacuum. These assemblies were heated for 2 days at 700°C and were slowly cooled to room temperature at a rate of 5°C·h^{-1}. Small amounts of $Cs_{24}In_{12}Nb_2As_{18}$ were found as impurities. $Cs_{13}In_6Nb_2As_{10}$ is dark-gray to black in color and very brittle. All manipulations were performed inside an Ar-filled glove box with moisture level below 1 ppm.

$Cs_{24}In_{12}Nb_2As_{18}$ also crystallizes in the triclinic structure with the lattice parameters $a = 951.9 \pm 0.4$, $b = 954.0 \pm 0.5$, and $c = 2516 \pm 1$ pm and $\alpha = 86.87 \pm 0.04°$, $\beta = 87.20 \pm 0.04°$, and $\gamma = 63.81 \pm 0.04°$ and a calculated density of 4.952 g·cm^{-3} (Gascoin and Sevov 2003a). To obtain this compound, stoichiometric mixtures of Cs, In, As, and Nb were loaded in Nb containers that were then sealed by arc-welding under Ar. The containers were enclosed in fused-silica ampoules, which were then flame-sealed under vacuum. These assemblies were heated for 2 days at 650°C and slowly cooled to room temperature at a rate of 5°C·h^{-1}. The title compound is dark-gray to black in color and very brittle. All manipulations were performed inside an Ar-filled glove box with moisture level below 1 ppm.

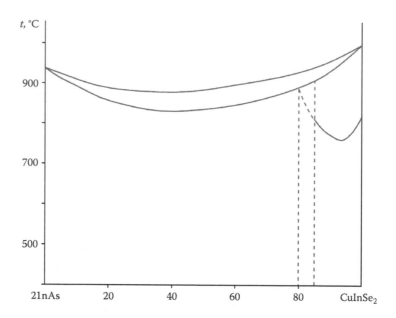

FIGURE 14.1 Phase relations in the $2InAs–CuInSe_2$ system. (Reprinted from *Sol. Energy Mater.*, 22 (1), Hasoon, F.S., and Al-Douri, A.A.J., Surface characterization and differential thermal analysis of the $(2InAs)_{1-x}(CuInSe_2)_x$ solid solution, 93–103, Copyright (191), with permission from Elsevier.)

14.16 INDIUM–COPPER–SELENIUM–ARSENIC

InAs–CuInSe$_2$. Phase relations in this system are shown in Figure 14.1 (Hasoon and Al-Douri 1991). A continuous solid solution exists in the $(2InAs)_{1-x}(CuInSe_2)_x$ system; samples crystallize with the zinc-blende-type structure from the InAs side up to $x = 0.8$ where no phase transition occurred (Al-Douri and Hasoon 1989; Hasoon and Al-Douri 1991). From $x = 0.85$ up to the ternary compound, they crystallize with the chalcopyrite-type structure, where a phase transition has been observed. The X-ray photoelectron spectroscopy and Auger electron spectroscopy data are assisted by electron probe micro-analysis to confirm that the samples are all uniform and that the stoichiometric composition observed in $(2InAs)_{1-x}(CuInSe_2)_x$ solid solution is within the experimental error limit. The X-ray powder photography analysis of specimens of composition $x = 0.85$ and 0.9 quenched from the melt in cold water shows that the phase transition *chalcopyrite* ⇔ *zinc blend* occurs. The dependence of lattice parameters on the composition shows that there is no obvious range, in which the transition *zinc blend* ⇔ *chalcopyrite* has taken place. In all samples in these solid solutions except for the compositions with $x = 0.5$ and $x = 0.6$ during the heating, a small kink occurred on the endothermal peak. These shoulders could refer to a solid–vapor reaction (arsenic evaporation), where a small amount of arsenic vapor condensates at the top of the differential thermal analysis (DTA) ampoules.

14.17 INDIUM–COPPER–CHLORINE–ARSENIC

InAs–CuCl. Indium arsenide chemically interacts with copper chloride according to the next scheme: $InAs + CuCl = InCl + Cu + As$ (Koppel et al. 1967).

14.18 INDIUM–SILVER–CHLORINE–ARSENIC

InAs–AgCl. Indium arsenide chemically interacts with silver chloride according to the next scheme: InAs + AgCl = InCl + Ag + As (Koppel et al. 1967).

14.19 INDIUM–SILVER–IODINE–ARSENIC

InAs–AgI. Indium arsenide chemically interacts with silver iodide according to the next scheme: InAs + AgI = InI + Ag + As (Koppel et al. 1967).

14.20 INDIUM–MAGNESIUM–CHLORINE–ARSENIC

InAs–MgCl₂. Indium arsenide chemically interacts with magnesium chloride (Koppel et al. 1967).

14.21 INDIUM–CALCIUM–CHLORINE–ARSENIC

InAs–CaCl₂. After standing at 1000°C a melt containing 16.5 mass% InAs and 83.5 mass% CaCl₂, a delamination was observed (Koppel et al. 1967). The salt layer contains 1–3.5 mass% In and ~0.1 mass% As. The solubility of InAs in CaCl₂ is practically absent.

14.22 INDIUM–STRONTIUM–CHLORINE–ARSENIC

InAs–SrCl₂. After standing at 1000°C a melt containing 16.5 mass% InAs and 83.5 mass% SrCl₂, a delamination was observed (Koppel et al. 1967). The salt layer contains 1–3.5 mass% In and ~0.1 mass% As. The solubility of InAs in SrCl₂ is practically absent.

14.23 INDIUM–BARIUM–CHLORINE–ARSENIC

InAs–BaCl₂. After standing at 1000°C a melt containing 16.5 mass% InAs and 83.5 mass% BaCl₂, a delamination was observed (Koppel et al. 1967). The salt layer contains 1–3.5 mass% In and ~0.1 mass% As. The solubility of InAs in BaCl₂ is practically absent.

14.24 INDIUM–ZINC–GERMANIUM–ARSENIC

InAs–ZnGeAs₂. The solid solutions based on InAs contained up to 80 mol% ZnGeAs₂ and crystallize in the cubic structure (Giesecke and Pfister 1961; Voitsehovskiy and Goryunova 1962). The solid solutions with higher concentration of ZnGeAs₂ exist in the chalcopyrite-type structure. The lattice parameters for both solid solutions obey Vegard's law.

14.25 INDIUM–ZINC–TIN–ARSENIC

InAs–ZnSnAs₂. The continuous series of the solid solutions are formed in this system (Giesecke and Pfister 1961; Voitsehovskiy and Goryunova 1962). The lattice parameter changes linearly with composition according to Vegard's law.

14.26 INDIUM–ZINC–SULFUR–ARSENIC

InAs–ZnS. The phase diagram is not constructed. The solubility of Zn + S in InAs at Zn/S = 3:1, 1:1, and 1:3 was investigated (Table 14.1) (Glazov et al. 1979a). Maximum solubility takes place at the equimolar ratio of doping elements. This system was investigated through

TABLE 14.1   Solubility of Zn + S (cm^{-3}) in InAs

t, °C	Zn/S = 3:1	Zn/S = 1:1	Zn/S = 1:3
700	2.0×10^{20}	6.1×10^{20}	3.2×10^{20}
800	2.3×10^{20}	7.9×10^{20}	4.3×10^{20}
850	3.5×10^{20}	15.8×10^{20}	9.4×10^{20}
900	5.8×10^{20}	22×10^{20}	11.5×10^{20}

Source: Glazov, V.M. et al., *Izv. AN SSSR. Neorgan. Mater.*, 15(3), 390–394, 1979. With permission.

metallography and measuring of microhardness. The ingots were annealed at 700°C, 800°C, 850°C, and 900°C for 1200, 1000, 800, and 500 h, respectively.

14.27 INDIUM–ZINC–SELENIUM–ARSENIC

InAs–ZnIn$_2$Se$_4$. The solid solutions based on InAs are formed in this system up to the composition (InAs)$_{0.4}$(ZnIn$_2$Se$_4$)$_{0.6}$ (Drobyazko and Kuznetsova 1975). The lattice parameter of these solid solutions changes near linearly. This system was investigated through X-ray diffraction (XRD), metallography, and measuring of the microhardness.

InAs–ZnSe. The phase diagram is not constructed. The solubility of Zn + Se in InAs at Zn/Se = 3:1, 1:1, and 1:3 (Table 14.2) (Glazov et al. 1979a) was investigated. Maximum solubility takes place at the equimolar ratio of doping elements. This system was investigated through metallography and measuring of microhardness. The ingots were annealed at 700°C, 800°C, 850°C, and 900°C for 1200, 1000, 800, and 500 h, respectively.

14.28 INDIUM–ZINC–TELLURIUM–ARSENIC

InAs–ZnIn$_2$Te$_4$. The continuous series of the solid solutions are formed in this system (Drobyazko and Kuznetsova 1975). The lattice parameter of these solid solutions changes near linearly. This system was investigated through XRD, metallography, and measuring of the microhardness.

InAs–ZnTe. The phase diagram is a eutectic type (Figure 14.2) (Shumilin et al. 1974). The eutectic composition and temperature are 94 mol% InAs and 920°C, respectively. The solubility of ZnTe in InAs is not higher than 1 mol% [2 mol% (Glazov et al. 1975b)], and the solubility of InAs in ZnTe reaches 30 mol% (Shumilin et al. 1974). The regions of the solid solutions in the Zn–Te–InAs quasiternary system based on InAs at the different temperatures are shown in Figure 14.3 (Glazov et al. 1975b).

TABLE 14.2   Solubility of Zn + Se (cm^{-3}) in InAs

t, °C	Zn/Se = 3:1	Zn/Se = 1:1	Zn/Se= 1:3
700	1.9×10^{20}	5.4×10^{20}	2.8×10^{20}
800	2.4×10^{20}	7.6×10^{20}	3.7×10^{20}
850	3.1×10^{20}	14.5×10^{20}	8.6×10^{20}
900	14.7×10^{20}	19×10^{20}	11.2×10^{20}

Source: Glazov, V.M. et al., *Izv. AN SSSR. Neorgan. Mater.*, 15(3), 390–394, 1979. With permission.

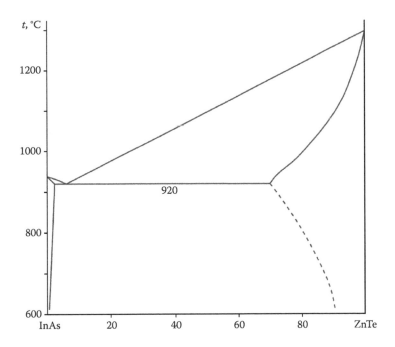

FIGURE 14.2   InAs–ZnTe phase diagram. (From Shumilin, V.P., *Izv. AN SSSR. Neorgan. Mater.*, 10 (8), 1414–1417, 1974. With permission.)

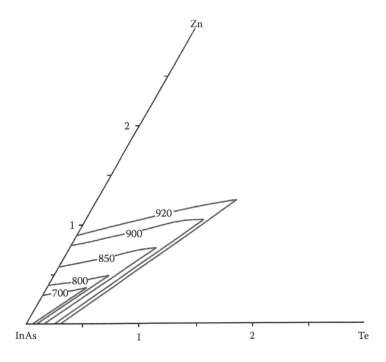

FIGURE 14.3   Solubility isotherms in the InAs–Zn–Te quasiternary system from the InAs-rich side. (From Glazov, V.M., *Izv. AN SSSR. Neorgan. Mater.*, 11(7), 1181–1183, 1975. With permission.)

This system was investigated through DTA, metallography, local XRD, and measuring of microhardness. The ingots were annealed at 750°C, 800°C, and 850°C for 350 h (Shumilin et al. 1974) [at 700°C, 800°C, 850°C, and 900°C for 1400, 1050, 700, and 350 h (Glazov et al. 1975b)].

14.29 INDIUM–ZINC–CHLORINE–ARSENIC

InAs–ZnCl$_2$. After standing a melt containing 16.5 mass% InAs and 83.5 mass% ZnCl$_2$, a delamination was observed (Koppel et al. 1967). Indium partially moves into the salt layer, and arsenic content in this layer is negligible. The next equilibrium $2InAs + ZnCl_2 \Leftrightarrow 2InCl + ZnAs_2$ takes place between InAs and ZnCl$_2$.

14.30 INDIUM–ZINC–IODINE–ARSENIC

InAs–ZnI$_2$. After standing a melt containing 16.5 mass% InAs and 83.5 mass% ZnI$_2$, a delamination was observed (Koppel et al. 1967). Indium partially moves into the salt layer, and arsenic content in this layer is negligible. The next equilibrium $2InAs + ZnI_2 \Leftrightarrow 2InI + ZnAs_2$ takes place between InAs and ZnI$_2$.

14.31 INDIUM–CADMIUM–TIN–ARSENIC

InAs–CdSnAs$_2$. The continuous series of the solid solutions are formed in this system (Goryunova and Prochukhan 1960; Baranov et al. 1965). This system was investigated through DTA, metallography, XRD, and microhardness measuring.

14.32 INDIUM–CADMIUM–SULFUR–ARSENIC

InAs–CdS. The phase diagram of this system belongs to the eutectic type (Figure 14.4) (Drobyazko and Kuznetsova 1983). The eutectic temperature and composition are 912°C and 20 mol% CdS, respectively. It is necessary to note that the melting temperature of InAs is 937.9°C (according to the date taken from this diagram, this temperature is equal to 953°C).

Using XRD, it was determined that the ingots containing up to 25 mol% CdS crystallize in the sphalerite structure, and at the more CdS contents, a two-phase region (sphalerite + wurtzite) exists in this system (Voitsehivskiy and Drobyazko 1967; Voitsehovskiy et al. 1968). According to the data of metallography, solubility of CdS in InAs reaches 18 mol% (Voitsehivskiy and Drobyazko 1967; Voitsehovskiy et al. 1968; Drobyazko and Kuznetsova 1983). Lattice parameters of forming solid solutions change nearly linearly with composition.

This system was investigated through DTA, XRD, metallography, and microhardness measuring (Voitsehivskiy and Drobyazko 1967; Voitsehovskiy et al. 1968; Drobyazko and Kuznetsova 1983). The ingots were annealing at 830°C, 780°C, and 730°C during 120, 240, and 480 h, respectively (Drobyazko and Kuznetsova 1983).

14.33 INDIUM–CADMIUM–SELENIUM–ARSENIC

InAs–CdIn$_2$Se$_4$. The solid solutions based on InAs are formed in this system up to the composition $(InAs)_{0.3}(CdIn_2Se_4)_{0.7}$ (Drobyazko and Kuznetsova 1975). The lattice parameter of these solid solutions changes near linearly. This system was investigated through XRD, metallography, and measuring of the microhardness.

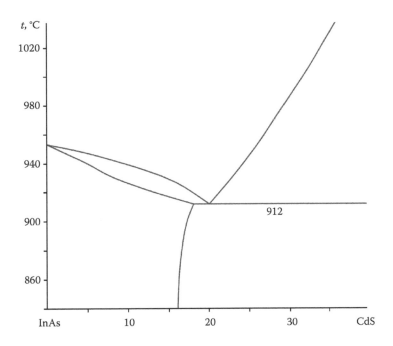

FIGURE 14.4   Part of the InAs–CdS phase diagram. (From Drobyazko, V.P., and Kuznetsova, S.T., *Zhurn. Neorgan. Khimii*, 28(11), 2929–2933, 1983. With permission.)

InAs–CdSe. The phase diagram of this system belongs to the eutectic type (Figure 14.5) (Drobyazko and Kuznetsova 1983). The eutectic temperature and composition are 897°C and 26 mol% CdSe, respectively. Liquidus and solidus temperatures within the concentration range up to 20 mol% CdSe were also determined by Voitsehovskiy et al. (1968). The solubility of CdSe in InAs reaches 25 mol% [30 mol% (Voitsehovskiy et al. 1968)], and the solubility of InAs in CdSe is equal to 15 mol% (Drobyazko and Kuznetsova 1983).

This system was investigated through DTA, XRD, metallography, and measuring of microhardness (Voitsehovskiy et al. 1968; Drobyazko and Kuznetsova 1983). The ingots were annealed at 830°C, 780°C, and 730°C for 120, 240, and 480 h, respectively (Drobyazko and Kuznetsova 1983).

14.34 INDIUM–CADMIUM–TELLURIUM–ARSENIC

InAs–CdIn$_2$Te$_4$. The continuous series of the solid solutions are formed in this system (Figure 14.6) (Drobyazko and Kuznetsova 1975). It should be noted that the phase diagram was slightly changed since the solidus temperatures on the CdInTe$_4$ side should be slightly higher than the melting temperature of this compound. The lattice parameter of these solid solutions changes near linearly. This system was investigated through XRD, metallography, and measuring of the microhardness.

InAs–CdTe. The phase diagram is a eutectic type (Figure 14.7) (Morozov and Chernov 1979). The eutectic composition and temperature are 33 ± 2 mol% CdTe and 874 ± 3°C (Buzevich 1972; Bazhenova et al. 1974; Balagurova et al. 1975; Morozov and Chernov 1979) [882°C (Drobyazko and Kuznetsova 1983); 870°C (Stuckes and Chasmar 1964)], respectively.

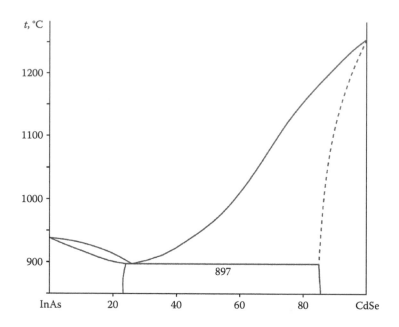

FIGURE 14.5 InAs–CdSe phase diagram. (From Drobyazko, V.P., and Kuznetsova, S.T., *Zhurn. Neorgan. Khimii*, 28(11), 2929–2933, 1983. With permission.)

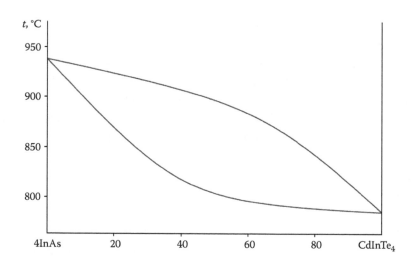

FIGURE 14.6 4InAs–CdInTe$_4$ phase diagram. (From Drobyazko, V.P., and Kuznetsova, S.T., *Zhurn. Neorgan. Khimii*, 20(11), 3061–3064, 1975. With permission.)

The solubility of CdTe in InAs at the eutectic temperature reaches 30 mol% (Voitsehovskiy and Goryunova 1962; Kleshchinski et al. 1964; Bazhenova et al. 1974; Balagurova et al. 1975, 1976) [26 mol% (Drobyazko and Kuznetsova 1983)], and the solubility of InAs in CdTe at the same temperature is equal to 5 mol% (Balagurova et al. 1975, 1976) [35 mol% (Drobyazko and Kuznetsova 1983)].

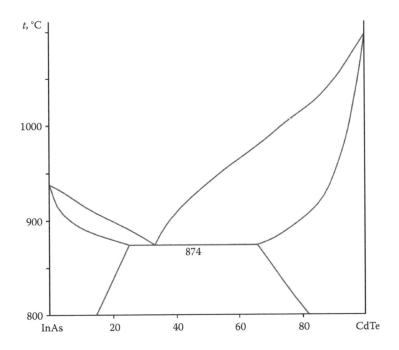

FIGURE 14.7   InAs–CdTe phase diagram. (From Morozov, V.N., and Chernov, V.G., *Izv. AN SSSR. Neorgan. Mater.*, 15(8), 1324–1329, 1979. With permission.)

Temperature dependence of CdTe solubility in InAs within the interval of 780–874°C is satisfactorily described by the following equation (Balagurova et al. 1975): $\ln x_{CdTe} = -13{,}080/T + 9.02$. The solubility of CdTe in InAs saturated by In is 1.47, 2.75, 4.30, 7.90, 13.50, and 16.00 mol% at 400°C, 450°C, 500°C, 550°C, 600°C, and 650°C, respectively (Buzevich 1972). Within the interval of 0–1 mol%, CdTe, InTe, and CdAs complexes are included in the crystal matrix, which are formed as the result of the following reaction: CdTe + InAs ⇔ CdAs + InTe. Thus, this system is nonquasibinary in the region of small CdTe contents and turns into an InTe–InAs–CdAs system.

The coefficient of CdTe distribution in the CdTe–InAs solid solutions obtained by the Bridgman method is equal to 0.25 (Kalashnikova et al. 1975).

This system was investigated by DTA, XRD, emission spectrum analysis, and measuring of microhardness (Buzevich 1972; Bazhenova et al. 1974; Balagurova et al. 1975, 1976; Morozov and Chernov 1979; Drobyazko and Kuznetsova 1983). The ingots were annealing at 830°C, 780°C, and 730°C during 120, 240, and 480 h, respectively (Drobyazko and Kuznetsova 1983).

InAs–CdTe–In. The liquidus surface of this quasiternary system (Figure 14.8) was constructed using DTA and mathematical simulation of experiment (Riazantsev and Telegina 1978).

14.35 INDIUM–CADMIUM–CHLORINE–ARSENIC

InAs–CdCl₂. After standing a melt containing 16.5 mass% InAs and 83.5 mass% CdCl₂, a delamination was observed (Koppel et al. 1967). Indium partially moves into the salt layer,

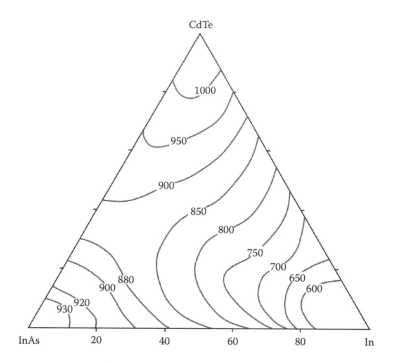

FIGURE 14.8   Liquidus surface of the InAs–CdTe–In quasiternary system. (From Riazantsev, A.A., and Telegina, M.P., *Zhurn. Neorgan. Khimii*, 23(8), 2211–2216, 1978. With permission.)

and arsenic content in this layer is negligible. The next equilibrium $2InAs + CdCl_2 \Leftrightarrow 2InCl + CdAs_2$ takes place between InAs and $CdCl_2$.

14.36 INDIUM–CADMIUM–BROMINE–ARSENIC

InAs–CdBr$_2$. After standing a melt containing 16.5 mass% InAs and 83.5 mass% $CdBr_2$, a delamination was observed (Koppel et al. 1967). Indium partially moves into the salt layer, and arsenic content in this layer is negligible. The next equilibrium $2InAs + CdBr_2 \Leftrightarrow 2InBr + CdAs_2$ takes place between InAs and $CdCl_2$.

14.37 INDIUM–CADMIUM–IODINE–ARSENIC

InAs–CdI$_2$. The polythermal section of this system was constructed by Luzhnaya et al. (1965) and Lushnaja et al. (1966). The next equilibrium $2InAs + CdI_2 \Leftrightarrow 2InI + CdAs_2$ which takes place between indium arsenide and cadmium iodide was established. This system was investigated through DTA, metallography, XRD, and chemical analysis.

14.38 INDIUM–MERCURY–TELLURIUM–ARSENIC

InAs–HgTe. The phase diagram is shown in Figure 14.9 (Kuz'mina 1976). Solid solutions over the entire range of concentrations are formed in this system (Goryunova et al. 1962; Inyutkin et al. 1964; Kuz'mina 1976). Small quantities of inclusions were found from the InAs-rich side (Goryunova et al. 1962). Lattice parameters of forming solid solutions change linearly with concentration. The region of possible solid solution degradation was found near the liquidus. The two-phase region was determined at 500°C by the XRD of ingots annealed for 250 h.

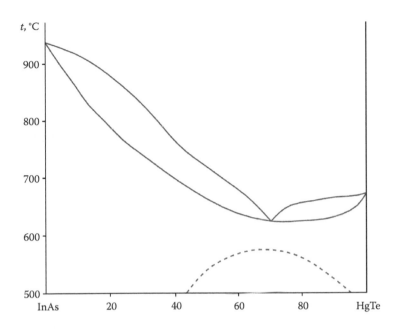

FIGURE 14.9 InAs–HgTe phase diagram. (From Kuz'mina, G.A., *Izv. AN SSSR. Neorgan. Mater.*, 12(6), 1121–1122, 1976. With permission.)

This system was investigated by DTA, metallography, and measuring of electro-conductivity (Goryunova et al. 1962; Inyutkin et al. 1964; Kuz'mina 1976). The ingots were annealed at 610°C for 250 h and then at 650–850°C (Kuz'mina 1976) [at 570–600°C for 550–600 h (Goryunova et al. 1962)].

14.39 INDIUM–MERCURY–BROMINE–ARSENIC

The **[Hg$_4$As$_2$](InBr$_{3.5}$As$_{0.5}$)** quaternary compound, which crystallizes in the hexagonal structure with the lattice parameters $a = 774.08 \pm 0.06$ and $c = 1253.50 \pm 0.19$ pm, a calculated density of 7.067 g·cm^{-3}, and an energy gap 1.71 eV, is formed in the In–Hg–Br–As system (Zou et al. 2011). Single crystals of the title compound were prepared by the solid-state reaction of a mixture of Hg$_2$Br$_2$ (1 mM), In (0.3 mM), and As (0.5 mM) at 320°C. The starting materials were ground into fine powders and pressed into a pellet, followed by being loaded into an evacuated Pyrex tube and flame-sealed. The tube was placed into a computer-controlled furnace, heated from room temperature to 250°C at a rate of 50°C·h^{-1}, and kept at this temperature for 4 h. The tube was then heated to 400°C at 25°C·h^{-1}, kept at 400°C for 120 h, and then slowly cooled down to 100°C at a rate of 2°C·h^{-1}. Finally, it was cooled down to room temperature over 10 h. A crop of dark-red crystals of [Hg$_4$As$_2$](InBr$_{3.5}$As$_{0.5}$) that are stable in air were obtained.

14.40 INDIUM–THALLIUM–OXYGEN–ARSENIC

The **TlInAs$_2$O$_7$** quaternary compound, which crystallizes in the triclinic structure with the lattice parameters $a = 782.7 \pm 0.2$, $b = 862.5 \pm 0.2$, and $c = 1049.4 \pm 0.2$ pm and $\alpha = 88.83 \pm 0.03$, $\beta = 89.98 \pm 0.03°$, and $\gamma = 74.38 \pm 0.03°$ and a calculated density of 5.66 g·cm^{-3}, is formed in the In–Tl–O–As system (Schwendtner 2006). The title compound was synthesized under hydrothermal conditions in Teflon-lined stainless steel autoclave at 220°C under autogeneous pressure. Small, colorless, pseudo-orthorhombic, non-merohedrically twinned crystals were prepared from a mixture of Tl_2CO_3, In_2O_3, H_3AsO_4, and distilled water. The autoclave was filled to about 60–80% of its inner volume and was then heated to 220°C, kept at this temperature for 7 days, and slowly cooled to room temperature overnight. Initial and final pH values were about 1.5 and 1, respectively. The reaction products were washed thoroughly with distilled water, filtered, and slowly dried at room temperature. This compound was accompanied by small X-ray amorphous, colorless to whitish plates of tetragonal outline and yellow, tabular, translucent crystals with an indistinct outline of a presently unidentified Tl-compound.

14.41 INDIUM–THALLIUM–SELENIUM–ARSENIC

The joint solubility of Tl and Se in InAs at the temperature region from 500°C to 800°C was determined by Mavlonov and Makhmatkulov (1983), and it was shown that this solubility is retrograde.

14.42 INDIUM–TIN–CHLORINE–ARSENIC

InAs–SnCl₂. After standing a melt containing 16.5 mass% InAs and 83.5 mass% $SnCl_2$, a delamination was observed (Koppel et al. 1967). A significant quantity of indium moves into the salt layer. Indium arsenide chemically interacts with tin(II) chloride according to the next scheme: $2InAs + SnCl_2 = 2InCl + Sn + 2As$.

14.43 INDIUM–LEAD–ANTIMONY–ARSENIC

InAs–InSb–Pb. The liquidus surface of the InAs–InSb–Pb quasiternary system near the Pb corner is given in Figure 14.10 (Grebenyuk et al. 1990). The fields of primary crystallization of Pb and $InAs_xSb_{1-x}$ solid solution exist in this system. The first is situated near the Pb corner, and the second occupies the biggest part of this system.

14.44 INDIUM–LEAD–CHLORINE–ARSENIC

InAs–PbCl₂. After standing a melt containing 16.5 mass% InAs and 83.5 mass% $PbCl_2$, a delamination was observed (Koppel et al. 1967). A significant quantity of indium moves into the salt layer. Indium arsenide chemically interacts with lead chloride according to the next scheme: $2InAs + PbCl_2 = 2InCl + Pb + 2As$.

14.45 INDIUM–LEAD–IODINE–ARSENIC

InAs–PbI₂. After standing a melt containing 16.5 mass% InAs and 83.5 mass% PbI_2, a delamination was observed (Koppel et al. 1967). A significant quantity of indium moves into the salt layer. Indium arsenide chemically interacts with lead iodide according to the next scheme: $2InAs + PbI_2 = 2InI + Pb + 2As$.

14.46 INDIUM–ANTIMONY–BISMUTH–ARSENIC

High-quality crystals of **$InBi_xSb_yAs_{1-x-y}$** solid solutions with $x \leq 0.017$ and $y \leq 0.096$ have been grown on InAs (110)-oriented substrates by atmospheric pressure metal–organic vapor-phase epitaxy using the precursors Me_3In, Me_3Bi, Me_3Sb, and AsH_3 at 400°C (Ma et al. 1989).

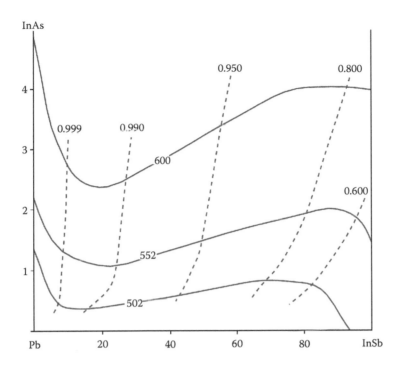

FIGURE 14.10 Liquidus surface of the InAs–InSb–Pb quasiternary system near the Pb corner: dashed lines correspond to the content of InAs in the InAs$_x$Sb$_{1-x}$ solid solution in molar fraction. (From Grebenyuk, A.M. et al., *Zhurn. Neorgan. Khimii*, 35(11), 2941–2944, 1990. With permission.)

14.47 INDIUM–SULFUR–SELENIUM–ARSENIC

The **InAs$_2$S$_3$Se**, **In$_3$As$_2$S$_3$Se$_3$**, and **In$_6$As$_4$S$_3$Se$_9$** quaternary compounds are formed in the In–S–Se–As system. InAs$_2$S$_3$Se and In$_3$As$_2$S$_3$Se$_3$ melt congruently at 370°C and 745°C, respectively (Rustamov et al. 1980). InAs$_2$S$_3$Se crystallizes in the monoclinic structure with the lattice parameters $a = 1231$, $b = 1143$, $c = 662$ pm, and $\beta = 107.6°$ and the calculated and experimental densities of 4.93 and 4.66 g·cm^{-3}.

In$_6$As$_4$S$_3$Se$_9$ melts congruently at 820°C and has an experimental density of 5.15 g·cm^{-3} (Aliev et al. 2009).

14.48 INDIUM–SULFUR–TELLURIUM–ARSENIC

The **InAs$_2$S$_3$Te**, **InAs$_6$S$_7$Te$_3$**, and **In$_3$As$_2$S$_3$Te$_3$** quaternary compounds, which melt incongruently at 325°C, 377°C, and 425°C, respectively, are formed in the In–S–Te–As system (Aliev et al. 1987; Safarov et al. 1992).

CHAPTER 15

Systems Based on InSb

15.1 INDIUM–SODIUM–POTASSIUM–ANTIMONY

The **NaK$_2$InSb$_2$** quaternary compound, which crystallizes in the orthorhombic structure with the lattice parameters $a = 1403.2 \pm 0.2$, $b = 1639.9 \pm 0.3$, and $c = 700.9 \pm 0.1$ pm and a calculated density of 3.784 g·cm^{-3}, is formed in the In–Na–K–Sb system (Carrillo-Cabrera et al. 1993). This compound was synthesized in sealed Nb ampoules from the stoichiometric mixture of elements, which were heated at 630°C for 2 days and then slowly cooled to room temperature (6°C·h^{-1}). NaK$_2$InSb$_2$ forms platelike crystals with silver metallic luster, which are very unstable in air and moisture. The title compound begins to decompose at about 670°C in high vacuum.

15.2 INDIUM–SODIUM–CALCIUM–ANTIMONY

The **Na$_{5.06\pm0.02}$Ca$_{0.94}$In$_{1.58\pm0.02}$Sb$_2$** or **Na$_5$CaIn$_{1.66}$Sb$_2$** quaternary compound, which crystallizes in the orthorhombic structure with the lattice parameters $a = 476.09 \pm 0.05$, $b = 916.80 \pm 0.05$, and $c = 788.98 \pm 0.08$ pm and a calculated density of 4.02 g·cm^{-3} at 200 K, is formed in the In–Na–Ca–Sb system (Wang et al. 2015). Single crystals of this compound were first identified from exploratory reactions of Na, Ca, In, and Sb in the molar ratio of 3:3:1:4. The metals were weighed in the glove box (total weight circa 500 mg) and loaded into Nb tube, which was then closed by arc welding under high-purity Ar gas. The sealed Nb container was subsequently jacketed within evacuated fused silica tube. The reaction mixture was heated to 900°C at a rate of 100°C·h^{-1} and equilibrated at this temperature for 24 h. Following a cooling step to 500°C at a rate of 5°C·h^{-1}, the product was brought to room temperature over 12 h. After the structure and chemical compositions were established by single-crystal work, new reaction with the proper stoichiometry was loaded. It yielded this compound as a major product. The produced sample contained small amounts of Na$_3$InSb$_2$, but the bulk consisted of small, shiny crystals with a dark metallic luster that were the targeted phase. The obtained compound is brittle and air sensitive. All manipulations were carried out inside an argon-field glove box with controlled atmosphere or under vacuum.

15.3 INDIUM–SODIUM–STRONTIUM–ANTIMONY

The $Na_{3.45\pm0.03}Sr_{2.55}InSb_4$ quaternary compound, which crystallizes in the hexagonal structure with the lattice parameters $a = 1026.10 \pm 0.14$ and $c = 799.5 \pm 0.2$ pm and a calculated density of 4.12 g·cm^{-3} at 200 K, is formed in the In–Na–Sr–Sb system (Wang et al. 2017a). The title compound was obtained from the elements. The starting materials were weighed in the glove box and loaded into Nb tube, which was then closed by arc-welding under high-purity argon gas. The sealed Nb container was subsequently jacketed within evacuated fused silica tube. The reaction mixture was heated in a muffle furnace to 900°C at a rate of 100°C·h^{-1} and equilibrated at this temperature for 24 h. After that, the sample were cooled to 500°C at a rate of 5°C·h^{-1}, at which point the cooling rate to room temperature was changed to 40°C·h^{-1}. All manipulations were carried out inside an argon-field glove box with controlled atmosphere or under vacuum.

15.4 INDIUM–POTASSIUM–GERMANIUM–ANTIMONY

The $K_9In_9GeSb_{22}$ quaternary compound, which crystallizes in the triclinic structure with the lattice parameters $a = 1162 \pm 1$, $b = 1304 \pm 3$, and $c = 1127 \pm 2$ pm; $\alpha = 113.5 \pm 0.2°$, $\beta = 91.7 \pm 0.1°$, and $\gamma = 112.3 \pm 0.1°$; and a calculated density of 4.851 g·cm^{-3}, is formed in the In–Na–Ge–Sb system (Shreeve-Keyer et al. 1996). The single crystals of the title compound were prepared by heating a mixture of K_3Sb_7, Sb, In, and Ge (5:7:12:18 molar ratio) in an evacuated sealed quartz tubes. The samples were heated to 650°C over 12 h, where the temperature was held for an additional 2 h. The samples were cooled at a linear rate to 500°C over 48 h and then cooled to ambient temperature over 12 h. The obtained compound is extremely air and moisture sensitive; therefore, all manipulations were performed using standard inert atmosphere techniques.

15.5 INDIUM–SILVER–SELENIUM–ANTIMONY

The $Ag_{1.75}InSb_{5.75}Se_{11}$ quaternary compound, which congruently melts at around 581°C and crystallizes in the monoclinic structure with the lattice parameters $a = 1341.9 \pm 0.1$, $b = 408.4 \pm 0.1$, $c = 1916.5 \pm 0.2$ pm, and $\beta = 105.83°$ at 153 ± 2 K; a calculated density of 6.152 g·cm^{-3}; and energy gap of 0.94 ± 0.02 eV, is formed in the In–Ag–Se–Sb system (Hao et al. 2014). Crystals of this compound were initially obtained from a reaction between binary starting materials Ag_2Se, In_2Se_3, and Sb_2Se_3. Reaction mixtures of Ag_2Se (1.75 mM), In_2Se_3 (1 mM), and Sb_2Se_3 (5.75 mM) were ground and loaded into fused silica tubes under an Ar atmosphere in a glove box. The tubes were flame sealed under a high vacuum and then placed in a computer-controlled furnace. The samples were heated to 950°C within 15 h, kept for 48 h, followed by slow cooling to 320°C at the rate of 4°C·h^{-1}, and finally cooled to room temperature by switching off the furnace. Many dark gray block-shaped crystals of $Ag_{1.75}InSb_{5.75}Se_{11}$ were obtained.

The polycrystalline samples of this compound were synthesized by the stoichiometric reaction of the binary materials in the molar ratio of $Ag_2Se/In_2Se_3/Sb_2Se_3 = 1.75:1:5.75$ at high temperature in sealed silica tubes evacuated to 10^{-3} Pa (Hao et al. 2014). The reaction mixtures were placed in a computer-controlled furnace and then heated to 550°C within 15 h and kept at this temperature for 48 h, and the furnace was then turned off.

15.6 INDIUM–STRONTIUM–OXYGEN–ANTIMONY

The **Sr$_2$InSbO$_6$** quaternary compound, which crystallizes in the cubic structure with the lattice parameter a − 809.8 pm (Wittmann et al. 1981), is formed in the In–Sr–O–Sb system. To obtain this compound, the starting materials SrCO$_3$ or Sr(NO$_3$)$_2$, In$_2$O$_3$, and Sb$_2$O$_3$ were finely ground and heated up to 1000–1100°C at the rate of 100°C·h^{-1} with the next annealing at these temperatures from 2 to 6 h in corundum crucibles in air. Then, the mixture was annealed at 1200°C for 12 h and at 1300–1350°C two times for 1 day.

15.7 INDIUM–BARIUM–OXYGEN–ANTIMONY

The **Ba$_2$InSbO$_6$** quaternary compound, which crystallizes in the cubic structure with the lattice parameter $a = 413.6$ pm (Wittmann et al. 1981) [$a = 826.9$ pm (Sleight and Ward 1964); $a = 413$ pm (Blasse 1965)] is formed in the In–Ba–O–Sb system. To prepare this compound, the starting materials BaCO$_3$ or Ba(NO$_3$)$_2$, In$_2$O$_3$, and Sb$_2$O$_3$ were finely ground and heated up to 1000–1100°C at the rate of 100°C·h^{-1} with the next annealing at these temperatures from 2 to 6 h in corundum crucibles in air (Wittmann et al. 1981). Then, the mixture was annealed at 1200°C for 12 h and at 1300–1350°C from two to four times for one day. The title compound could also be prepared at the sintering the mixture of BaO, In$_2$O$_3$, and Sb$_2$O$_3$ at 1100°C (Sleight and Ward 1964).

15.8 INDIUM–BARIUM–SELENIUM–ANTIMONY

The **Ba$_2$InSbSe$_5$** quaternary compound, which crystallizes in the orthorhombic structure with the lattice parameters $a = 430.5 \pm 0.1$, $b = 1896.8 \pm 0.4$, and $c = 1303.4 \pm 0.3$ pm; a calculated density of 5.65 g·cm^{-3} at 153 ± 2 K; and energy gap of 1.92 ± 0.5 eV is formed in the In–Ba–Se–Sb system (Hao et al. 2013). To prepare this compound, the reaction mixture of BaSe (2 mM), Sb$_2$Se$_3$ (0.5 mM), and In$_2$Se$_3$ (0.5 mM) were ground and loaded into fused silica tube under an Ar atmosphere in a glove box. The tube was flame sealed under a high vacuum (10^{-3} Pa) and then placed in a computer-controlled furnace. The samples were heated to 1000°C within 15 h, kept for 48 h, then slowly cooled to 320°C at the rate of 4°C·h^{-1}, and finally cooled to room temperature by switching off the furnace.

After the structural identification, the polycrystalline samples of Ba$_2$InSbSe$_5$ were synthesized by the stoichiometric reaction of the binary materials in the molar ratio of BaSe/Sb$_2$Se$_3$/In$_2$Se$_3$ = 4:1:1 at high temperature in sealed silica tubes evacuated to 10^{-3} Pa. The reaction mixtures were placed in a computer-controlled furnace and then heated to 750°C in 15 h and kept at that temperature for 48 h, and then the furnace was turned off.

15.9 INDIUM–BARIUM–TELLURIUM–ANTIMONY

The **Ba$_2$InSbTe$_5$** quaternary compound, which crystallizes in the orthorhombic structure with the lattice parameters $a = 1378.2 \pm 0.3$, $b = 457.3 \pm 0.1$, and $c = 2018.4 \pm 0.4$ pm and a calculated density of 6.00 g·cm^{-3} at 153 ± 2 K, is formed in the In–Ba–Te–Sb system

(Hao et al. 2013). The title compound was obtained by the same way as $In_2GaSbSe_5$ was synthesized using BaTe, Sb_2Te_3, and In_2Te_3 instead of BaSe, Sb_2Se_3, and In_2Se_3, respectively.

15.10 INDIUM–BARIUM–COBALT–ANTIMONY

Double-filled skutterudite $In_xBa_yCo_4Sb_{12}$ ($x + y$ =0.5) have been synthesized by the high-pressure high-temperature method (Deng et al. 2017). The In, Ba, Sb, and Co powders were mixed and milled in Ar as the protective gas, and then they were compressed into a thick cylindrical sample by hydraulic press. After that, the samples were prepared by a cubic anvil high-pressure apparatus at about 630°C, and the synthesized pressure was 1 GPa. $In_{0.4}Ba_{0.1}Co_4Sb_{12}$ sample showed a very low thermal conductivity, and the maximum ZT value of 0.97 has been obtained for an In and Ba double-filled sample at 450°C.

15.11 INDIUM–ZINC–TIN–ANTIMONY

InSb–ZnSnSb$_2$. This system is not quasibinary in character as the variation of the lattice parameter with composition was observed for a range of composition where a second phase was clearly present (Woolley and Williams 1964). According to the data of X-ray diffraction (XRD), the solubility of $ZnSnSb_2$ in 2InSb is equal to 3 mol%, and the results of differential thermal analysis (DTA) indicated that 10 mol% $ZnSnSb_2$ could be dissolved in 2InSb. Voitsehovskiy and Goryunova (1962) noted that $ZnSnSb_2$ dissolves in 2InSb up to 80 mol%. Solid solutions based on InSb were obtained up to 30 mol% $ZnSnSb_2$ by zone recrystallization (Khalilov and Aliev 1968).

15.12 INDIUM–ZINC–BISMUTH–ANTIMONY

The results of the determination of the homogeneity region of Bi and Zn solid solution in InSb at 400°C in the corresponding ternary systems of the In–Zn–Bi–Sb quaternary system show (Figure 15.1) that the solubility of both Bi and Zn when they are simultaneously introduced into InSb increases (Ufimtsev et al. 1987). In this case, the solubility of Bi and Zn decreases when passing from the sections containing Bi as one component (another is Zn), to a section containing InBi, and reaches a minimum value for the section containing In_2Bi. The homogeneity regions of the Bi and Zn solid solution in InSb in the InSb–Zn_3Sb_2–InBi and InSb–Zn_3Sb_2–In_2Bi are almost identical.

The ingots were annealed at 400°C for 300 h, and the system was investigated through XRD, metallography, and measuring of microhardness.

InSb–Zn$_3$Sb$_2$–InBi. The liquidus surface of this quasiternary system is presented in Figure 15.2 (Lapkina et al. 1986). The ternary eutectic crystallizes at 110°C and is situated near InBi. Quaternary compounds were not found in this system. The field of Zn_3Sb_2 primary crystallization occupies the biggest part of the concentration triangle and the field of InBi primary crystallization is degenerated. Homogeneity regions based on InSb at 380°C and 500°C are shown in Figure 15.3. It is seen that these regions are elongated to the Zn_3Sb_2 side.

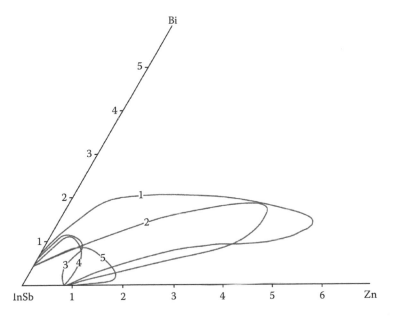

FIGURE 15.1 Solubility isotherms (400°C) of Bi and Zn in InSb in the next ternary systems of the In–Zn–Bi–Sb quaternary system: (1) InSb–Bi–Zn; (2) InSb–InBi–Zn; (3) InSb–In$_2$Bi–Zn; (4) InSb–InBi–Zn$_3$Sb$_2$; (5) InSb–In$_2$Bi–Zn$_3$Sb$_2$. (From Ufimtsev, V.B. et al., *Izv. AN SSSR. Neorgan. Mater.*, 23 (11), 1784–1787, 1987. With permission.)

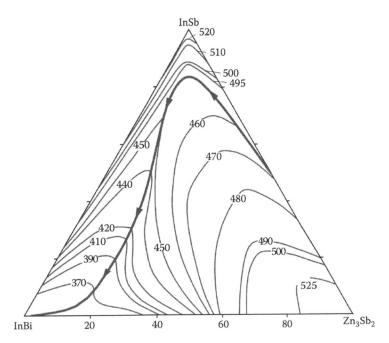

FIGURE 15.2 Liquidus surface of the InSb–InBi–Zn$_3$Sb$_2$ system. (From Lapkina, I.A. et al., *Zhurn. Neorgan. Khimii*, 31(1), 206–209, 1986. With permission.)

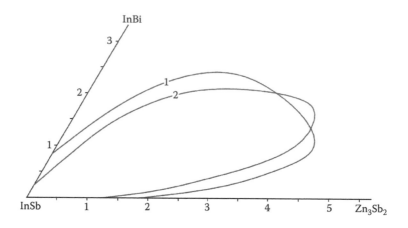

FIGURE 15.3   Homogeneity region of the solid solution based on InSb at (1) 380°C and (2) 500°C in the InSb–Zn₃Sb₂–InBi quasiternary system. (From Lapkina, I.A. et al., *Zhurn. Neorgan. Khimii*, 31 (1), 206–209, 1986. With permission.)

InSb–Bi–Zn. The part of the liquidus surface of this system in the field of InSb crystallization was constructed through DTA, XRD, metallography, and measuring of microhardness by Lapkina et al. (1983) and is presented in Figure 15.4.

The homogeneity regions of the solid solution based on InSb in the InSb–Zn–Bi system at 400°C, 480°C, and 500°C are given in Figure 15.5 (Yevgen'ev et al. 1985).

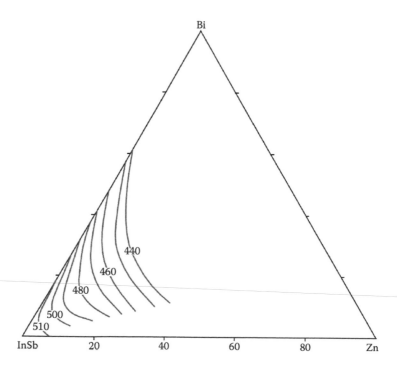

FIGURE 15.4   Part of the liquidus surface of the InSb–Zn–Bi system. (From Lapkina, I.A. et al., *Zhurn. Neorgan. Khimii*, 28(10), 2622–2626, 1983. With permission.)

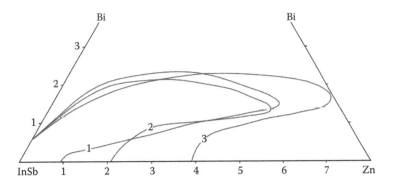

FIGURE 15.5 Homogeneity region of the solid solution based on InSb in the InSb–Zn–Bi quasiternary system at (1) 500°C, (2) 400°C, and (3) 400°C. (From Yevgen'ev, S.B. et al., *Izv. AN SSSR. Neorgan. Mater.*, 21(3), 490–492, 1985. With permission.)

The temperature dependence of the Bi solubility in InSb, doped with 5 at% Zn, Zn solubility in InSb, doped with 1.5 at% Bi, as well as the joint solubility of Bi and Zn at a molar ratio of 1:1 is characteristic of retrograde solubility (Figure 15.6) (Yevgen'ev et al. 1985). Maximum solubility of Bi and Zn in InSb takes place at 480°C and is equal to 2.1 at% Bi and 6.2 at% Zn.

InBi–Zn₃Sb₂. The phase diagram of this system is a eutectic type (Figure 15.7) (Lapkina et al. 1986). The eutectic crystallizes at 110°C and is degenerated from the InBi side.

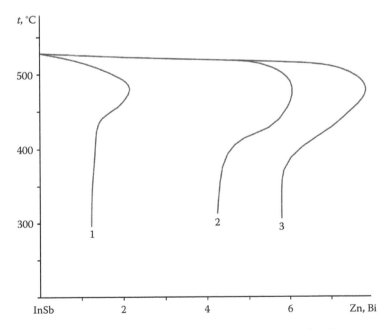

FIGURE 15.6 Temperature dependence of Bi solubility in InSb, doped with 5 at% Zn (1), Zn solubility in InSb (2), doped with 1.5 at% Bi, and the joint solubility of Bi and Zn (3) at a molar ratio of 1:1. (From Yevgen'ev, S.B. et al., *Izv. AN SSSR. Neorgan. Mater.*, 21(3), 490–492, 1985. With permission.)

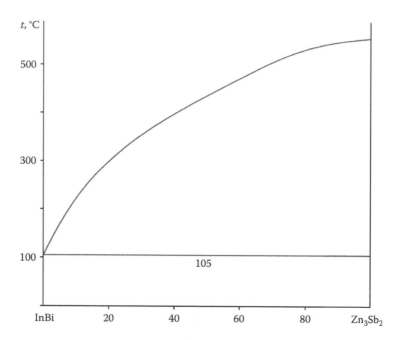

FIGURE 15.7   InBi–Zn₃Sb₂ phase diagram. (From Lapkina, I.A. et al., *Zhurn. Neorgan. Khimii*, 31 (1), 206–209, 1986. With permission.)

15.13 INDIUM–ZINC–SELENIUM–ANTIMONY

InSb–ZnSe. The phase diagram is not constructed. According to the data of XRD and band-gap measuring, solid solutions based on InSb contain up to 30 mol% ZnSe (Kirovskaya et al. 2002).

15.14 INDIUM–ZINC–TELLURIUM–ANTIMONY

InSb–ZnTe. The phase diagram is a eutectic type (Figure 15.8) (Puris et al. 1970). The eutectic composition and temperature are 2.4 mol% (2 mass%) ZnTe and 520°C, respectively. The maximum solubility of ZnTe in InSb at 450°C is equal to 2.0 mol% (1.6 mass%). The solubility of InSb in ZnTe is not determined. According to the data of Kirovskaya et al. (2002), solid solution based on InSb with sphalerite structure contains up to 10 mol% ZnTe [up to 20 mol% ZnTe (Kirovskaya et al. 2007b)]. This system was investigated through DTA, metallography, and XRD (Puris et al. 1970, Kirovskaya et al. 2002).

15.15 INDIUM–CADMIUM–EUROPIUM–ANTIMONY

Eu₅Cd$_x$In$_{2-x}$Sb₆ solid solutions exist in the In–Cd–Eu–Sb system (Lv et al. 2017). The polycrystalline samples at x = 0.02, 0.06, 0.08, and 0.1 were prepared by element combination reaction. Starting materials Eu, Sb, In, and Cd were weighed according to the formula Eu₅Cd$_x$In$_{2.1-x}$Sb₆ in an Ar-filled glove box. The mixtures were placed into BN crucibles and then sealed in quartz ampoules. Subsequently, all the samples were heated to 900°C for 24 h with a heating rate of 1°C·min⁻¹, followed by annealing process at 700°C for

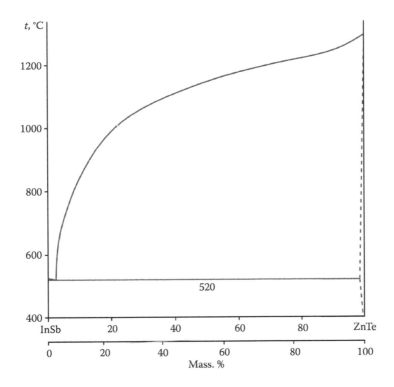

FIGURE 15.8   InSb–ZnTe phase diagram. (From Puris, T.E. et al., *Izv. AN SSSR. Neorgan. Mater.*, 6 (10), 1811–1815, 1970. With permission.)

72 h, and cooled to room temperature. Cd substituting In leads to the enhancement of the power factor and allows for the transition from nondegenerate to degenerate semiconductor.

15.16 INDIUM–CADMIUM–TIN–ANTIMONY

InSb–CdSnSb₂. The phase equilibria in this system are presented in Figure 15.9 (Abrikosov et al. 1966). This system is nonquasibinary section of the In–Cd–Sn–Sb quaternary system. InSb crystallizes within the concentration interval from 30 to 100 mol% InSb, and antimony crystallizes from the CdSnSb₂ side. The nonvariant equilibrium L + Sb ⇔ InSb + CdSb + β-SnSb takes place at 345°C, and at 300°C, the next interaction L + β-SnSb ⇔ β′-SnSb + InSb + CdSb apparently exists. InSb dissolves less than 1 mol% CdSnSb₂. Woolley and Williams (1964) indicated that the solubility of CdSnSb₂ in 2InSb is within the interval from 5 to 10 mol%, and according to the data by Goryunova and Prochukhan (1960), the region of the solid solution reaches up to 50 mol% CdSnSb₂.

This system was investigated through DTA, XRD, metallography, and microhardness measurement (Goryunova and Prochukhan 1960; Woolley and Williams 1964; Abrikosov et al. 1966).

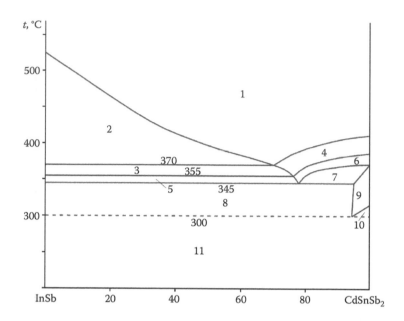

FIGURE 15.9 Phase equilibria in the InSb–CdSnSb$_2$ system; 1, L; 2, L + InSb; 3, L + InSb + Sb; 4, L + Sb; 5, L + InSb + Sb + CdSb; 6, L + Sb + β; 7, L + Sb + β + CdSb; 8, L + InSb + CdSb + β; 9, L + β + CdSb; 10, L + β′ + β + CdSb; 11, InSb + CdSb + β′, wherein β and β′ are the binary phases of the Sn–Sb system. (From Abrikosov, N.Kh., *Izv. AN SSSR. Neorgan. Mater.*, 2(8), 1416–1428, 1966. With permission.)

15.17 INDIUM–CADMIUM–BISMUTH–ANTIMONY

InSb–Cd–Bi. The liquidus surface of this system in the field of InSb crystallization is given in Figure 15.10 (Lapkina et al. 1990b). The region of the solid solution based on InSb at 400°C is presented in Figure 15.11 (Lapkina et al. 1990a). This system was investigated through DTA, XRD, metallography, and microhardness measurement.

InBi–CdSb. The phase diagram of this system is a eutectic type (Lapkina et al. 1990b). Ternary compound was not found in the InBi–CdSb system.

15.18 INDIUM–CADMIUM–SULFUR–ANTIMONY

InSb–CdS. The phase diagram is not constructed. The solubility of CdS in InSb is equal to 3–4 mol% (Kirovskaya and Filatova 2007).

15.19 INDIUM–CADMIUM–SELENIUM–ANTIMONY

InSb–Cd–Se. The joint solubility of Cd and Se in InSb was determined through the microhardness measuring (Figure 15.12) (Glazov and Smirnova 1984). Maximum solubility takes place along the InSb–CdSe section. The samples were annealed at 400°C, 450°C, and 490°C for 600 h.

InSb–CdSe. The phase diagram is not constructed. According to the data of Kirovskaya et al. (2002), the solid solution based on InSb with the sphalerite structure contains up to 10 mol% CdSe.

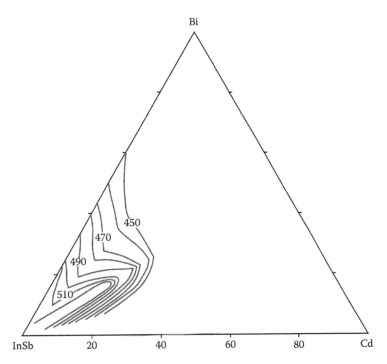

FIGURE 15.10 Part of the liquidus surface of the InSb–Cd–Bi system in the field of InSb crystallization. (From Lapkina, I.A. et al., *Izv. AN SSSR. Neorgan. Mater.*, 35(12), 3189–3193, 1990. With permission.)

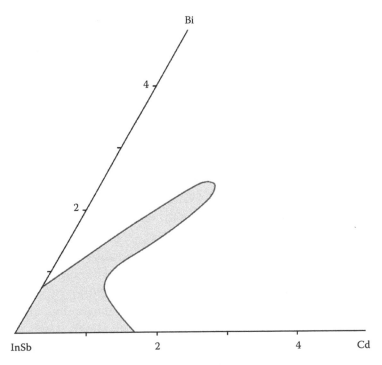

FIGURE 15.11 Solid solution based on InSb in the InSb–Cd–Bi system at 400°C. (From Lapkina, I.A. et al., *Izv. AN SSSR. Neorgan. Mater.*, 26(7), 1550–1551, 1990. With permission.)

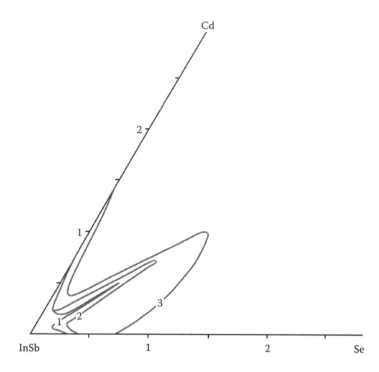

FIGURE 15.12   Joint solubility of Cd and Se in InSb at (1) 400°C, (2) 450°C, and (3) 490°C. (From Glazov, V.M., and Smirnova, E.B., *Zhurn. Fiz. Khimii*, 58(12), 2991–2995, 1984. With permission.)

15.20 INDIUM–CADMIUM–TELLURIUM–ANTIMONY

InSb–CdTe. The phase diagram of this system is a eutectic type (Figure 15.13) (Morozov et al. 1974). The eutectic composition and temperature are 6 ± 1 mol% CdTe and 510 ± 1°C, respectively. The solubility of CdTe in InSb reaches 5 mol% (Goryunova et al. 1961; Khabarov and Sharavskiy 1963, 1964; Khabarova et al. 1963; Inyutkin et al. 1964; Morozov et al. 1974, Kirovskaya and Mironova 2006, 2007; Kirovskaya et al. 2007b), and the solubility of InSb in CdTe at eutectic temperature is equal to 6 mol% and decreases to 3 mol% at 730°C (Riazantsev et al. 1980; Kirovskaya and Mironova 2006, 2007). The coefficient of CdTe distribution in the InAs–CdTe solid solutions obtained by the Bridgman method is equal to 0.3 (Kalashnikova et al. 1975).

Some maximums exist on the dependence of the lattice parameter from the composition within the interval of 0–5 mol% CdTe (Brodovoy et al. 1997). An opinion was expressed about the formation of donor–acceptor complexes in the solid solutions based on InSb. The band-gap of solid solution based on InSb is characterized by a minimum (0.17 eV) at 3 mol% CdTe (Kirovskaya and Mironova 2006; Kirovskaya et al. 2007b).

This system was investigated by DTA, XRD, metallography, and measuring of microhardness (Goryunova et al. 1961; Khabarov and Sharavskiy 1963, 1964; Khabarova et al. 1963; Inyutkin et al. 1964; Morozov et al. 1974; Kirovskaya and Mironova 2006; Riazantsev et al. 1980; Kirovskaya et al. 2007b). The ingots were annealed at 510°C for 350 h

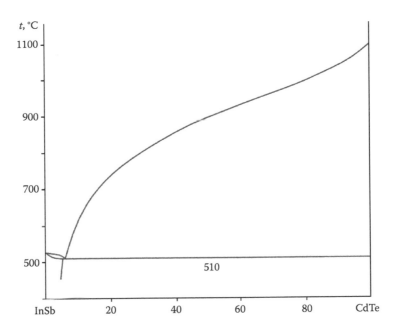

FIGURE 15.13   InSb–CdTe phase diagram. (From Morozov, V.N., *Izv. Sib. otd. AN SSSR. Ser. Khim. Nauk*, 4(9), 52–56, 1974. With permission.)

(Morozov et al. 1974). The melts were homogenized at 1100°C for 100 h, and then the ingots were annealed at 500°C for 100–150 h (Brodovoy et al. 1997).

InSb–Cd–CdTe. The liquidus surface of this quasiternary system (Figure 15.14) was constructed using DTA and mathematical simulation of experiment (Riazantsev and Telegina 1978).

InSb–CdTe–In. The liquidus surface of this quasiternary system (Figure 15.15) was constructed using DTA and mathematical simulation of experiment (Riazantsev and Telegina 1978).

15.21 INDIUM–MERCURY–TELLURIUM–ANTIMONY

InSb–HgTe. The phase diagram is a eutectic type (Figure 15.16) (Pashaev et al. 2006). The eutectic composition and temperature are 27 mol% HgTe and 350°C, respectively. The solubility of HgTe in InSb is equal to 10 mol% at room temperature. According to the data of Belotskiy et al. (1978), the solubility of HgTe in InSb at 500°C reaches 10 mol%, and the solubility of InSb in HgTe at the same temperature is equal to 5 mol%. The ingots within the interval of 10–80 mol% InSb contain elemental mercury.

This system was investigated by DTA, XRD, metallography, and measuring of microhardness and density (Pashaev et al. 2006). The ingots were annealed at 340°C for 500 h and the ingots containing 8, 9, 10, 12, and 15 mol% HgTe were additionally annealed at 100°C, 200°C, and 300°C during 250 h with the next quenching.

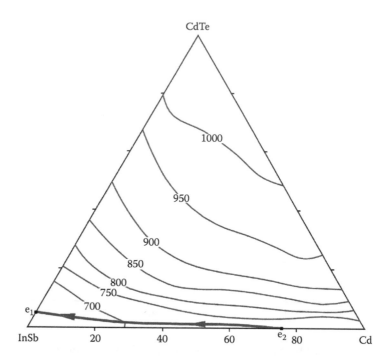

FIGURE 15.14 Liquidus surface of the InSb–Cd–CdTe quasiternary system. (From Riazantsev, A.A., and Telegina, M.P., *Zhurn. Neorgan. Khimii*, 23(8), 2211–2216, 1978. With permission.)

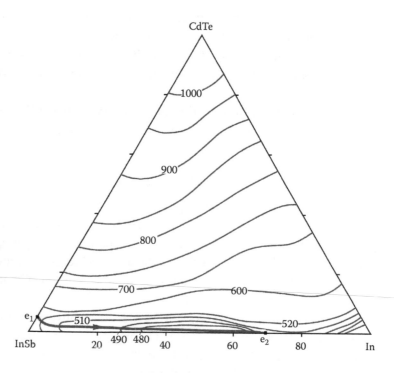

FIGURE 15.15 Liquidus surface of the InSb–CdTe–In quasiternary system. (From Riazantsev, A.A., and Telegina, M.P., *Zhurn. Neorgan. Khimii*, 23(8), 2211–2216, 1978. With permission.)

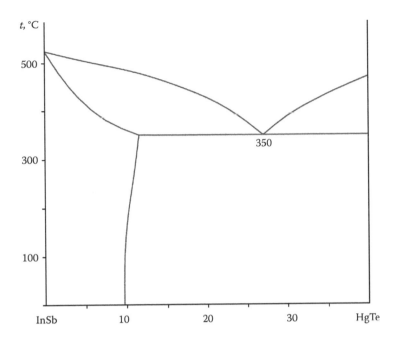

FIGURE 15.16 Part of the InSb–HgTe phase diagram. (From Pashaev, A. M. et al., *Zhurn. Khim. Problem*, (1), 119, 2006. With permission.)

15.22 INDIUM–MERCURY–BROMINE–ANTIMONY

The [**Hg₆Sb₄**](**InBr₆**)**Br** quaternary compound, which crystallizes in the cubic structure with the lattice parameter $a = 1285.42 \pm 0.03$ pm, a calculated density of 7.395 g·cm⁻³, and energy gap of 1.63 eV is formed in the In–Hg–Br–Sb system (Zou et al. 2009). Single crystals of this compound were synthesized from the solid-state reaction of HgBr₂ (1 mM), Sb (1 mM), and In (0.5 mM). The starting materials were ground into fine powders and pressed into a pellet, followed by being loaded into a Pyrex tube, evacuated to 0.01 Pa, and flame-sealed. The tube was placed into a computer-controlled furnace, heated from room temperature to 200°C at a rate of 30°C·h⁻¹, kept at this temperature for 4 h, then heated to 450°C at 20°C·h⁻¹, kept at this temperature for 120 h, slowly cooled to 100°C at a rate of 2°C·h⁻¹, and finally cooled to room temperature over 5 h. A crop of red dark crystals of [Hg₆Sb₄](InBr₆)Br that are stable in air was obtained.

15.23 INDIUM–LANTHANUM–SULFUR–ANTIMONY

The **La₄InSbS₉** quaternary compound, which crystallizes in the tetragonal structure, exhibits excellent thermal stability, shows no obvious mass loss up to 765°C, and has an energy gap of 2.07 eV, is formed in the In–La–S–Sb system (Zhao et al. 2012b). This compound was synthesized by the solid-state reactions.

15.24 INDIUM–LANTHANUM–SELENIUM–ANTIMONY

The **La₄InSbSe₉** quaternary compound, which crystallizes in the tetragonal structure with the lattice parameters $a = 1064.99 \pm 0.04$ and $c = 2947.90 \pm 0.12$ pm, a calculated density of

5.971 g·cm^{-3}, and an energy gap of 1.76 ± 0.02 eV, is formed in the In–La–Se–Sb system (Yin et al. 2017a). This compound was synthesized by the solid-state reactions. Its crystals were initially identified in reactions of mixtures of La, In_2Se_3, Sb_2Se_3, and Se in a molar ratio of 8:1:1:12 and have a total mass of ~0.3 g, which were pressed into pellets and loaded into fused-silica tubes, which were evacuated, sealed, and placed in a computer-controlled furnace. The tubes were heated to 850°C over 30 h, kept at that temperature for 72 h, cooled to 300°C over 72 h, and then slowly cooled to room temperature by shutting off the furnace. Small black block-shaped air-stable crystals were obtained, but in low yields.

Polycrystalline samples of the title compound were obtained in an optimized synthesis using a slightly different heat treatment. As before, mixtures of La, In_2Se_3, Sb_2Se_3, and Se in a molar ratio of 8:1:1:12 were placed in evacuated and sealed fused silica tubes. The tubes were heated to 800°C over 24 h and kept there for 72 h, after which the furnace was turned off. The samples were reground, loaded into new tubes, and reheated at 750°C for 96 h.

15.25 INDIUM–CERIUM–SELENIUM–ANTIMONY

The **Ce$_4$InSbSe$_9$** quaternary compound, which crystallizes in the tetragonal structure with the lattice parameters $a = 1057.25 \pm 0.04$ and $c = 2929.89 \pm 0.11$ pm, a calculated density of 6.116 g·cm^{-3}, and an energy gap of 1.69 eV, is formed in the In–Ce–Se–Sb system (Yin et al. 2017a). This compound was synthesized by the same way as La$_4$InSbSe$_9$ was prepared using Ce instead of La.

15.26 INDIUM–PRASEODYMIUM–SULFUR–ANTIMONY

The **Pr$_4$InSbS$_9$** quaternary compound, which crystallizes in the tetragonal structure and has an energy gap of 2.09 eV, is formed in the In–Pr–S–Sb system (Zhao et al. 2012b). This compound was synthesized by the solid-state reactions.

15.27 INDIUM–PRASEODYMIUM–SELENIUM–ANTIMONY

The **Pr$_4$InSbSe$_9$** quaternary compound, which crystallizes in the tetragonal structure with the lattice parameters $a = 1052.08 \pm 0.07$ and $c = 2915.2 \pm 0.2$ pm, a calculated density of 6.220 g·cm^{-3}, and an energy gap of 1.89 eV, is formed in the In–Pr–Se–Sb system (Yin et al. 2017a). This compound was synthesized by the same way as La$_4$InSbSe$_9$ was prepared using Pr instead of La.

15.28 INDIUM–NEODYMIUM–SULFUR–ANTIMONY

The **Nd$_4$InSbS$_9$** quaternary compound, which crystallizes in the tetragonal structure and has an energy gap of 2.12 eV, is formed in the In–Nd–S–Sb system (Zhao et al. 2012b). This compound was synthesized by the solid-state reactions.

15.29 INDIUM–NEODYMIUM–SELENIUM–ANTIMONY

The **Nd$_4$InSbSe$_9$** quaternary compound, which crystallizes in the tetragonal structure with the lattice parameters $a = 1047.77 \pm 0.04$ and $c = 2903.84 \pm 0.11$ pm, a calculated density of 6.351 g·cm^{-3}, and an energy gap of 1.57 eV, is formed in the In–Nd–Se–Sb system

(Yin et al. 2017a). This compound was synthesized by the same way as $La_4InSbSe_9$ was prepared using Nd instead of La.

15.30 INDIUM–SAMARIUM–SULFUR–ANTIMONY

The Sm_4InSbS_9 quaternary compound, which crystallizes in the tetragonal structure with the lattice parameters $a = 1000.42 \pm 0.04$ and $c = 2768.10 \pm 0.15$ pm, a calculated density of 5.402 g·cm^{-3}, and energy gap 2.13 eV, is formed in the In–Sm–S–Sb system (Zhao et al. 2016b). This compound was synthesized by the solid-state reactions. To prepare it, the stoichiometric mixtures of Sm, In, Sb, and S were accordingly weighed with an overall loading of about 300 mg and loaded into a silica tube, which was evacuated and sealed. The sample was heated to 950°C for 24 h and kept at that temperature for 4 d and subsequently cooled at 2.5°C·h^{-1} to 200°C before the furnace was turned off. Yellow block crystals were obtained. All manipulations were performed inside the glove box.

15.31 INDIUM–GERMANIUM–TELLURIUM–ANTIMONY

InSb–In$_2$GeTe. The specimens with compositions in the range 0–50 mol% In$_2$GeTe were investigated by Woolley and Williams (1964). It was shown that for the quenched from 800°C materials, solid solubility extends out to about 33 mol% In$_2$GeTe, but this was reduced to 12 mol% for alloys annealed at 400°C for 14 days. Alloys containing 30 and 40 mol% In$_2$GeTe showed signs of partial melting when annealed at 450°C, indicating the presence of a eutectic or peritectic horizontal in the vicinity of 450°C. The energy gap of the solid solutions increases to a maximum value of 0.45 eV at 0.5 mol% In$_2$GeTe and then decreases with increased In$_2$GeTe content. According to the data by Aliev et al. (1981), the solubility of In$_2$GeTe in 2InSb reaches 12 mol% for the alloys annealed at 400°C for 340 h. This system was investigated through DTA, XRD, and metallography (Woolley and Williams 1964; Aliev et al. 1981).

15.32 INDIUM–TIN–LEAD–ANTIMONY

InSb–Sn–Pb. The vertical section InSb–(80 mass% Sn + 20 mass% Pb) of this system is presented in Figure 15.17 (Kuznetsov and Bobrov 1971). In most alloys of this section, InSb crystallizes primarily.

15.33 INDIUM–TIN–SELENIUM–ANTIMONY

InSb–In$_2$SnSe. Specimens with compositions in the range of 0–20 mol% In$_2$SnSe were investigated by Woolley and Williams (1964). It was shown that for the specimens quenched from 800°C, the solid solubility extends out to about 5 mol% In$_2$SnSe. The energy gap of the solid solutions increases to a maximum value of 0.42 eV at 1.5 mol% In$_2$SnSe and then decreases with increasing In$_2$SnSe content.

15.34 INDIUM–TIN–TELLURIUM–ANTIMONY

InSb–In$_2$SnTe. Specimens with compositions in the range 0–50 mol% In$_2$SnTe were investigated by Woolley and Williams (1964). It was shown that for the quenched from

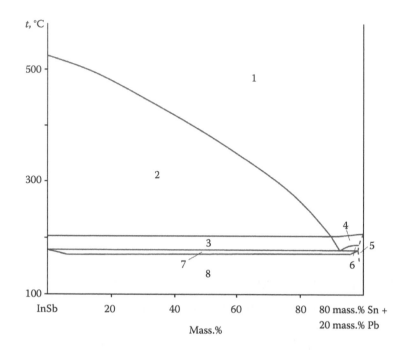

FIGURE 15.17 Vertical section InSb–(80 mass% Sn + 20 mass% Pb) of the InSb–Sn–Pb system: 1, L; 2, L + InSb; 3, L + InSb + Sn; 4, L + Sn; 5, Sn, 6, L + Sn + Pb; 7, L + InSb + Sn + Pb; 8, InSb + Sn + Pb. (From Kuznetsov, G.M., and Bobrov, A.P., *Izv. AN SSSR. Neorgan. Mater.*, 7(5), 766–768, 1971. With permission.)

800°C materials, solid solubility extends out to about 26 mol% In_2SnTe, but this was reduced to 7 mol% for alloys annealed at 400°C for 6 weeks, with a similar value for annealing temperatures up to 450°C. Alloys containing 30 and 40 mol% In_2SnTe were found to melt at 500°C and showed signs of melting when annealed at 450°C. The energy gap of the solid solutions increases to a maximum value of 0.47 eV at 0.5 mol% In_2SnTe, and then it remains constant. This system was investigated through XRD and metallography

15.35 INDIUM–LEAD–OXYGEN–ANTIMONY

The $Pb_2In_{0.5}Sb_{3.5}O_{6.5}$ quaternary compound, which crystallizes in the cubic structure with the lattice parameter a = 1058.92 ± 0.01 pm and a calculated density of 8.48 g·cm^{-3}, is formed in the In–Pb–O–Sb system (Cascales and Rasines 1985). Orange–yellow powder of the title compound was prepared from a mixture of stoichiometric amounts of PbO, In_2O_3, and Sb_2O_3, which, after being ground in an agate mortar, was heated in air at 680°C, 800°C, 850°C, and 950°C. After each thermal treatment for 24 h, the materials were quenched, weighed, and ground.

15.36 INDIUM–LEAD–TELLURIUM–ANTIMONY

$InSb–PbSb_2Te_4$. The phase equilibria in this system are presented in Figure 15.18 (Vassilev et al. 2011). $(2 ± \delta)InSb·PbSb_2Te_4$ quaternary compound of variable composition, which

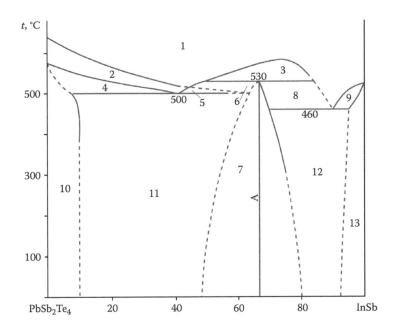

FIGURE 15.18 Phase equilibria in the InSb–PbSb$_2$Te$_4$ system: (1, L; 2, PbTe + L; 3, L$_1$ + L$_2$; 4, PbSb$_2$Te$_4$ + PbTe; 5, PbTe + L + PbSb$_2$Te$_4$; 6, L + A; 7, A; 8, A + L$_2$; 9, L$_2$ + InSb; 10, PbSb$_2$Te$_4$; 11, PbSb$_2$Te$_4$ + A; 12, A + InSb; 13, InSb, wherein A = (2 ± δ) InSb·PbSb$_2$Te$_4$). (Reprinted from *Thermochim. Acta*, 520(1–2), Vassilev, V. et al., Phase equilibria in the PbSb2Te4–InSb system, 80 83, Copyright (2011), with permission from Elsevier.)

incongruently melts at 530 ± 5°C and crystallizes in the orthorhombic structure with the lattice parameters a = 776.40, b = 641.44, and c = 360.10 pm (for InSb·PbSb$_2$Te$_4$ composition), is formed in this system. The title compound has an asymmetric area of homogeneity shifted to the side rich in InSb. The system also contains InSb- and PbSb$_2$Te$_4$-based boundary solid solutions extended from 0 to 10 mol% and from 95 to 100 mol% InSb, respectively, at room temperature. There are three nonvariant equilibria in this system: syntectic equilibrium (66.7 mol% InSb and 530 ± 5°C) and two eutectic equilibria (90 mol% InSb and 460 ± 5°C, and 40 m% InSb and 500 ± 5°C).

This system was investigated through DTA, XRD, and microhardness and density measurements.

15.37 INDIUM–SELENIUM–TELLURIUM–ANTIMONY

In$_3$Sb$_2$Se$_3$Te$_3$ and **In$_6$Sb$_4$Se$_9$Te$_6$** quaternary compounds are formed in the In–Se–Te–Sb system. In$_3$Sb$_2$Se$_3$Te$_3$ incongruently melts at 550°C (Alieva et al. 1978b).

In$_6$Sb$_4$Se$_9$Te$_6$ incongruently melts at 572–575°C and crystallizes in the hexagonal structure with the lattice parameters a = 404 and c = 2935 pm (Belotskiy and Legeta 1972a, 1972b; Alieva et al. 1978a).

References

Abrikosov N.Kh., Skudnova E.V., Poretskaya L.V., Pavlova N.G. Investigation of the In–Sb–Cd–Sn quaternary system for the determination of the phase equilibria in the InSb–(CdSnSb$_2$) section [in Russian], *Izv. AN SSSR. Neorgan. Mater.*, 2(8), 1416–1428 (1966).

Acharya S., Hajra J.P. Thermodynamic modelling of phase equilibria in Al–Ga–P–As system, *Bull. Mater. Sci.*, 28(2), 179–185 (2005).

Addamiano A. Some observations on the system ZnS–AlP, *J. Electrochem. Soc.*, 107(12), 1006–1007 (1960).

Ahmed E., Breternitz J., Groh M.F., Isaeva A., Ruck M. [Sb$_7$Se$_8$Br$_2$]$^{3+}$ and [Sb$_{13}$Se$_{16}$Br$_2$]$^{5+}$—Double and quadruple spiro cubanes from ionic liquidus, *Eur. J. Inorg. Chem.*, (19), 3037–3042 (2014).

Ahmed E., Isaeva A., Fiedler A., Haft M., Ruck M. [Sb$_{10}$Se$_{10}$]$^{2+}$, a heteronuclear polycyclic polycation from a room-temperature ionic liquid, *Chem. Eur. J.*, 17(24), 6847–6852 (2011).

Ai L., Zhou S., Qi M., Xu A., Wang S. InGaAsBi materials grown by gas source molecular beam epitaxy, *J. Cryst. Growth*, 477, 135–138 (2017).

Al-Douri A.A.J., Hasoon F.S. Physical properties of the (2InAs)$_{1-x}$(CuInSe$_2$)$_x$ solid solution, *Sol. Energy Mater.*, 19(3–5), 181–185 (1989).

Alieva Z.G., Aliev A.F., Eybatova Sh.E., Gasanova F.A., Zaidova R.R. Investigation of the In$_2$Te$_3$–Sb$_2$Se$_3$ system [in Russian], *Azerb. Khim. Zhurn.*, (5), 124–128 (1978a).

Alieva Z.G., Shakhguseynova T.F., Eybatova Sh.E., Gasanova F.A. InTe–Sb$_2$Se$_3$ phase diagram [in Russian], *Uch. zap. Azerb. un-t. Ser. Khim. n.*, (4), 33–37 (1978b).

Aliev I.I., Magammedragimova R.S., Farzaliev A.A., Veliev Dzh. Phase diagram of the In$_3$As$_2$Se$_6$–In$_3$As$_2$S$_3$Se$_3$ system, *Rus. J. Inorg. Chem.*, 54(4) 634–637 (2009).

Aliev I.I., Rustamov P.G., Maksudova T.F., Aliev F.G. Interaction in the As$_2$S$_3$–InTe system [in Russian], *Zhurn. Neorgan.. Khim.*, 32(8), 1991–1993 (1987).

Aliev M.I., Suleymanov Z.I., Arasly D.G., Ragimov R.N. Investigation of the InSb–In$_2$GeTe system [in Russian], *Izv. AN SSSR. Neorgan. Mater.*, 17(10), 1763–1766 (1981).

Ali T., Bauer E., Hilscher G., Michor H. Structural, superconducting and magnetic properties of La$_{3-x}$R$_x$Ni$_2$B$_2$N$_{3-\delta}$ with R = Ce, Pr, Nd, *Solid State Phenom.*, 170, 165–169 (2011a).

Ali T., Bauer E., Hilscher G., Michor H. Anderson lattice in the intermediate valence compound Ce$_3$Ni$_2$B$_2$N$_{3-\delta}$, *Phys. Rev. B*, 83(11), 115131_1–115131_7 (2011b).

Ali T., Rupprecht C., Khan R.T., Bauer E., Hilscher G., Michor H. The effect of nitrogen vacancies in La$_3$Ni$_2$B$_2$N$_{3-\delta}$, *J. Phys.: Conf. Ser.*, 200(1), 012004_1–012004_4 (2010).

Ali T., Steiner S., Ritter C., Michor H. Neutron diffraction study of superconducting La$_3$Ni$_2$^{11}B$_2$N$_{3-\delta}$, *J. Alloys Compd.*, 716, 251–258 (2017).

Allen V.T., Fahey J.J., Axelrod J.M. Mansfieldite, a new arsenate, the aluminum analogue of scorodite, and the mansfieldite-scorodite series, *Amer. Mineralog.*, 33(3–4), 122–134 (1948).

Ambros V.P., Burdiyan I.I. Investigation of mercury telluride solubility in gallium antimonide [in Russian], *Uchen. zap. Tirasp. ped. in-ta*, (21, Pt. 1), 31–38 (1970).

Anishchenko V.A., Voitsehovskiy A.V., Pashun A.D. Some physico-chemical properties of the ingots of the GaAs–ZnTe system [in Russian], *Izv. AN SSSR. Neorgan. Mater.*, 16(2), 354–355 (1980).

Anisimova N., Bork M., Hoppe R., Meisel M. Crystal structure of a new acentric $CsGaHP_3O_{10}$ form III, *Z. anorg. allg. Chem.*, 621(6), 1069–1074 (1995).

Ansara I., Dutartre D. Thermodynamic study of the Al–Ga–As–Ge system, *Calphad*, 8(4), 323–342 (1984).

Anya C.C., Hendry A. Stoichiometry and crystal structure of X-phase sialon, *J. Eur. Ceram. Soc.*, 10 (2), 65–74 (1992).

Arakcheeva A., Chapuis G., Petříček V., Dušek M., Schönleber A. The incommensurate structure of $K_3In(PO_4)_2$, *Acta Crystallogr.*, B59(1), 17–27 (2003).

Araki T., Finney J.J., Zoltai T. The crystal structure of augelite, *Amer. Mineralog.*, 53(7–8), 1096–1103 (1968).

Araki T., Zoltai T. The crystal structure of wavellite, *Z. Kristallogr.*, 127(1–4), 21–33 (1968).

Arbenina V.V., Buff G., Lill' T. Investigation of the possibilities and solid solutions obtaining conditions in the $GaSb–CdSnSb_2$ system [in Russian], In: *Kristalliz. i svoystva kristallov*, Novocherkassk: Publish. House of Novocherkassk Politekh. In-t, 150–156 (1989).

Arbenina V.V., Pak Cher, Skakovskiy S.I., Tarasov A.V. Investigation of equilibria in the GaSb–AlSb–Sn system [in Russian], *Neorgan. Mater.*, 28(3), 513–517 (1992).

Asahia H., Fushida M., Yamamoto K., Iwata K., Koh H., Asami K., Gonda S., Oe K. New semiconductors TlInGaP and their gas source MBE growth, *J. Cryst. Growth*, 175/176(Pt. 2), 1195–1199 (1997).

Asahia H., Yamamoto K., Iwata K., Gonda S.-I., Oe K. New III-V compound semiconductors TlInGaP for 0.9 μm to over 10 μm wavelength range laser diodes and their first successful growth, *Jpn. J. Appl. Phys.*, 35(7B), Pt. 2, L876–L879 (1996).

Asomoza R., Elyukhin V.A., Peña-Sierra R. Spinodal decomposition in the $A^{III}_xB^{III}_{1-x}C^V_yD^V_{1-y}$ quaternary alloys, *J. Cryst. Growth*, 222(1–2), 58–63 (2001a).

Asomoza R., Elyukhin V.A., Peña-Sierra R. Spinodal decomposition in the $B_xGa_yIn_{1-x-y}As$ alloys, *Appl. Phys. Lett.*, 78(17), 2494–2496 (2001b).

Asomoza R., Elyukhin V.A., Peña-Sierra R. Spinodal decomposition range of $In_xGa_{1-x}N_yAs_{1-y}$ alloys, *Appl. Phys. Lett.*, 81(10), 1785–1787 (2002).

Avaliani M.A., Chudinova N.N., Tananaev I.V. Synthesis of condensed indium phosphates in polyphosphoric acid melts [in Russian], *Izv. AN SSSR. Neorgan. Mater.*, 15(9), 1688–1689 (1979).

Avent A.G., Hitchcock P.B., Lappert M.F., Liu D.-S., Mignani G., Richard C., Roche E. Preparation, spectra and X-ray structure of an archetypal coordination compound $[BCl_3(NH_3)]$ and its thermolysis, *J. Chem. Soc., Chem. Commun.*, (8), 855–856 (1995).

Avignon-Poquillon L., Verdier P., Laurent Y., Rebouillon P., Gambino M., Bros H., Castanet R. Mesure des capacites calorifiques et des enthalpies de la solution solide AlN–SiC, *J. Therm. Anal.*, 36(7–8), 2623–2633 (1990).

Bagirov Z.B., Mamedov A.N., Bakhtiyarov I.B., Kurbanov Z.B. Thermal analysis of phase equilibria in the mutual systems Ga,In‖As,Sb and Ga,La‖O,S, *Thermochim. Acta*, 93, 717–720 (1985).

Bailey M.S., Shen D.Y., McGuire M.A., Fredrickson D.C., Toby B.H., DiSalvo F.J., Yamane H., Sasaki S., Shimada M. The indium subnitrides $Ae_6In_4(In_xLi_y)N_{3-z}$ (Ae = Sr and Ba), *Inorg. Chem.*, 44 (19), 6680–6690 (2005).

Bailey M.S., DiSalvo F.J. Synthesis and crystal structure of $LiCaGaN_2$, *J. Alloys Compd.*, 417, 50–54 (2006).

Balagurova E.A., Bazhenova G.I., Riazantsev A.A., Khabarov E.N. *T-x*-projection of InAs–CdTe phase diagram [in Russian], In: *Protsessy rosta i sinteza poluprovodnikovyh kristallov i plenok*. Novosibirsk: Nauka, Pt. 2, pp. 236–239 (1975).

Balagurova E.A., Morozov V.N., Khabarov E.N. Microhardness dependence of the solid solutions crystals from components interaction nature [in Russian], *Izv. AN SSSR. Neorgan. Mater.*, 12 (1), 105–107 (1976).

Bando Y., Mitomo M., Kitami Y., Izumi F. Structure and composition analysis of silicon aluminium oxynitride polytypes by combined use of structure imaging and microanalysis, *J. Microsc.*, 142, Pt. 2, 235–246 (1986).

Banno H., Asaka T., Fukuda K. Electron density distribution and crystal structure of 27R-SiAlON, $Si_{3-x}Al_{6+x}O_xN_{10-x}$ ($x \sim 1.9$), *J. Ceram. Soc. Jpn.*, 122(1424), 281–287 (2014a).

Banno H., Asaka T., Fukuda K. Electron density distribution and disordered crystal structure of 8H-SiAlON, $Si_{3-x}Al_{1+x}O_xN_{5-x}$ ($x \sim 2.2$), *J. Solid State Chem.*, 213, 169–175 (2014b).

Banno H., Hanai T., Asaka T., Kimoto K., Fukuda K. Electron density distribution and disordered crystal structure of 15R-SiAlON, $SiAl_4O_2N_4$, *J. Solid State Chem.*, 211, 124–129 (2014c).

Banno H., Hanai T., Asaka T., Kimoto K., Nakano H., Fukuda K. Electron density distribution and disordered crystal structure of 12H-SiAlON, $SiAl_5O_2N_5$, *Powder Diffr.*, 29(4), 318–324 (2014d).

Baranov A.N., Dzhurtanov B.E., Litvak A.M., Syavris S.V., Charykov N.A. Phase equilibria melt–solid in the Al–Ga–As–Sb system [in Russian], *Zhurn. Neorgan. Khim.*, 35(4), 1020–1023 (1990).

Baranov A.N., Litvak A.M., Moiseev K.D., Sherstrnev V.V., Yakovlev Yu.P. Obtaining of the In–Ga–As–Sb/GaSb and In–Ga–As–Sb/InAs solid solution in the region near InAs [in Russian], *Zhurn. Prikl. Khim.*, 67(12), 1951–1956 (1994).

Baranov B.V., Prochukhan V.D., Goryunova N.A. Thermal analysis of some solid solutions [in Russian], *Izv. AN LatvSSR. Ser. Khim.*, (3), 301–308 (1965).

Barsoum M.W., Ali M., El-Raghy T. Processing and characterization of Ti_2AlC, Ti_2AlN and $Ti_2AlC_{0.5}N_{0.5}$, *Metall. Trans. A*, 31(7), 1857–1865 (2000).

Bass J.D., Sclar C.B. The stability of trolleite and the Al_2O_3–$AlPO_4$–H_2O phase diagram, *Amer. Mineralog.*, 64(11–12), 1175–1183 (1979).

Bauer H. Über eine Reihe isotyper Erdalkaloboratphosphate und -arsenate vom Typus 2MeO·X_2O_5·B_2O_3 II. Die Verbindungen 2BaO·P_2O_5·B_2O_3 und 2BaO·As_2O_5·B_2O_3, *Z. anorg. allg. Chem.*, 345(5–6), 225–229 (1966).

Bauer H. Über eine Reihe isotyper Erdalkaloboratphosphate und -arsenate vom Typus 2MeO·X_2O_5·B_2O_3, *Z. anorg. allg. Chem.*, 337(3–4), 183–190 (1965).

Bayer G., Mocellin A. Displacement reaction sintering in the ZrO_2–AlN, *Rev. Chim. Minér.*, 23(1), 80–87 (1986).

Baykal A., Gözel G., Kizilyalli M., Torpak M., Kniep R. X-ray powder diraction and IR study of calcium borophosphate, $CaBPO_5$, *Turk. J. Chem.*, 24(4), 381–388 (2000a).

Baykal A., Kizilyalli M., Gözel G., Kniep R. Synthesis of strontium borophosphate, $SrBPO_5$ by solid state and hydrothermal methods and characterisation, *Cryst. Res. Technol.*, 35(3), 247–254 (2000b).

Bazhenova G.I., Balagurova E.A., Riazantsev A.A., Khabarov E.N. Solid solutions in the InAs–CdTe system [in Russian], *Izv. AN SSSR. Neorgan. Mater.*, 10(10), 1770–1773 (1974).

Beck J., Schlüter S. Synthesis and crystal structure of $(Sb_2Te_2)[AlCl_4]$, a compound containing a heteroatomic polymeric cation $(Sb_2Te^+)_n$, *Z. anorg. allg. Chem.*, 631(2–3), 569–574 (2005).

Beck J., Schlüter S., Zotov N. The polycationic main group element clusters $(As_3S_5)^+$ and $(As_3Se_4)^+$— Syntheses from chloroaluminate melts, crystal structures, and vibrational spectra, *Z. anorg. allg. Chem.*, 631(12), 2450–2456 (2005).

Belik A.A., Izumi F., Ikeda T., Okui M., Malakho A.P., Morozov V.A., Lazoryak B.I. Whitlockite-related phosphates $Sr_9A(PO_4)_7$ (A = Sc, Cr, Fe, Ga, and In): Structure refinement of $Sr_9In(PO_4)_7$ with synchrotron X-ray powder diffraction data, *J. Solid State Chem.*, 168(1), 237–244 (2002).

Belokoneva E.L., Ruchkina E.A., Dimitrova O.V., Stefanovich S.Yu. Synthesis and crystal structure of new lead borophosphate $Pb_6[PO_4][B(PO_4)_4]$ [in Russian], *Zhurn. Neorgan. Khim.*, 46(2), 226–232 (2001).

Belotskiy D.P., Babiuk P.F., Cherveniuk G.I. Investigation of solid solutions in the InSb–HgTe system [in Russian], *Izv. AN SSSR. Neorgan. Mater.*, 14(3), 589–590 (1978).

Belotskiy D.P., Legeta L.V. In_2Se_3–Sb_2Te_3 system [in Russian], *Izv. AN SSSR. Neorgan. Mater.*, 8(6), 1160–1162 (1972a).

Belotskiy D.P., Legeta L.V. Crystal structure and electrical properties of the alloys of the In_2Se_3–Sb_2Te_3 system [in Russian], *Izv. AN SSSR. Neorgan. Mater.*, 8(11), 1908–1912 (1972b).

Ben Ali A., Antic-Fidancev E., Viana B., Aschehoug P., Taibi M., Aride J., Boukhari A. Crystal structure of $ABPO_5$ and optical study of Pr^{3+} embedded in these compounds, *J. Phys. Condens. Matter*, 13(42), 9663–9672 (2001).

Benko E., Barr T.L., Bernasik A., Hardcastle S., Hoppe E., Bielańska E., Klimczyk P. Experimantal and calculated phase equilibria in the cubic BN–Ta–C system, *Ceram. Intern.*, 30(1), 31–40 (2004).

Bennarndt W., Boehm G., Amann M.-C. Domains of molecular beam epitaxial growth of Ga(In)AsBi on GaAs and InP substrates, *J. Cryst. Growth*, 436, 56–61 (2016).

Bergman B., Ekström T., Micski A. The Si–Al–O–N system at temperatures of 1700–1775°C, *J. Eur. Ceram. Soc.*, 8(3), 141–151 (1991).

Bernhardt E., Bernhardt-Pitchougina V., Willner H., Ignatiev N. Umpolung at boron by reduction of $[B(CN)_4]^-$ and formation of the dianion $[B(CN)_3]^{2-}$, *Angew. Chem. Int. Ed.*, 50(50), 12085–12088 (2011a).

Bernhardt E., Bernhardt-Pitchougina V., Willner H., Ignatiev N. Umpolung von Bor im $[B(CN)_4]^-$ Anion durch Reduktion zum Dianion $[B(CN)_3]^{2-}$, *Angew. Chem.*, 123(50), 12291–12294 (2011b).

Bernhardt E., Finze M., Willner H. Eine effiziente Synthese von Tetracyanoboraten durch Sinterprozesse, *Z. anorg. allg. Chem.*, 629(7–8), 1229–1234 (2003).

Bernhardt E., Henkel G., Willner H. Die Tetracyanoborate $M[B(CN)_4]$, M = $[Bu_4N]^+$, Ag^+, K^+, *Z. anorg. allg. Chem.*, 626(2), 560–568 (2000).

Berul' S.I., Voskresenskaya N.K. Interaction of sodium metaphosphate with alumina [in Russian], *Zhurn. Neorgan. Khim.*, 13(2), 422–457 (1968).

Bessler E. Darstellung und Eigenschaften von $AgB(CN)_4$ und $CuB(CN)_4$, *Z. anorg. allg. Chem.*, 430(1), 38–42 (1977).

Bhansali A.S., Sinclair R., Morgan A.E. A thermodynamic approach for interpreting metallization layer stability and thin-film reactions involving four elements: Application to integrated circuit contact metallurg, *J. Appl. Phys.*, 68(3), 1043–1049 (1990).

Blaschkowski B., Meyer H.-J. Electronic conditions of diatomic (BN) anions in the structure of CaNiBN, *Z. anorg. allg. Chem.*, 628(6), 1249–1254 (2002).

Blaschkowski B., Meyer H.-J. X-Ray single crystal refinement and superconductivity of $La_3Ni_2B_2N_3$, *Z. anorg. allg. Chem.*, 629(1), 129–132 (2003).

Blasse G. New compounds with eulytine structure: Crystal chemistry and luminescence, *J. Solid State Chem.*, 2(1), 27–30 (1970).

Blasse G. New compounds with perovskite-like structures, *J. Inorg. Nucl. Chem.*, 27(5), 993–1003 (1965).

Bluhm K., Park C.H. Die Synthese und Kristallstruktur des Borat-Phosphats: α-$Zn_3(BO_3)(PO_4)$, *Z. Naturforsch.*, 52B(1), 102–106 (1997).

Bontchev R.P., Sevov S.C. $Co_5BP_3O_{14}$: The first borophosphate with planar BO_3 groups connected to PO_4 tetrahedra, *Inorg. Chem.*, 35(24), 6910–6911 (1996).

Bondarenko V.P., Khalepa A.P., Cherepinina E.S. Investigation of cubic boron nitride interaction with transition metals and their carbides [in Russian], *Sintetich. almazy*, (4), 22–25 (1978).

Botelho N.F., Roger G., d'Yvoire F., Moëlo Y., Volfinger M. Yanomamite, $InAsO_4{\cdot}2H_2O$, a new indium mineral from topaz-bearing greisen in the Goiás Tin Province, Brazil, *Eur. J. Mineral.*, 6(2), 245–254 (1994).

Boughzala H., Driss A., Jouini T. Structure de $RbAlAs_2O_7$, *Acta Crystallogr.*, C49(3), 425–427 (1993).

Boughzala H., Jouini T. $KAlAs_2O_7$, *Acta Crystallogr.*, C51(2), 179–181 (1995).

Bowden M.E., Barris G.C., Brown W.M., Jefferson D.A. A new, low-temperature polymorph of O'-SiALON, *J. Amer. Ceram. Soc.*, 81(8), 2188–2190 (1998).

Brennan G.L., Dahl G.H., Schaeffer R. Studies of boron–nitrogen compounds: II. Preparation and reactions of B-trichloroborazole, *J. Amer. Chem. Soc.*, 82(24), 6248–6250 (1960).

Brodovoy V.A., Vialyi N.G., Knorozok L.M., Markiv I.Ya. Peculiarities of lattice parameter changing of the $(ZnSb)_{1-x}(CdTe)_x$ solid solutions [in Russian], *Neorgan. Mater.*, 33(3), 303–304 (1997).

Brosheer J.C., Lenfesty F.A., Anderson, J.F.Jr. Solubility in the system aluminum phosphate–phosphoric acid–water. *J. Amer. Chem. Soc.*, 76(23), 5951–5956 (1954).

Brousseau L.C., Williams D., Kouvetakis J., O'Keeffe M. Synthetic routes to $Ga(CN)_3$ and $MGa(CN)_4$ (M = Li, Cu) framework structures, *J. Amer. Chem. Soc.*, 119(27), 6292–6296 (1997).

Brovkin V.N., Kazakov A.I., Kishmar I.N., Presnov V.A. Analysis of the phase equilibria in the M–Si–Ga–As systems [in Russian], *Izv. AN SSSR. Neorgan. Mater.*, 17(3), 407–411 (1981).

Brown C.A., Laubengayer A.W. B-Trichloroborazole, *J. Amer. Chem. Soc.*, 77(14), 3699–3700 (1955).

Brownlow C.E.A., Salmon J.E., Wall J.G.L. Indium phosphates: Phase-diagram, ion-exchange, and pH titration studies, *J. Chem. Soc.*, 2452–2457 (1960).

Brühne B., Jansen M. Synthese und Kristallstruktur von $LiAl[(P_2O_6)_2]$, *Z. anorg. allg. Chem.*, 620(8), 1409–1412 (1994).

Brunet F., Morineau D., Schmid-Beurmann P. Heat capacity of lazulite, $MgAl_2(PO_4)_2(OH)_2$, from 35 to 298 K and a (S–V) value for P_2O_5 to estimate phosphate entropy, *Mineralog. Mag.*, 68(1), 123–134 (2004).

Bunin V.A., Karpov A.V., Senkovenko M.Yu. Fabrication, structure, and properties of TiB_2–AlN ceramics, *Inorg. Mater.*, 38(7), 746–749 (2002).

Burdiyan I.I. The solid solutions of gallium antimonide with cadmium, zinc and mercury tellurides [in Russian], In: *Nekot. voprosy Khim. i fiz. poluprovodn. slozhn. sostava.* Uzhgorod: Publish. House of Uzhgorod. un-t, 190–195 (1970).

Burdiyan I.I. Investigation of solid solutions of GaSb with cadmium, zinc and mercury tellurides [in Russian], *Izv. AN SSSR. Neorgan. Mater.*, 7(3), 414–416 (1971).

Burdiyan I.I., Georgitse E.I. About solubility of aluminium antimonide in mercury telluride [in Russian], *Uchen. zap. Tirasp. ped. in-ta*, (21), P. 1, 3–5 (1970).

Burdiyan I.I., Korolevski B.P. About possibility of solid solution formation in the GaSb–ZnTe system [in Russian], *Uchen. zap. Tirasp. ped. in-ta*, 16, 127–128 (1966).

Burdiyan I.I., Makeichik A.I. Solid solutions in the GaSb–CdTe system [in Russian], *Uchen. zap. Tirasp. ped. in-ta*, 16, 125–126 (1966).

Bu W., Xu J., Qiu T., Li X. Investigation of reaction synthesis of AlN–SiC solid solution, *J. Mater. Sci. Lett.*, 21(9), 730–732 (2002).

Bu X., Feng P., Stucky G.D. Large-cage zeolite structures with multidimensional 12-ring channels, *Science*, 278(5346), 2080–2085 (1997).

Buzevich G.I. Investigation of growth and physical properties of epytaxial layers of InAs–CdTe solid solutions [in Russian], *Avtoref. dis. … kand. fiz.-mat. nauk*, Irkutsk, (1972).

Cai Y., Shen Z., Grins J., Esmaeilzadeh S. Self-reinforced nitrogen-rich calcium-α-SiALON ceramics, *J. Amer. Ceram. Soc.*, 90(2), 608–613 (2007).

Cámara F., Gatta G.D., Belakovskiy D. New mineral names, *Amer. Mineralog.*, 99(2–3), 551–555 (2014).

Cannard P.J., Ekström T., Tilley R.J.D. The formation of phases in the AlN–rich corner of the Si–Al–O–N system, *J. Eur. Ceram. Soc.*, 8(6), 375–382 (1991).

Cao C.C., Wang Y.G., Zhu L., Meng Y., Dai Y.D., Chen J.K. Evolution of structural and magnetic properties of the FeCuBP amorphous alloy during annealing, *J. Alloys Compd.*, 722, 394–399 (2017).

Carim A.H. Identification and characterization of $(Ti,Cu,Al)_6N$, a new η nitride phase, *J. Mater Res.*, 4(6), 1456–1461 (1989).

Caron M., Gagnon G., Fortin V., Currie J.F., Ouellet L., Tremblay Y., Biberger M., Reynolds R. Calculation of a Al–Ti–O–N quaternary isotherm diagram for the prediction of stable phases in TiN/Al alloy contact metallization, *J. Appl. Phys.*, 79(8), 4468–4470 (1996).

Carrillo-Cabrera W., Caroca-Canales N., Schnering von H.G. Dipotassium sodium diantimonidoindate, $K_2Na[InSb_2]$, a compound with the polyanion $^2_\infty[In_2Sb_2Sb_{4/2}]^{6-}$, *Z. anorg. allg. Chem.*, 619(10), 1717–1720 (1993).

Carrillo-Heian E.M., Xue H., Ohyanagi M., Munir Z.A. Reactive synthesis and phase stability investigations in the aluminum nitride–silicon carbide system, *J. Amer. Ceram. Soc.*, 83(5), 1103–1107 (2000).

Cascales C., Rasines I., García Casado P., Vega J. The new pyrochlores $Pb_2(M_{0.5}Sb_{1.5})O_{6.5}$ (M = Al, Sc, Cr, Fe, Ga, Rh), *Mater. Res. Bull.*, 20(11), 1359–1365 (1985).

Cascales C., Rasines I. New pyrochlore $Pb^{II}_2[In_{0.5}Sb_{1.5}]O_{6.5}$, *Z. anorg. allg. Chem.*, 529(10), 229–234 (1985).

Cava R.J., Zandbergen H.W., Batlogg B., Eisaki H., Takagi H., Krajewskii J.J., Peck C.OMMAJ.R.X.X.X W.F., Gyorgy E.M., Uchida S. Superconductivity in lanthanum nickel boro-nitride, *Nature*, 372(6503), 245–247 (1994).

Cemič L., Schmid-Beurmann P. Lazulite stability relations in the system $Al_2O_3–AlPO_4–Mg_3(PO_4)_2–H_2O$, *Eur. J. Mineralog.*, 7(4), 921–929 (1995).

Chen H.-H., Ge M.-H., Yang X.-X., Mi J.-X., Zhao J.-T. Synthesis and crystal structure of $Fe_2BP_3O_{12}$ [in Chinese], *J. Inorg. Mater.*, 19(2), 429–432 (2004).

Chen J., Tian Q., Virkar A.V. Effect of coherency strains on phase separation in the $AlN–Al_2OC$ pseudobinary system, *J. Amer. Ceram. Soc.*, 76(10), 2419–2432 (1993).

Chen M.-C., Li L.-H., Chen Y.-B., Chen L. In-phase alignments of asymmetric building units in Ln_4GaSbS_9 (Ln = Pr, Nd, Sm, Gd-Ho) and their strong nonlinear optical responses in middle IR, *J. Amer. Chem. Soc.*, 133(12), 4617–4624 (2011).

Chen M.-C., Wu L.-M., Lin H., Zhou L.-J., Chen L. Disconnection enhances the second harmonic generation response: Synthesis and characterization of $Ba_{23}Ga_8Sb_2S_{38}$, *J. Amer. Chem. Soc.*, 134 (14), 6058–6060 (2012).

Chen S., Hoffmann S., Carrillo-Cabrera W., Akselrud L.G., Prots Yu., Schwarz U., Zhao J.-T., Kniep R. $Sr_{10}[(PO_4)_{5.5}(BO_4)_{0.5}](BO_2)$: Growth and crystal structure of a strontium phosphate orthoborate metaborate closely related to the apatite-type crystal structure, *J. Solid State Chem.*, 183(3), 658–661 (2010).

Chen Z.-X., Weng L.-H., Zhou Y.-M., Zhang H.-Y., Zhao D.-Y. Synthesis and structure of a new three-dimensional microporous indium arsenate, *Acta Chim. Sinica*, 60(2), 305–309 (2002).

Cheviré F., Pallu A., Ray E., Tessier F. Characterization of Nd_2AlO_3N and Sm_2AlO_3N oxynitrides synthesized by carbothermal reduction and nitridation, *J. Alloys Compd.*, 509(19), 5839–5842 (2011).

Chondroudis K., Chakrabarty D., Axtell E.A., Kanatzidis M.G. Synthesis of the one-dimensional compound $(Ph_4P)[In(P_2Se_6)]$ in a Ph_4P^+-containing selenophosphate flux, and structure of $[In(P_2Se_6)_2]^{5-}$—A discrete molecular fragment of the $[In(P_2Se_6)]_n^{n-}$ chain. *Z. anorg. allg. Chem.*, 624(6), 975–979 (1998).

Chondroudis K., Kanatzidis M.G. $K_4In_2(PSe_5)_2(P_2Se_6)$ and $Rb_3Sn(PSe_5)(P_2Se_6)$: One-dimensional compounds with mixed selenophosphate anions, *J. Solid State Chem.*, 136(1), 79–86 (1998).

Chudinova N.N., Avaliani M.A., Guzeeva K.S., Tananaev I.V. Reactions in the $K_2O–Ga_2O_3–P_2O_5–H_2O$ system at 150–500°C [in Russian], *Izv. AN SSSR. Neorgan. Mater.*, 14(11), 2054–2060 (1978).

Chudinova N.N., Avaliani M.A., Guzeeva K.S., Tananaev I.V. Synthesis of binary condensed phosphates of gallium and alkali metals [in Russian], *Izv. AN SSSR. Neorgan. Mater.*, 15(12), 2176–2179 (1979).

Chudinova N.N., Grunze I., Guzeeva L.S., Avaliani M.A. Synthesis of binary condensed cesium-gallium phosphates [in Russian], *Izv. AN SSSR. Neorgan. Mater.*, 23(4), 604–609 (1987).

Chudinova N.N., Tananaev I.V., Avaliani M.A. Synthesis of binary gallium–potassium polyphosphates in polyphosphoric acid melts [in Russian], *Izv. AN SSSR. Neorgan. Mater.*, 13 (12), 2234–2235 (1977).

Clarke S.J., DiSalvo F.J. Synthesis and structure of the subnitrides Ba_2GeGaN and $(Ba_xSr_{1-x})_3Ge_2N_2$; $x \approx 0.7$, *J. Alloys Compd.*, 259(1–2), 158–162 (1997).

Cocksedge H.E. Boron thiocyanate, *J. Chem. Soc., Trans.*, 93, 2177–2179 (1908).

Cox C.A., Toberer E.S., Levchenko A.A., Brown S.R., Snyder G.J., Navrotsky A., Kauzlarich S.M. Structure, heat capacity, and high-temperature thermal properties of $Yb_{14}Mn_{1-x}Al_xSb_{11}$, *Chem. Mater.*, 21(7), 1354–1360 (2009).

Curda J., Herterich U., Peters K., Somer M., Schnering von H.G. Crystal structure of lithium tetraeuropium tris(dinitridoborate), $LiEu_4(BN_2)_3$, *Z. Kristallogr.*, 209(7), 618 (1994a).

Curda J., Herterich U., Peters K., Somer M., Schnering von H.G. Crystal structure of tetrabarium bis (dinitridoborate) monoxide, $Ba_4(BN_2)_2O$, *Z. Kristallogr.*, 209(2), 181 (1994b).

Cutler I.B., Miller P.D., Rafaniello W., Park H.K., Thopson D.P., Jack K.H. New materials in the Si-C-Al-O-N and related systems, *Nature*, 275(5679), 434–435 (1978).

Dąbrowska G. Synthesis and some properties of a new compound in the Al–Sb–V–O system, *J. Therm. Anal. Calorim.*, 109(2), 745–749 (2012).

Dai Z., Zhu Z., Ma W., Peng S., Yin F. The Zn-rich corner of the Zn–Al–V–Sb quaternary system at 450 and 600°C, *J. Phase Equilibr. Dif.*, 37(5), 574–580 (2016).

Daly S.R., Bellott B.J., Kim D.Y., Girolami G.S. Synthesis of the long-sought unsubstituted aminodiboranate $Na(H_3B-NH_2-BH_3)$ and its *N*-alkyl analogs, *J. Amer. Chem. Soc.*, 132(21), 7254–7255 (2010).

Deichman T.N., Ezhova Zh.A., Tananaev I.V., Kharitonov Yu.Ya. Obtaining and investigation of some properties of indium arsenate [in Russian], *Zhurn. Neorgan. Khim.*, 19(1), 35–58 (1974).

Deichman E.N., Tananaev I.V., Ezhova Zh.A. On the acid phosphates of indium [in Russian], *Izv. AN SSSR. Neorgan. Mater.*, 5(8), 1397–1401 (1969).

Deichman E.N., Tananaev I.V., Ezhova Zh.A. On the interaction of lithium diphosphate and indium chloride in the aqueous solution [in Russian], *Izv. AN SSSR. Neorgan. Mater.*, 3(5), 900–902 (1967a).

Deichman E.N., Tananaev I.V., Ezhova Zh.A. On the potassium indium diphosphates [in Russian], *Izv. AN SSSR. Neorgan. Mater.*, 3(10), 1946–1947 (1967b).

Deichman T.N., Tananaev I.V., Ezhova Zh.A., Palkina K.K. A study of indium phosphates [in Russian], *Izv. AN SSSR. Neorgan. Mater.*, 6(9), 1645–1649 (1970).

Deng L., Ni J., Wang L.B., Jia X.P., Qin J.M., Liu B.W. Structure and thermoelectric properties of $In_xBa_yCo_4Sb_{12}$ samples prepared by HPHT, *J. Alloys Compd.*, 712, 477–481 (2017).

Députier S., Guérin R., Ballini Y., Guivarc'h A. Solid state phase equilibria in the Ni–Al–As system, *J. Alloys Compd.*, 217(1), 13–21 (1995).

Devenson J., Pačebutas V., Butkuté R., Baranov A., Krotkus A. Structure and optical properties of InGaAsBi with up to 7% bismuth, *Appl. Phys. Expr.*, 5(1), 015503_1–015503_3 (2012).

Devi R.N., Vidyasagar K. Synthesis and characterisation of novel layered compounds, $Cs_2MP_3O_{10}$ (M = Al or Ga), containing triphosphate groups, *J. Chem. Soc., Dalton Trans.*, (10), 1605–1608 (2000).

Deyneko D.V., Morozov V.A., Stefanovich S.Yu., Belik A.A., Lazoryak B.I., Lebedev O.I. Structural changes in $Sr_9In(PO_4)_7$ during antiferroelectric phase transition, *Inorg. Mater.*, 52(2), 176–185 (2016).

Djoudi L., Lachebi A., Merabet B., Abid H. First-principles investigation of structural and electronic properties of the $B_xGa_{1-x}N$, $B_xAl_{1-x}N$, $Al_xGa_{1-x}N$ and $B_xAl_yGa_{1-x-y}N$ compounds, *Acta Phys. Polon. A*, 122(4), 748–753 (2012).

Dolginov L.M., Eliseev P.G., Lapshin A.N., Milvidskii M.G. A study of phase equilibria and heterojunctions in Ga–In–As–Sb quaternary system, *Krist. Techn.*, 13(6), 631–638 (1978).

Dörner P., Gauckler L.J., Krieg H., Lukas H.L., Petzow G., Weiss J. Calculation of heterogeneous phase equilibria in the SiAlON system, *J. Mater. Sci.*, 16(4), 935–943 (1981).

Drew P., Lewis M.H. The microstructures of silicon nitride/alumina ceramics, *J. Mater. Sci.*, 9(11), 1833–1838 (1974).

Driss A., Jouini T. Structure crystalline de $NaAlAs_2O_7$, *J. Solid State Chem.*, 112(2), 277–280 (1994).

Driss A., Jouini T. Structure de $Na(Al_{1.5}As_{0.5})(As_2O_7)_2$, *Acta Crystallogr.*, C45(3), 356–360 (1989).

Drobyazko V.P., Kuznetsova S.T. Interaction of indium arsenide with $A^2B^3_2C^6_4$ compounds [in Russian], *Zhurn. Neorgan. Khim.*, 20(11), 3061–3064 (1975).

Drobyazko V.P., Kuznetsova S.T. Interaction of indium arsenide with cadmium chalcogenides [in Russian], *Zhurn. Neorgan. Khim.*, 28(11), 2929–2933 (1983).

Dronskowski R. $In_{2.24}(NCN)_3$ and $NaIn(NCN)_2$: Synthesis and crystal structures of new main group metal cyanamides, *Z. Naturforsch.*, 50B(7), 1245–1251 (1995).

Duggan M.B., Jones M.T., Richards D.N.G., Kamprad J.L. Phosphate minerals in altered andesite from Mount Perry, Queensland, Australia, *Canad. Mineralog.*, 28(1), 125–131 (1990).

Dumitrescu L., Sundman B. A thermodynamic reassessment of the Si–Al–O–N system, *J. Eur. Ceram. Soc.*, 15(3), 239–247 (1995).

Dunn P.J., Chao G.Y., Fleischer M., Ferraiolo J.A., Langley R.H., Pabst A., Zilczer J.A. New mineral names, *Amer. Mineralog.*, 70(1–2), 214–221 (1985).

Durlu N., Gruber U., Pietzka M.A., Schmidt H., Schuster J.C. Phases and phase equilibria in the quaternary system Ti–Cu–Al–N at 850°C, *Z. Metallkde*, 88(5), 390–400 (1997).

Dutczak D., Wurst K.M., Ströbele M., Enseling D., Jüstel T., Meyer H.-J. Defect-related luminescence in nitridoborate nitride, $Mg_3Ga(BN_2)N_2$, *Eur. J. Inorg. Chem.*, (6), 861–866 (2016).

d'Yvoire F., Bretey E., Collin G. Crystal structure, non-stoichiometry and conductivity of II-$Na_3M_2(AsO_4)_3$ (M = Al, Ga, Cr, Fe), *Solid State Ionics*, 28–30, Pt. 2, 1259–1264 (1988).

d'Yvoire F., Pintard-Screpel M., Bretey E. Polymorphism and cation transport properties in arsenate $Na_3M_2(AsO_4)_3$ (M = Al, Cr, Fe, Ga), *Solid State Ionics*, 18–19, Pt. 1, 502–506 (1986).

Egorysheva A.V., Milenov T.I., Rafailov P.M., Gaitko O.M., Avdeev G.V., Dudkina T.D. Optical and vibrational spectra of $Bi_{1.8}Fe_{1.2(1-x)}Ga_{1.2x}SbO_7$ solid solutions with pyrochlore-type structure, *Russ. J. Inorg. Chem.*, 62(7), 960–963 (2017).

Ehrenberg H., Laubach S., Schmidt P.C., McSweeney R., Knapp M., Mishra K.C. Investigation of crystal structure and associated electronic structure of $Sr_6BP_5O_{20}$, *J. Solid State Chem.*, 179(4), 968–973 (2006).

Eich A., Bredow T., Beck J. Mixed pentele-chalcogen cationic chains from aluminum and gallium halide melts, *Inorg. Chem.*, 54(2), 484–491 (2015).

Eich A., Hoffbauer W., Schnakenburg G., Bredow T., Daniels J., Beck J. Double-cube-shaped mixed chalcogen/pentele clusters from $GaCl_3$ melts, *Eur. J. Inorg. Chem.*, (19), 3043–3052 (2014).

Eich A., Schlüter S., Schnakenburg G., Beck J. $(Sb_7Te_8)^{5+}$—A double cube shaped polycationic cluster, *Z. anorg. allg. Chem.*, 639(2), 375–383 (2013).

Ekström T., Käll P.O., Nygren M., Olsson P.O. Dense single-phase β-sialon ceramics by glass-encapsulated hot isostatic pressing, *J. Mater. Sci.*, 24(5), 1853–1861 (1989).

El Haj Hassan F., Postnikov A.V., Pagès O. Structural, electronic, optical and thermal properties of $Al_xGa_{1-x}As_ySb_{1-y}$ quaternary alloys: First-principles study, *J. Alloys Compd.*, 504(2), 559–565 (2010).

Ellert O.G., Egorysheva A.V., Maksimov Yu.V., Gajtko O.M., Efimov N.N., Svetogorov R.D. Isomorphism in the $Bi_{1.8}Fe_{1.2(1-x)}Ga_{1.2x}SbO_7$ pyrochlores with spin glass transition, *J. Alloys Compd.*, 8, Pt. A, 1–7 (2016).

Ensslin F., Dreyer H., Lessman O. Zur Chemie des Indiums. XI Verbindungen des Indiums mit den Sauerstoffsäuren des Phosphors, *Z. anorg. allg. Chem.*, 254(5–6), 315–318 (1947).

Ewald B., Prots Yu., Kniep R. Refinement of the crystal structures of praseodymium- and samariumoxoborate-bis(oxophosphate)-oxide, $Ln_7O_6[BO_3][PO_4]_2$, (Ln = Pr, Sm), Z. Kristallogr., New Cryst. Str., 219(3), 213–215 (2004).

Ezhova Zh.A., Deichman T.N., Tananaev I.V., Kharitonov Yu.Ya. On the acidic indium arsenates [in Russian], Zhurn. Neorgan. Khim., 21(2), 395–340 (1976).

Fabrichnaya O., Pavlyuchkov D., Neher R., Herrmann M., Seifert H.J. Liquid phase formation in the system $AlN–Al_2O_3–Y_2O_3$: Part II. Thermodynamic assessment, J. Eur. Ceram. Soc., 33(13–14), 2457–2463 (2013).

Feng G., Oe K., Yoshimoto M. Temperature dependence of Bi behavior in MBE growth of InGaAs/InP, J. Cryst. Growth, 301–302, 121–124 (2007).

Feng G., Yoshimoto M., Oe K., Chayahara A., Horino Y. New III–V semiconductor InGaAsBi alloy grown by molecular beam epitaxy, Jpn. J. Appl. Phys., 44(37), L1161–L1163 (2005).

Feng P., Bu X., Stucky G.D. Hydrothermal syntheses and structural characterization of zeolite analogue compounds based on cobalt phosphate, Nature, 388(6644), 735–740, (1997).

Fleischer M., Cabri L.J., Nickel E.H., Pabst A. New mineral names, Amer. Mineralog., 62(5–6), 593–600 (1977).

Fleischer M., Chao G.Y., Pabst A. New mineral names, Amer. Mineralog., 64(3–4), 464–467 (1979).

Fransolet A.-M. La vantasselite, $Al_4(PO_4)_3(OH)_3·9H_2O$, une nouvelle espèce minérale du Massif de Stavelot, Belgique, Bull. Minéral., 110(6), 647–656 (1987).

Gamondès J.-P., d'Yvoire F., Boullé A. Préparartion et étude radiocristallographique de diphosphates $M^IM^{III}P_2O_7$ (M^I = Na, K; M^{III} = Fe, Al) et $Na_{1-x}Fe_{1-y}H_{x+3y}P_2O_7$, C. r. Acad. Sci., Sér. C, 272(1), 49–52 (1971).

García-Guinea, J., Chagoyen A.M., Nickel E.H. A re-investigation of bolivarite and evansite, Canad. Mineralog., 33(1), 59–65 (1995).

Garvie L.A.J., Hubert H., Rez P., McMillan P.F., Buseck P.R. $BN_{0.5}O_{0.4}C_{0.1}$: Carbon- and oxygen-substituted hexagonal BN, J. Alloys Compd., 290(1–2), 34–40 (1999).

Gascoin F., Sevov S.C. Cubane with a handle: $[\{In_3As_4Nb\}–As]^{7-}$ in $Cs_7NbIn_3As_5$, Angew. Chem. Int. Ed., 41(7), 1232–1234 (2002a).

Gascoin F., Sevov S.C. Cubane with a handle: $[\{In_3As_4Nb\}–As]^{7-}$ in $Cs_7NbIn_3As_5$, Angew. Chem., 114(7), 1280–1282 (2002b).

Gascoin F., Sevov S.C. KBa_2InAs_3 with coexisting monomers of $[In_2As_7]^{13-}$ and their one-dimensional polymers, Inorg. Chem., 41(8), 2292–2295 (2002c).

Gascoin F., Sevov S.C. Synthesis and characterization of transition-metal Zintl phases: $Cs_{24}Nb_2In_{12}As_{18}$ and $Cs_{13}Nb_2In_6As_{10}$ with isolated complex anions, Inorg. Chem., 42(25), 8567–8571 (2003a).

Gascoin F., Sevov S.C. Synthesis and characterization of transition-metal Zintl phases: $K_{10}NbInAs_6$ and $K_9Nb_2As_6$, Inorg. Chem., 42(3), 904–907 (2003b).

Gauckler L.J., Lukas H.L., Petzow G. Contribution to the phase diagram $Si_3N_4–AlN–Al_2O_3–SiO_2$, J. Amer. Ceram. Soc., 58(7–8), 346–347 (1975).

Geisz J.F., Friedman D.J., Kurtz S., Olson J.M., Swartzlander A.B., Reedy R.C., Norman A.G. Epitaxial growth of BGaAs and BGaInAs by MOCVD, J. Cryst. Growth, 225(2–4), 372–376 (2001).

Genkina E.A., Muradyan L.A., Maksimov B.A., Merinov B.V., Sigarev S.E. Crystal structure of Li_3In_2 $(PO_4)_3$ [in Russian], Kristallografiya, 32(1), 74–78 (1987).

Giesecke G., Pfister H. Mischkristalle des Systems $ZnSnAs_2–InAs$ und des Systems $ZnGeAs_2–InAs$, Acta Crystallogr., 14(12), 1289 (1961).

Gilev V.G. IR spectra and phases structure of the Si–Al–O–N system obtained by carbothermic reduction and simultaneous nitriding of kaolin [in Russian], Neorgan. Mater., 37(10), 1224–1229 (2001).

Gillott L., Cowlam N., Bacon G.E. A neutron diffraction investigation of some β'-sialons, J. Mater. Sci., 16(8), 2263–2268 (1981).

Glaser J., Mori T., Meyer H.-J. Crystal structure of $Ce_3Ni_2(BN)_2N$ and magnetic behavior of $(CeNi(BN))_2(CeN)_x$ with x = 0, 1, *Z. anorg. allg. Chem.*, 634(6–7), 1067–1070 (2008).

Glazov V.M., Chernyaev V.N., Nagiev V.A., Ufimtsev V.B., Abdullaev A.A., Rzaev F.R. Investigation of separate and joint solubility of zinc, cadmium and tellurium in indium phosphide [in Russian], *Izv. AN SSSR. Neorgan. Mater.*, 10(1), 144–145 (1974).

Glazov V.M., Kiselev A.N., Shvedkov E.I. Solubility and donor-acceptor interaction in InAs, doped with S, Se, and Zn [in Russian], *Izv. AN SSSR. Neorgan. Mater.*, 15(3), 390–394 (1979a).

Glazov V.M., Krestovnikov A.N., Nagiev V.A., Rzaev F.R. Investigation of phase equilibria in the InP–ZnTe and InP–CdTe quasibinary system [in Russian], *Elektron. tehnika. Ser. 6: Materialy*, (4), 127–129 (1972).

Glazov V.M., Krestovnikov A.N., Nagiev V.A., Rzaev F.R. Investigation of phase equilibrium and analysis of intermolecular interaction in the quasibinary systems [in Russian], In: *Termodin. svoistva metallich. splavov*, Baku: Elm, 380–386 (1975a).

Glazov V.M., Krestovnikov A.N., Nagiev V.A., Rzaev F.R. Phase equilibria in the InP–ZnTe and InP–CdTe quasibinary system [in Russian], *Izv. AN SSSR. Neorgan. Mater.*, 9(11), 1883–1889 (1973).

Glazov V.M., Nagiev V.A., Glagoleva N.N. Separated and combined solubility of Zn, Cd and Te in InAs [in Russian], *Izv. AN SSSR. Neorgan. Mater.*, 11(7), 1181–1183 (1975b).

Glazov V.M., Pavlova L.M., Griazeva N.L. Investigation of phase equilibria and analysis of intermolecular interaction in the GaSb–Zn(Cd)Te quasibinary systems [in Russian], In: *Termodinamicheskiye svoistva metallicheskih splavov*, Baku: Elm, 368–371 (1975c).

Glazov V.M., Pavlova L.M., Griazeva N.L. Phase equilibria and analysis of intermolecular interaction in the GaSb–Zn(Cd)Te quasibinary systems [in Russian], *Izv. AN SSSR. Neorgan. Mater.*, 11(3), 418–423 (1975d).

Glazov V.M., Pavlova L.M., Lebedeva L.V. Thermodynamic analysis of interaction of gallium arsenide with zinc and cadmium tellurides [in Russian], In: *Termodinamicheskiye svoistva metallicheskih splavov*, Baku: Elm, 372–375 (1975e).

Glazov V.M., Pavlova L.M., Perederiy L.I. Analysis of intermolecular interaction and thermodynamic properties of the GaAs–CdSe melts [in Russian], In: *Termodinamicheskie svoistva metallicheskih rasplavov: Materialy 4-go Vsesoyuz. oveshch. po termodinamike metal. splavov (rasplavov)*. Alma-Ata: Nauka, Pt. 2, 26–29 (1979b).

Glazov V.M., Pavlova L.M., Perederiy L.I. Joint solubility and donor-acceptor interaction of selenium and cadmium in gallium arsenide [in Russian], *Zhurn. Fiz. Khim.*, 59(1), 32–36 (1985).

Glazov V.M., Pavlova L.M., Perederiy L.I. Phase equilibria and analysis of the intermolecular interaction in the GaAs–CdSe system [in Russian], *Izv. AN SSSR. Neorgan. Mater.*, 19(2), 193–196 (1983).

Glazov V.M., Smirnova E.B. Donor-acceptor interaction in the solid solutions of indium antimonide doped with cadmium and selenium [in Russian], *Zhurn. Fiz. Khim.*, 58(12), 2991–2995 (1984).

Goggin P.L., McColm I.J., Shore R. Indium tricyanide and indium trithiocyanate, *J. Chem. Soc. A: Inorg., Phys., Theor.*, 1314–1317 (1966).

Golubev V.N., Viting B.N., Dogadin O.B., Lazoryak B.I., Aziev R.G. About binary phosphates $Ca_9M(PO_4)_7$ (M = Al, Fe, Cr, Ga, Sc, Sb, In) [in Russian], *Zhurn. Neorgan. Khim.*, 35(12), 3037–3041 (1990).

Gordon S.G. Crystallographic data on wavellite from Llallagua, Bolivia and on cacoxenite from Hellertown, Pennsylvania, *Amer. Mineralog.*, 35(1–2), 132 (1950).

Goryunova N.A., Averkieva G.K., Sharavskiy P.V., Tovpentsev Yu.K. Investigation of quaternary alloys based on indium antimonide and cadmium telluride [in Russian]. In: *Fizika i khimia*. L.: Publish. of Leningr. inzh.-stroit. in-t, p. 10 (1961).

Goryunova N.A., Baranov B.V., Valov Yu.A., Prochukhan V.D. About solubility of boron phosphide in various melts [in Russian], In: *Fizika*, Leningrad, Publish. House of Lening. Inst. Civil Eng., 15–19 (1964).

Goryunova N.A., Fedorova N.N. About solid solutions in the ZnSe–GaAs system [in Russian], *Fiz. tv. tela*, 1(2), 344–345 (1959).

Goryunova N.A., Grigor'eva V.S., Sharavskiy P.V., Osnach L.A. Solid solutions in the InAs–HgTe system [in Russian], In: *Fizika*, L.: Publish. of Leningr. inzh.-stroit. in-t, pp. 7–10 (1962).

Goryunova N.A., Prochukhan V.D. Solid solutions in the quaternary systems based on InAs and InSb [in Russian], *Fiz. tv. tela*, 2(1), 176–178 (1960).

Goryunova N.A., Sokolova V.I. About complex phosphide [in Russian], *Izv. Mold. fil. AN SSSR*, 3 (69), 31–35 (1960).

Gottschalch V., Leibiger G., Benndorf G. MOVPE growth of $B_xGa_{1-x}As$, $B_xGa_{1-x-y}In_yAs$, and $B_xAl_{1-x}As$ alloys on (001) GaAs, *J. Cryst. Growth*, 248, 468–473 (2003).

Gözel G., Baykal A., Kizilyalli M., Kniep R. Solid-state synthesis, X-ray powder investigation and IR study of α-$Mg_3[BPO_7]$, *J. Eur. Ceram. Soc.*, 18(14), 2241–2246 (1998).

Granon A., Goeuriot P., Thevenot F., Guyader J., L'Haridon P., Laurent Y. Reactivity in the Al_2O_3–AlN–MgO system. The MgAlON spinel phase, *J. Eur. Ceram. Soc.*, 13(4), 365–370 (1994).

Grebenyuk A.M., Charykov N.A., Puchkov L.V. Phase equilibria melt—Solid in the Pb–InSb–GaSb and Pb–InAs–GaAs systems [in Russian], *Zhurn. Neorgan. Khim.*, 37(1), 201–203 (1992).

Grebenyuk A.M., Litvak A.M., Charykov N.A., Yakovlev Yu.P. Phase equilibria melt—Solid in the Pb–InAs–InSb system [in Russian], *Zhurn. Neorgan. Khim.*, 35(11), 2941–2944 (1990).

Grebenyuk A.M., Litvak A.M., Popov A.A., Charykov N.A., Sherstnev V.V., Yakovlev Yu.P. Phase equilibria melt—Solid in the Pb–GaAs–GaSb system [in Russian], *Zhurn. Neorgan. Khim.*, 36 (4), 1067–1071 (1991).

Grey I.E., Kampf A.R., Price J.R., MacRae C.M. Bettertonite, $[Al_6(AsO_4)_3(OH)_9(H_2O)_5]\cdot11H_2O$, a new mineral from the Penberthy Croft mine, St. Hilary, Cornwall, UK, with a structure based on polyoxometalate clusters, *Mineralog. Mag.*, 79(7), 1849–1858 (2015a).

Grey I.E., Kampf A.R., Price J.R., MacRae C.M. Bettertonite, IMA 2014-074. CNMNC Newsletter No. 23, February 2015, page 55, *Mineralog. Mag.*, 79(1), 51–58 (2015b).

Groen W.A., Hal van P.F., Kraan M.J., Sweegers N., With de G., Preparation and properties of ALCON ($Al_{28}C_6O_{21}N_6$) ceramics, *J. Mater. Sci.*, 30(19), 4775–4780 (1995a).

Groen W.A., Kraan M.J., Hal van P.F., De Veirman A.E.M. A new diamond-related compound in the system Al_2O_3–Al_4C_3–AlN, *J. Solid State Chem.*, 120(2), 211–215 (1995b).

Grunze I., Palkina K.K., Chudinova N.N., Guzeeva L.S., Avaliani M.A., Maksimova S.I. The structure and thermal transformations of binary cesium–gallium phosphates [in Russian], *Izv. AN SSSR. Neorgan. Mater.*, 23(4), 610–615 (1987).

Gruß M., Glaum R. Beiträge zur Kristallchemie und zum thermischen Verhalten von wasserfeien Phosphaten. XXIV. Darstellung, Kristallstruktur und Eigenschaften des Kupfer(II)–Indium(III)–Orthophosphates $Cu_3In_2[PO_4]_4$, *Z. anorg. allg. Chem.*, 627(6), 1377–1382 (2001).

Gusarov V.V., Mikirticheva G.A., Shitova V.I., Grabovenko L.Yu., Kuchaeva S.K. Phase relationships in the $NaPO_3$–Al_2O_3 glass-forming system, *Glass Phys. Chem.*, 28(5), 309–316 (2002).

Guzairov R.S., Leytsin V.A., Grekov S.D. Solubility and solubility product of indium pyrophosphates [in Russian], *Zhurn. Neorgan. Khim.*, 9(1), 20–24 (1964).

Gvozdetskyi V., German N., Gladyshevskii R. Phase equilibria in the systems Gd–Fe–{Ga,Ge}–Sb at 500°C. Crystallographic parameters of the compounds $GdFe_2Ge_2$ and $GdFe_{0.52}Ge_2$ [in Ukrainian], *Visnyk L'viv un-tu. Ser. Khim.*, (53), 12–19 (2012).

Györi B., Emri J., Fehér I. Preparation and properties of novel cyano and isocyano derivatives of borane and the tetrahydroborate anion, *J. Organomet. Chem.*, 255(1), 17–28 (1983).

Häberlen M., Glaser J., Meyer H.-J. Zwei neue Nitridoborate des Calciums, *Z. anorg. allg. Chem.*, 628(9–10), 2169 (2002).

Hamdoune S., Tran Qui D., Schouler E.J.L. Ionic conductivity and crystal structure of $Li_{1+x}Ti_{2-x}In_xP_3O_{12}$, *Solid State Ionics*, 18–19, Pt. 1, 587–589 (1986).

Han B.Y., Singh S.P., Sohn K.-S. Photoluminescent and structural properties of $MgAlSiN_3:Eu^{2+}$ phosphors, *J. Electrochem. Soc.*, 158(2), J32–J35 (2011).

Hao W., Han Y., Huang R., Feng K., Yin W., Yao J., Wu Y. $Ag_{1.75}InSb_{5.75}Se_{11}$: A new non-centrosymmetric compound with congruent-melting behavior, *J. Solid State Chem.*, 218, 196–201 (2014).

Hao W., Mei D., Yin W., Feng K., Yao J., Wu Y. Synthesis, structural characterization and optical properties of new compounds: Centrosymmetric Ba_2GaMQ_5 (M = Sb, Bi; Q = Se, Te), $Ba_2InSbTe_5$ and noncentrosymmetric $Ba_2InSbSe_5$, *J. Solid State Chem.*, 198, 81–86 (2013).

Harata M., Imai A., Ohta T., Sugaike S. Formation and ionic conductivity of nitrogen-doped β-alumina ceramics, *Solid State Ionics*, 3–4, 409–412 (1981).

Harrison W.T.A. Synthetic mansfieldite, $AlAsO_4{\cdot}2H_2O$, *Acta Crystallogr.*, C56(10), e421 (2000).

Hasegawa T., Yamane H. Crystal structure of $Li_2B_3PO_8$ with 2D-linkage of BO_3, BO_4 and PO_4 groups, *Dalton Trans.*, 43(39), 14525–14528 (2014a).

Hasegawa T., Yamane H. Synthesis, crystal structure and lithium ion conduction of $Li_3BP_2O_8$, *Dalton Trans.*, 43(5), 2294–2300 (2014b).

Hasoon F.S., Al-Douri A.A.J. Surface characterization and differential thermal analysis of the $(2InAs)_{1-x}(CuInSe_2)_x$ solid solution, *Sol. Energy Mater.*, 22(1), 93–103 (1991).

Hauf C., Friedrich T., Kniep R. Crystal structure of pentasodium catena-(diborato-triphosphate), $Na_5[B_2P_3O_{13}]$, *Z. Kristallogr.*, 210(6), 446 (1995).

Hauf C., Yilmaz A., Kizilyalli M., Kniep R. Borophosphates: Hydrothermal and microwave-assisted synthesis of $Na_5[B_2P_3O_{13}]$, *J. Solid State Chem.*, 140(1), 154–156 (1998).

Hecht C., Stadler F., Schmidt P.J., Schmedt auf der Günne J., Baumann V., Schnick W. $SrAlSi_4N_7$: Eu^{2+}—A nitridoalumosilicate phosphor for warm white light (pc)LEDs with edge-sharing tetrahedra, *Chem. Mater.*, 21(8), 1595–1601 (2009).

He H., Stoyko S., Bobev S. New insights into the application of the valence rules in Zintl phases: Crystal and electronic structures of $Ba_7Ga_4P_9$, $Ba_7Ga_4As_9$, $Ba_7Al_4Sb_9$, $Ba_6CaAl_4Sb_9$, and $Ba_6CaGa_4Sb_9$, *J. Solid State Chem.*, 236, 116–122 (2016).

He H., Tyson C., Bobev S. Synthesis and crystal structures of the quaternary Zintl phases $RbNa_8Ga_3Pn_6$ (Pn = P, As) and $Na_{10}NbGaAs_6$, *Crystals*, 2(2), 213–223 (2012).

Hendsbee A.D., Pye C.C., Masuda J.D. Hexaaquagallium(III) trinitrate trihydrate, *Acta Crystallogr.*, E65(8), i65 (2009).

Herterich U., Curda J., Peters K., Somer M., Schnering von H.G. Crystal structure of lithium magnesium dinitridoborate, $LiMgBN_2$, *Z. Kristallogr.*, 209(7), 617 (1994).

Hesse K.-F., Cemič L. Crystal structure of $FeAlPO_5$, *Z. Kristallogr.*, 209(4), 346–347 (1994a).

Hesse K.-F., Cemič L. Crystal structure of $MgAlPO_5$, *Z. Kristallogr.*, 209(8), 660–661 (1994b).

Hillert M., Jonsson S. Thermodynamic calculation of the Si–Al–N–O system *Z. Metallkde*, 83(10), 720–728 (1992).

Hinteregger E., Wurst K., Perfler L., Kraus F., Huppertz H. High-pressure synthesis and characterization of the actinide borate phosphate $U_2[BO_4][PO_4]$, *Eur. J. Inorg. Chem.*, (30), 5247–5252 (2013).

Hirosaki N., Takeda T., Funahashi S., Xie R.-J. Discovery of new nitridosilicate phosphors for solid state lighting by the single-particle-diagnosis approach, *Chem. Mater.*, 26(14), 4280–4288 (2014).

Hoard J.L., Geller S., Cashin W.M. Structures of molecular addition compounds. III. Ammonia-boron trifluoride, $H_3N–BF_3$, *Acta Crystallogr.*, 4(5), 396–398 (1951).

Hochleitner R., Fehr K.T., Kaliwoda M., Günther A., Rewitzer C., Schmahl W.W., Park S. Hydroniumpharmacoalumite, $(H_3O)Al_4[(OH)_4(AsO_4)_3]{\cdot}4–5\ H_2O$, a new mineral of the pharmacosiderite supergroup from Rodalquilar, Spain, *Neues Jahrb. Mineral. Abh.*, 192(2), 169–176 (2015).

Hochleitner R., Fehr K.T., Kaliwoda M., Günther A., Schmahl W.W., Park S. Hydro-niumpharmacoalumite, IMA 2012-050. CNMNC Newsletter No. 16, August 2013, page 2699, *Mineralog. Mag.*, 77(6), 2695–2709 (2013).

Höhn P., Prots Yu., Kokal I., Somer M. Crystal structure of dieuropium(II) dinitridoborate bromide, $Eu_2[BN_2]Br$, *Z. Kristallogr., New Cryst. Str.*, 224(3), 379–380 (2009).

Huang Q., Chakoumakos B.C., Santoro A., Cava R.J., Krajewski J.J., Peck, W.F.Jr. Neutron powder diffraction study of the 12 K superconductor $La_3Ni_2B_2N_{3-x}$, *Physica C*, 244(1–2), 101–105 (1995).

Huang Q., Hwu S.-J. $Cs_2Al_2P_2O_9$: An exception to Löwenstein's rule: Synthesis and characterization of a novel layered aluminophosphate containing linear Al–O–Al linkages, *Chem. Commun.*, (23), 2343–2344 (1999).

Huang W., Oe K., Feng G., Yoshimoto M. Molecular-beam epitaxy and characteristics of $GaN_yAs_{1-x-y}Bi_x$, *J. Appl. Phys.*, 98(5), 053505_1–053505_6 (2005).

Huang W., Yoshimoto M., Takehara Y., Saraie J., Oe K. $GaN_yAs_{1-x-y}Bi_x$ alloy lattice matched to GaAs with 1.3 μm photoluminescence emission, *Jpn. J. Appl. Phys.*, 43(10B), L1350–L1352 (2004).

Huang Y.-X., Liu J.-Y., Mi J.-X., Zhao J.-T. $(Ga_{0.71}B_{0.29})PO_4$ with a high-cristobalite-type structure refined from powder data, *Acta Crystallogr.*, E66(2), i4 (2010).

Huang Z.K., Yan D.S. (Yen T.S.), Tien T.Y. Compound formation and melting behavior in the *AB* compound and rare earth oxide systems, *J. Solid State Chem.*, 85(1), 51–55 (1990).

Iizuki E. Process for producing boron nitride of cubic system, *Pat. USA*, 4 551 316 (1985).

Ilegems M., Panish M.B. Phase equilibria in III–V quaternary systems: Application to Al–Ga–P–As, *J. Phys. Chem. Solids*, 35(3), 409–420 (1974).

Il'yasov T.M. Phase diagram of the As_2Te_3–GaS system [in Russian], *Izv. AN SSSR. Neorgan. Mater.*, 19(3), 380–383 (1983).

Il'yasov T.M., Rustamov P.G. Chemical interaction and glass formation in the As_2X_3–GaX chalcogenide systems [in Russian], *Zhurn. Neorgan. Khim.*, 27(10), 2651–2654 (1982).

Il'yasov T.M., Rustamov P.G. Liquidus surface projection of the Ga,As ‖ S,Se system [in Russian], *Zhurn. Neorgan. Khim.*, 28(8), 2087–2090 (1983).

Il'yasov T.M., Safarov M.G., Rustamov P.G. As_2Se_3–GaS system [in Russian], *Zhurn. Neorgan. Khim.*, 22(9), 2539–2543 (1977).

Imamura N., Mizoguchi H., Hosono H. Superconductivity in LaT_MBN and $La_3T_{M2}B_2N_3$ (T_M = transition metal) synthesized under high pressure, *J. Amer. Chem. Soc.*, 134(5), 2516–2519 (2012).

Inoue A., Park R.E. Soft magnetic properties and wide supercooled liquid region of Fe–P–B–Si base amorphous alloys, *Mater. Trans., JIM*, 37(11), 1715–1721 (1996).

Inoue A., Shinohara Y., Cook J.S. Thermal and magnetic properties of bulk Fe-based glassy alloys prepared by copper mold casting, *Mater. Trans., JIM*, 36(12), 1427–1433 (1995).

Inuzuka H., Kaga M., Urushihara D., Nakano H., Asaka T., Fukuda K. Synthesis and structural characterization of a new aluminum oxycarbonitride, $Al_5(O,C,N)_4$, *J. Solid State Chem.*, 183 (11), 2570–2575 (2010).

Inyutkin A., Kolosov E., Osnach L., Khabarova V., Khabarov E., Sharavskiy P. Some investigations of solid solutions based on III–V and II–VI compounds [in Russian], *Izv. AN SSSR. Ser. Fiz.*, 28(6), 1010–1016 (1964).

Ischenko V., Kienle L., Jansen M. Formation and structure of $LiSi_2N_3$–AlN solid solutions, *J. Mater. Sci.*, 37(24), 5305–5317 (2002).

Ishida K., Tokunaga H., Ohtani H., Nishizawa T. Data base for calculating phase diagrams of III–V alloy semiconductors, *J. Cryst. Growth*, 98(1–2), 140–147 (1989).

Ismunandar, Kennedy B.J., Hunter B.A. Observation on pyrochlore oxide structures, *Mater. Res. Bull.*, 34(8), 1263–1274 (1999).

Ismunandar, Kennedy B.J., Hunter B.A. Structural and surface properties of $Bi_3(MSb_2)O_{11}$ (M = Al, Ga), *J. Solid State Chem.*, 127(2), 178–185 (1996).

Ito A., Aoki H., Akao M., Miura N., Otsuka R., Tsutsumi S. Flux growth and crystal structure of boron-containing apatite [in Japanese], *J. Ceram. Soc. Jpn.*, 96(3) 305–309 (1988)

Jack K.H. Sialons and related nitrogen ceramics, *J. Mater. Sci.*, 11(6), 1135–1158 (1976).

Jack K.H., Wilson W.I. Ceramics based on the Si–Al–O–N and related systems, *Nature Phys. Sci.*, 238(80), 28–29 (1972).

Jambor J.L., Bladh K.W., Ercit T.S., Grice J.D., Grew E.S. New mineral names, *Amer. Mineralog.*, 73 (7–8), 927–935 (1988).

Jambor J.L., Roberts A.C. New mineral names, *Amer. Mineralog.*, 80(1–2), 184–188 (1995).

Jameson R.F., Salmon J.E. Aluminium phosphates: Phase-diagram and ion-exchange studies of the system aluminium oxide–phosphoric oxide–water at 25°, *J. Chem. Soc.*, 4013–4017 (1954).

Jansen S.R., Haan de J.W., Ven van de J.M., Hanssen R., Hintzen H.T., Metselaar R. Incorporation of nitrogen in alkaline-earth hexaaluminates with β-alumina- or a magnetoplumbite-type structure, *Chem. Mater.*, 9(7), 516–523 (1997a).

Jansen S.R., Hintzen H.T., Metselaar R. $BaAl_{11.54}O_{17.11}N_{0.80}$: A new oxynitride with the β-alumina structure, *J. Mater. Sci. Let.*, 15(9), 794–796 (1996).

Jansen S.R., Hintzen H.-T., Metselaar R. Phase relations in the $BaO–Al_2O_3–AlN$ system: Materials with the β-alumina structure, *J. Solid State Chem.*, 129(1), 66–73 (1997b).

Jäschke B., Jansen M. Synthese, Kristallstruktur und spektroskopische Charakterisierung der Phosphaniminato-Komplexe $[Cl_2BNPCl_2NPCl_3]_2$ und $[Br_2BNPCl_3]_2$, *Z. anorg. allg. Chem.*, 628(9–10), 2000–2004 (2002a).

Jäschke B., Jansen M. Synthese, Kristallstruktur und spektroskopische Charakterisierung der Phosphaniminato-Komplexes $[Cl_2AlNPCl_3]_2$ und des Phosphanimin-Komplexes $[Me_3SiNPCl_3.AlCl_3]$, *Z. Naturforsch.*, 57B(11), 1237–1243 (2002b).

Jing H., Blaschkowski B., Meyer H.-J. Eine Festkörper-Metathese-Route zur Synthese von Nitridoboraten des Lanthans, *Z. anorg. allg. Chem.*, 628(9–10), 1955–1958 (2002).

Jing H., Meyer H.-J. $La_6(BN_3)O_6$, ein Nitridoborat-Oxid des Lanthans, *Z. anorg. allg. Chem.*, 628(7), 1548–1551 (2002).

Johan Z. La sénégalite, $Al_2(PO_4)(OH)_3 \cdot H_2O$, un nouveau minéral, *Lithos*, 9(2), 165–171 (1976).

Johan Z., Slansky E., Povondra P. Vashegyite, a sheet aluminum phosphate: New data, *Canad. Mineralog.*, 21(3), 489–498 (1983).

Kalashnikova E.V., Korzhov V.I., Morozov V.N., Petrov A.A., Semikolenova N.A., Skorobogatova L.A., Finogenova V.K., Khabarova V.A., Khabarov E.N. Obtaining of the $A^{III}B^V–A^{II}B^{VI}$ solid solutions closed to intrinsic [in Russian]. In: *Processy rosta i sinteza poluprovodnikovyh kristallov i plenok*. Novosibirsk: Nauka, Pt. 2, pp. 232–236 (1975).

Kamler G., Weisbrod G., Podsiadło S. Formation and thermal decomposition of gallium oxynitride compounds, *J. Therm. Anal. Calorim.*, 61(3), 873–877 (2000).

Kannan K.K., Viswamitra M.A. Unit cell, space group and refractive indices of $Al(NO_3)_3 \cdot 9H_2O$ and $Cr(NO_3)_3 \cdot 9H_2O$, *Acta Crystallogr.*, 19(1), 151–152 (1965).

Karouta F., Marbeuf A., Joullié J.H. Low temperature phase diagram of the $Ga_{1-x}In_xAs_ySb_{1-y}$ system, *J. Cryst. Growth*, 79(1–3), Pt. 1, 445–450 (1986).

Kasper B. Phasengleichgewichte im System B–C–N–Si, *Thesis*, Max-Planck-Institut, Stuttgart, 1–225 (1996).

Katz G., Kedesdy H. A new synthetic hydrate of aluminum arsenate, *Amer. Mineralog.*, 39(11–12), 1005–1017 (1954).

Kaufman L. Calculation of quasibinary and quasiternary oxynitride systems—III, *CALPHAD*, 3(4), 275–291 (1979).

Kawaguchi M., Kawashima T., Nakajima T. Syntheses and structures of new graphite-like materials of composition BCN(H) and $BC_3N(H)$, *Chem. Mater.*, 8(6), 1197–1201 (1996).

Kechele J.A., Hecht C., Oeckler O., Schmedt auf der Günne J., Schmidt P.J., Schnick W. $Ba_2AlSi_5N_9$— A new host lattice for Eu^{2+}-doped luminescent materials comprising a nitridoalumosilicate framework with corner- and edge-sharing tetrahedra, *Chem. Mater.*, 21(7), 1288–1295 (2009).

Keegan T.D., Araki T., Moore P.B. Senegalite, $Al_2(OH)_3(H_2O)(PO_4)$, a novel structure type, *Amer. Mineralog.*, 64(11–12), 1243–1247 (1979).

Ketchum D.R., DeGraffenreid A.L., Niedenzu P.M., Shore S.G. Synthesis of amorphous boron nitride from the molecular precursor ammonia-monochloroborane, *J. Mater. Res.*, 14(5), 1934–1938 (1999).

Khabarova V.A., Khabarov E.N., Sharavskiy P.V. Determination of CdTe solubility limit in InSb [in Russian], *Izv. vuzov. Ser. Fizika*, (6), 62 64 (1963).

Khabarov E.N., Sharavskiy P.V. About interatomic constraint forces in the InSb–CdTe solid solutions [in Russian]. In: *Fizika. L.*: Publish. of Leningr. inzh.-stroit. in-t, p. 31 (1963).

Khabarov E.N., Sharavskiy P.V. Investigation of properties of InSb·CdTe limited solid solutions [in Russian], *Dokl. AN SSSR*, 155(3), 542–544 (1964).

Khalilov Kh.Ya., Aliev M.I. Investigation of the electrical propeties of the $InSb–ZnSnSb_2$ solid solutions [in Russian], *Izv. AN SSSR. Neorgan. Mater.*, 4(1), 149–151 (1968).

Khorari S., Rulmont A., Tarte P. Alluaudite-like structure of the arsenate $Na_3In_2(AsO_4)_3$. *J. Solid State Chem.*, 134(1), 31–37 (1997).

Kikkawa S., Nagasaka K., Takeda T., Bailey M., Sakurai T., Miyamoto Y. Preparation and lithium doping of gallium oxynitride by ammonia nitridation via a citrate precursor route, *J. Solid State Chem.*, 180(7), 1984–1989 (2007).

Kim N.H., Fun Q.D., Komeya K., Meguro T. Phase reaction and sintering behavior in the pseudoternary system $AlN–Y_2O_3–Al_2O_3$, *J. Amer. Ceram. Soc.*, 79(10), 2645–2651 (1996a).

Kim N.H., Komeya K., Meguro T. Effect of Al_2O_3 addition on phase reaction of the $AlN–Y_2O_3$ system, *J. Mater. Sci.*, 31(6), 1603–1608 (1996b).

Kirovskaya I.A., Azarova O.P., Shubenkova E.G., Dubina O.N. Synthesis and optical absorption of the solid solutions of the InSb-(II-VI) systems [in Russian], *Neorgan. Mater.*, 38(2), 135–138 (2002).

Kirovskaya I.A., Filatova T.N. New semiconductor InSb–CdS system [in Russian], *Sovrem. naukoemk. tekhnol.*, (8), 29–31 (2007).

Kirovskaya I.A., Mironova E.V. Obtaining and identification of substitutional solid solutions of the system InSb–CdTe [in Russian], *Zhurn. Neorgan. Khim.*, 51(4), 701–705 (2006).

Kirovskaya I.A., Mironova E.V. Synthesis and properties of new materials—Solid solutions $(InSb)_x(CdTe)_{1-x}$ [in Russian], *Dokl. AN Vysshey shkoly Rosii*, (1), 34–43 (2007).

Kirovskaya I.A., Mulikova G.M. The GaAs–ZnSe system [in Russian], *Izv. AN SSSR. Neorgan. Mater.*, 11(6), 1131–1132 (1975).

Kirovskaya I.A., Novgorodtseva L.V., Vasina M.V., Baranovskaya M.V., Chkalova A.L., Kuznetsova I.Yu. New materials based on the $GaSb-A^{II}B^{VI}$ systems [in Russian], *Sovrem. naukoemk. tekhnol.*, (6), 96–97 (2007a).

Kirovskaya I.A., Shubenkova E.G., Mironova E.V., Rud'ko T.A., Bykova E.I. Obtaining and properties of the solid solutions of the $InSb-A^{II}B^{VI}$ system [in Russian], *Sovremen. naukoemk. tekhnol.*, (8), 31–33 (2007b).

Kirovskaya I.A., Timoshenko O.T. Pecularities of the InP–CdS system and its obtaining [in Russian], *Sovrem. naukoemk. tekhnol.*, (6), 97–98 (2007).

Kirovskaya I.A., Zemtsov A.E. Chemical composition and acid-base properties of the surface of GaAs–CdS solid solutions [in Russian], *Zhurn. Fiz. Khim.*, 81(1), 101–106 (2007).

Kirovskaya I.A., Zemtsov A.E., Shedenko A.V. Solid solutions of heterovalent substitution based on the $GaAs-A^{II}B^{VI}$ system [in Russian], *Sovremen. naukoemk. tekhnol.*, (8), 33–35 (2007c).

Kleshchinskiy L.I., Habarov E.N., Sharavski P.V. Determination of solid solution limits in the InAs–CdTe system [in Russian]. In: *Fizika. L.*: Publish. of Leningr. inzh.-stroit. in-t, pp. 12–15 (1964).

Klyucharov Ya.V., Skoblo L.I. About $Al_2O_3·3P_2O_5$ phosphates and the interaction of $Al_4(P_4O_{12})_3$ with melted potassium chloride [in Russian], *Dokl. AN SSSR*, 154(3), 634–637 (1964).

Kniep R., Gözel G., Eisenmann B., Röhr C., Asbrand M., Kizilyalli M. Borophosphate—eine vernachlässigte Verbindungsklasse: Die Kristallstrukturen von $M^{II}[BPO_5]$ (M^{II} = Ca, Sr) und $Ba_3[BP_3O_{12}]$, *Angew. Chem.*, 106(7), 791–793 (1994a).

Kniep R., Gözel G., Eisenmann B., Röhr C., Asbrand M., Kizilyalli M. Borophosphates—A neglected class of compounds: Crystal structures of $M^{II}[BPO_5]$ (M^{II} = Ca, Sr) and $Ba_3[BP_3O_{12}]$, *Angew. Chem. Int. Ed. Engl.*, 33(7), 749–751 (1994b).

Kniep R., Mootz D. Metavariscite—A redetermination of its crystal structure, *Acta Crystallogr.*, B29 (10), 2292–2294 (1973).

Kniep R., Mootz D., Vegas A. Variscite, *Acta Crystallogr.*, B33(1), 263–265 (1977).

Knobloch G., Meier U., Butter E. Phase eqilibria in the $GaAs–HCl–H_2$ system, *J. Cryst. Growth*, 66 (2), 338–345 (1984).

Kokal I., Aydemir U., Prots Yu., Förster T., Sichelschmidt J., Yahyaoglu M., Auffermann G., Schnelle W., Schappacher F., Pöttgen R., Somer S. Synthesis, crystal structure and magnetic properties of $Li_{0.44}Eu_3[B_3N_6]$, *J. Solid. State Chem.*, 210(1), 96–101 (2014).

Kokal I., Aydemir U., Prots Yu., Schnelle W., Akselrud L., Höhn P., Somer M. Syntheses, crystal structures, magnetic properties and vibrational spectra of nitridoborate-halide compounds $Sr_2[BN_2]Br$ and $Eu_2[BN_2]X$ (X = Br, I) with isolated $[BN_2]^{3-}$ units, *Z. Kristallogr.*, 226 (8), 633–639 (2011a).

Kokal I., Somer M., Akselrud L., Höhn P., Carrillo-Cabrera W. Synthesis, crystal structure and vibrational spectra of $Mg_2[BN_2]Br$: A new magnesium nitridoborate halide with discrete $[N–B–N]^{3-}$ groups, *Z. anorg. allg. Chem.*, 637(7–8), 915–918 (2011b).

Kolitsch U., Schwendtner K. Octahedral-tetrahedral framework structures of $InAsO_4 \cdot H_2O$ and $PbIn(AsO_4)(AsO_3OH)$, *Acta Crystallogr.*, C61(9), i86–i89 (2005).

Kopnin E., Coste S., Jobic S., Evain M., Brec R. Synthesis and crystal structure determination of three layered-type thiophosphate compounds KMP_2S_7 (M = Cr, V, In), *Mater. Res. Bull.*, 35(9), 1401–1410 (2000).

Koppel Kh.D., Luzhnaya N.P., Medvedeva Z.S. Interaction of indium arsenide with some salts [in Russian], *Izv. AN SSSR. Neorgan. Mater.*, 3(8), 1354–1359 (1967).

Köster W., Thoma B. Aufbau ternärer Systeme von Metallen der dritten und fünften Gruppe des periodischen Systems, *Z. Metallkde*, 46(4), 293–297 (1955).

Kouvetakis J., O'Keefe M., Brouseau L., McMurran J., Williams D., Smith D.J. New pathways to heteroepitaxial GaN by inorganic CVD synthesis and characterization of related Ga–C–N novel systems, *Mater. Res. Soc. Symp. Proc.*, 449, 313–318 (1996).

Krivovichev S.V., Molchanov A.V., Filatov S.K. Crystal structure of urusovite $Cu[AlAsO_5]$: A new type of a tetrahedral aluminoarsenate polyanion, *Crystallogr. Rep.*, 45(5), 723–727 (2000).

Kuhn A., Eger R., Nuss J., Lotsch B.V. Synthesis and crystal structures of the alkali aluminium thiohypodiphosphates $M^IAlP_2S_6$ (M = Li, Na), *Z. anorg. allg. Chem.*, 639(7), 1087–1089 (2013).

Kulinich S.A., Sevast'yanova L.G., Burdina K.P., Leonova M.E., Sirotinkin S.P. Complex lithium-potassium boronitride $LiCa_4(BN_2)_3$ with cubic structure [in Russian], *Zhurn. Obshch. Khim.*, 66(7), 1067–1069 (1996).

Kulinich S.A., Sevast'yanova L.G., Burdina K.P., Semenenko K.N. New lithium–magnesium boronitride: Obtaining and properties [in Russian], *Zhurn. Obshch. Khim.*, 67(10), 1598–1601 (1997).

Kulinich S.A., Sevast'yanova L.G., Zhukov A.N., Burdina K.P. Boronitrides of alkali and alkaline earth metals containing N^{3-} anion [in Russian], *Zhurn. Obshch. Khim.*, 70(2), 190–196 (2000).

Kumar V., Bhardwaj N., Tomar N., Thakral V., Uma S. Novel lithium-containing honeycomb structures, *Inorg. Chem.*, 51(20), 10471–10473 (2012).

Kuo S.-Y., Virkar A.V. Morphology of phase separation in $AlN–Al_2OC$ and SiC–AlN ceramics, *J. Amer. Ceram. Soc.*, 1(9), 2640–2646 (1990).

Kuo S.-Y., Virkar A.V. Phase equilibria and phase transformation in the aluminum nitrided–aluminum oxycarbide pseudobinary system, *J. Amer. Ceram. Soc.*, 72(4), 540–550 (1989).

Kuo S.-Y., Virkar A.V., Rafaniello W. Modulated structures in SiC–A1N ceramics, *J. Amer. Ceram. Soc.*, 70(6), C-125–C-128 (1987).

Küppers T., Bernhardt E., Willner H., Rohm H.W., Köckerling M. Tetracyanoborate salts $M[B(CN)_4]$ with M = singly charged cations: Properties and structures, *Inorg. Chem.*, 44(4), 1015–1022 (2005).

Ku S.M., Bodi L.J. Synthesis and some properties of ZnSe:GaAs solid solutions, *J. Phys. Chem. Sol.*, 29(12), 2077–2082 (1968).

Kuz'mina G.A. Investigation of the AlSb–CdTe and InAs–HgTe phase diagram [in Russian], *Izv. AN SSSR. Neorgan. Mater.*, 12(6), 1121–1122 (1976).

Kuz'mina G.A., Khabarov E.N. Obtaining of the solid solutions in the AlSb–CdTe system [in Russian], *Izv. AN SSSR. Neorgan. Mater.*, 5(1), 30–32 (1969).

Kuznetsov G.M., Bobrov A.P. Investigation of the InSb interaction with metals [in Russian], *Izv. AN SSSR. Neorgan. Mater.*, 7(5), 766–768 (1971).

Kuznetsov V.V., Stus' N.M., Talalakin G.Sh., Rubtsov E.R. Interfacial interaction in the InPAsSb system [in Russian], *Kristallografiya*, 37(4), 998–1002 (1992).

Lahav A., Eizenberg M. Interfacial reactions of Ni–Ta thin films on GaAs, *Appl. Phys. Lett.*, 45(3), 256–258 (1984).

Lakeenkov V.M., Mil'vidskiy M.G., Pelevin O.V. Phase diagram of the GaAs–ZnSe system [in Russian], *Izv. AN SSSR. Neorgan. Mater.*, 11(7), 1311–1312 (1975).

Lakeenkov V.M., Mil'vidskiy M.G., Pelevin O.V. Physico-chemical investigation of the Ga–GaAs–ZnSe quasiternary system [in Russian], *Deposited in VINITI*, № 520–74Dep (1974).

Landmann J., Sprenger J.A.P., Bertermann R., Ignat'ev N., Bernhardt-Pitchougina V., Bernhardt E., Willner H., Finze M. Convenient access to the tricyanoborate dianion $B(CN)_3{}^{2-}$ and selected reactions as a boron-centred nucleophile, *Chem. Commun.*, 51(24), 4989–4992 (2015).

Land P.L., Wimmer J.M., Barns R.W., Choudhury N.S. Compounds and properties of the system Si–Al–O–N, *J. Amer. Ceram. Soc.*, 61(1–2), 56–60 (1978).

Łapinski Z., Podsiadło S. Formation and thermal decomposition of aluminium nitroxy compounds, *J. Therm. Anal.*, 32(1), 49–53 (1987).

Lapkina I.A., Sorokina O.V., Shcherbovskiy E.Ya. Joint bismuth and cadmium solubility in indium antimonide [in Russian], *Izv. AN SSSR. Neorgan. Mater.*, 26(7), 1550–1551 (1990a).

Lapkina I.A., Sorokina O.V., Ufimtsev V.B. Investigation of the indium antimonide crystallization regions in the In–Sb–Bi–Zn system [in Russian], *Zhurn. Neorgan. Khim.*, 28(10), 2622–2626 (1983).

Lapkina I.A., Sorokina O.V., Ufimtsev V.B. Phase equilibria in the InBi–InSb–Zn_3Sb_2 system [in Russian], *Zhurn. Neorgan. Khim.*, 31(1), 206–209 (1986).

Lapkina I.A., Sorokina O.V., Zak Yu.M. Investigation of the indium antimonide crystallization fields in the In–Sb–Bi–Cd system [in Russian], *Zhurn. Neorgan. Khim.*, 35(12), 3189–3193 (1990b).

Laubengayer A.W., Condike G.F. Donor-acceptor bonding: IV. Ammonia-boron trifluoride, *J. Amer. Chem. Soc.*, 70(6), 2274–2276 (1948).

Lazar D., Ribár B., Prelesnik B. Redetermination of the structure of hexaaquaaluminium(III) nitrate trihydrate, *Acta Crystallogr.*, C47(11), 2282–2285 (1991).

Leithe-Jasper A., Tanaka T., Bourgeois L., Mori T., Michiue Y. New quaternary carbon and nitrogen stabilized polyborides: $REB_{15.5}CN$ (RE: Sc, Y, Ho, Er, Tm, Lu), crystal structure and compound formation, *J. Solid State Chem.*, 177(2), 431–438 (2004).

Liebertz J., Stähr S. Zur Existenz und Einkristallzüchtung von Zn_3BPO_7 und Mg_3BPO_7, *Z. Kristallogr.*, 160(1–2), 135–137 (1982).

Liang J.-J., Navrotsky A., Leppert V.J., Paskowitz M.J., Risbud S.H., Ludwig T., Seifert H.J., Aldinger F., Mitomo M. Thermochemistry of $Si_{6-z}Al_zO_zN_{8-z}$ (z = 0 to 3.6) materials, *J. Mater. Res.*, 14(12), 4630–4636 (1999).

Li F.F., Zhang H.J., Zhang L.N. Redetermination of $Fe_2[BP_3O_{12}]$, *Acta Crystallogr.*, E66(9), i63 (2010).

Lii K.-H. Hydrothermal synthesis and crystal structure of $Na_3In(PO_4)_2$, *Eur. J. Solid State Inorg. Chem.* 33(6), 519–526 (1996).

Lii K.-H., Ye J. Hydrothermal synthesis and structures of $Na_3In_2(PO_4)_3$ and $Na_3In_2(AsO_4)_3$: Synthetic modifications of the mineral alluaudite, *J. Solid State Chem.*, 131(1), 131–137 (1997).

Li J., Watanabe T., Sakamoto N., Wada H., Setoyama T., Yoshimura M. Synthesis of a multinary nitride, Eu-doped $CaAlSiN_3$, from alloy at low temperatures, *Chem. Mater.*, 20(6), 2095–2105 (2008a).

Li J., Watanabe T., Wada H., Setoyama T., Yoshimura M. Low-temperature crystallization of Eu-doped red-emitting $CaAlSiN_3$ from alloy-derived ammonometallates, *Chem. Mater.*, 19(15), 3592–3594 (2007).

Lim C.S. Microstructure and toughening of SiC–AlN ceramics induced by seeding with α-SiC, *J. Ceram. Proc. Res.*, 11(6), 716–720 (2010).

Lindqvist O., Sjöberg J., Hull S., Pompe R. Structural changes in O′-sialons, $Si_{2-x}Al_xN_{2-x}O_{1+x}$, $0.04 \leq x \leq 0.40$, *Acta Crystallogr.*, B47(5), 672–678 (1991).

Lingam H.K., Wang C., Gallucci J.C., Chen X., Shore S.G. New syntheses and structural characterization of NH_3BH_2Cl and $(BH_2NH_2)_3$ and thermal decomposition behavior of NH_3BH_2Cl, *Inorg. Chem.*, 51(24), 13430–13436 (2012).

Lin K.-J., Lii K.-H. $KGaAs_2O_7$, *Acta Crystallogr.*, C52(10), 2387–2389 (1996a).

Lin K.-J., Lii K.-H. Synthesis and crystal structure of a novel galloarsenate containing the $As_3O_{10}^{5-}$ triarsenate anion: $Cs_2Ga_3As_5O_{18}$, *Chem. Commun.*, (10), 1137–1138 (1996b).

Lipp C., Burns P.C. $Th_2[BO_4][PO_4]$: A rare example of an actinide borate–phosphate, *Canad. Mineralog.*, 49(5), 1211–1220 (2011).

Litvak A.M., Moiseev K.D., Stepanov M.V., Sherstnev V.V., Yakovlev Yu.P. Preparation of the In–As–Sb–P solid solutions isoperiodic to InAs and GaSb near the immiscibility region [in Russian], *Zhurn. Prikl. Khim.*, 67(12), 1957–1960 (1994).

Litvak A.M., Nogovitsa D., Charykov N.A., Puchkov L.V., Yakovlev Yu.P., Klepikov V.V., Izotova S.G., Nikitin V.A., Zubkova M.Yu. Phase equilibria melt–solid in the In–Ga–As–P system [in Russian], *Zhurn. Neorgan. Khim.*, 44(4), 633–638 (1999).

Liu X.-X., Wang C.-X., Luo S.-M., Mi J.-X. The layered monodiphosphate $Li_9Ga_3(P_2O_7)_3(PO_4)_2$ refined from X-ray powder data, *Acta Crystallogr.*, E62(5), i112–i113 (2006).

Li Y.Q., Hirosaki N., Xie R.J., Takeda T., Mitomo M. Yellow-orange-emitting $CaAlSiN_3:Ce^{3+}$ phosphor: Structure, photoluminescence, and application in white LEDs, *Chem. Mater.*, 20(21), 6704–6714 (2008b).

Li Z., Wu Y., Fu P., Pan S., Lin Z., Chen C. Czochralski crystal growth and properties of $Na_5[B_2P_3O_{13}]$, *J. Cryst. Growth*, 255(1–2), 119–122 (2003).

Loong C.-K., Richardson, Jr. J.W., Sukuzi S., Ozawa M. Crystal phase and phonon densities of states of β′-SiAlON ceramics $Si_{6-z}Al_zO_zN_{8-z}$ ($0 \leq z \leq 4$), *J. Amer. Ceram Soc.*, 79(12), 3250–3256 (1996).

Lushnaja N.P., Medwedjewa Z.S., Koppel H.D. Löslichkeit einiger III-V-Verbindungen in Schmelzen, *Z. anorg. allg. Chem.*, 344(5–6), 323–328 (1966).

Lutsko V.A., Pap O.G., Ksenofontova N.M. An X-ray and spectroscopic investigations of binary triphosphates $M^{III}Cs_2P_3O_{10}$ where M^{III}–Al, Ga, Cr, Fe [in Russian], *Izv. AN SSSR. Neorgan. Mater.*, 22(8), 1373–1377 (1986).

Lux C., Wenski G., Mewis A. $Eu_2Pt_7AlP_{\sim3}$ und isotype Verbindungen: Eine neue Struktur aus $CaBe_2Ge_2$- und Cu_3Au-Einheiten, *Z. Naturforsch.*, 46B(8), 1035–1038 (1991).

Luzhnaya N.P., Medvedeva Z.S., Koppel Kh.D. On the interaction of indium arsenide with cadmium iodide [in Russian], *Zhurn. Neorgan. Khim.*, 10(10), 2320–2323 (1965).

Lv W., Yang C., Lin J., Hu X., Guo K., Yang X., Luo J., Zhao J.-T. Cd substitution in Zintl phase $Eu_5In_2Sb_6$ enhancing the thermoelectric performance, *J. Alloys Compd.*, 726, 618–622 (2017).

Ma H.W., Liang J.K., Wu L., Liu G.Y., Rao G.H., Chen X.L. Ab initio structure determination of new compound $Ba_3(BO_3)(PO_4)$, *J. Solid State Chem.*, 177(10), 3454–3459 (2004).

Makino A., Kubota T., Chang C., Makabe M., Inoue A. FeSiBP bulk metallic glasses with unusual combination of high magnetization and high glass-forming ability, *Mater. Trans., JIM*, 48(11), 3024–3027 (2007).

Makino A., Kubota T., Makabe M., Chang C.T., Inoue A. FeSiBP metallic glasses with high glass-forming ability and excellent magnetic properties, *Mater. Sci. Eng. B*, 148(1–3), 166–170 (2008).

Maksimova N.K., Poznyakov A.G., Kravtsov V.I., Krasil'nikova L.M., Yakubenya M.P., Yanovskiy V.P. Interphase interaction in the Au/V/GaAs contacts [in Russian], *Poverkhnost'. Fiz., Khim., Mekh.*, (10), 96–104 (1991).

Ma K.Y., Fang Z.M., Jaw D.H., Cohen R.M., Stringfellow G.B., Kosar W.P., Brown D.W. Organo-metallic vapor phase epitaxial growth and characterization of InAsBi and InAsSbBi, *Appl. Phys. Lett.*, 55(23), 2420–2422 (1989).

Mallinson P.M., Gál Z.A., Clarke S.J. Two new structurally related strontium gallium nitrides: Sr_4GaN_3O and $Sr_4GaN_3(CN)_2$, *Inorg. Chem.*, 45(1), 419–423 (2006).

Manoun B., Saxena S.K., Hug G., Ganguly A., Hoffman E.N., Barsoum M.W. Synthesis and com-pressibility of $Ti_3(Al,Sn_{0.2})C_2$ and $Ti_3Al(C_{0.5}N_{0.5})_2$, *J. Appl. Phys.*, 101(11), 113523_1–113523_7 (2007).

Mao H., Selleby M. Thermodynamic reassessment of the Si_3N_4–AlN–Al_2O_3–SiO_2 system— Modeling of the SiAlON and liquid phase, *CALPHAD: Comput. Coupling Phase Diagr. and Thermochem.*, 31(2), 269–280 (2007).

Marchand R. Oxynitrures à structure K_2NiF_4. Les composés Ln_2AlO_3N (Ln = La, Na, Sm), *C. r. Acad. Sci. Sér. C*, 282(7), 329–331 (1976).

Martin R., Duc-Maugé C., Guérin H. Sur les phosphates d'aluminium: Diagramme d'équilibre P_2O_5–Al_2O_3–H_2O à 60°, *Bull. Soc. Chim. Fr.*, (5), 851–856 (1960).

Masquelier C., d'Yvoire F., Bretey E., Berthet P., Peytour-Chansac C. A new family of sodium ion conductots: The diphosphates and diarsenites $Na_7M_3(X_2O_7)_4$; (M = Al, Ga, Cr, Fe; X = P, As), *Solid State Ionics*, 67(3–4), 183–189 (1994).

Masquelier C., d'Yvoire F., Collin G. Crystal structure of $Na_7Fe_4(AsO_4)_6$ and α-$Na_3Al_2(AsO_4)_3$, two sodium ion conductors structurally related to II-$Na_3Fe_2(AsO_4)_3$, *J. Solid State Chem.*, 118(1), 33–42 (1995).

Matsumoto H., Urata A., Yamada Y., Makino A. To enhance the efficiency of a power supply circuit by the use of Fe–P–B–Nb-type ultralow loss glassy metal core, *J. Appl. Phys.*, 105(7), 07A317_1–07A317_4 (2009).

Matsuoka T. Calculation of unstable mixing region in wurtzite $In_{1-x-y}Ga_xAl_yN$, *Appl. Phys. Lett.*, 71(1), 105–106 (1997).

Matsuoka T. Unstable mixing region in wurtzite $In_{1-x-y}Ga_xAl_yN$, *J. Cryst. Growth*, 189–190, 19–23 (1998).

Matsuoka T., Yoshimoto N., Sasaki T., Katsui A. Wide-gap semiconductor InGaN and InGaAlN grown by MOVPE, *J. Electron. Mater.*, 21(2), 157–163 (1992).

Mavlonov Sh., Makhmatkulov M. Study of separate and joint solubility of thallium and selenium in indium arsenide [in Russian], *Izv. AN TadzhSSR. Otd-nie fiz.-mat., khim. i geol. n.*, (4), 86–90 (1983).

McConnell D. Are vashegyite and kingite hydrous aluminum phyllophosphates with kaolinite-type structures?, *Mineralog. Mag.*, 39(307), 802–806 (1974).

McConnell D. Clinobarrandite and the isodimorphous series, variscite-metavariscite, *Amer. Mineralog.*, 25(11), 719–725 (1940).

Medraj M., Baik Y., Thompson W.T., Drew R.A.L. Understanding AlN sintering through compu-tational thermodynamics combined with experimental investigation, *J. Mater. Proc. Technol.*, 161(3), 415–422 (2005).

Medraj M., Hammond R., Thompson W.T., Drew R.A.L. High-temperature neutron diffraction of the AlN–Al_2O_3–Y_2O_3 system, *J. Amer. Ceram. Soc.*, 86(4), 717–726 (2003a).

Medraj M., Hammond R., Thompson W.T., Drew R.A.L. Phase equilibria, thermodynamic modeling and neutron diffraction of the $AlN–Al_2O_3–Y_2O_3$, *Canad. Metallurg. Quart.*, 42(4), 495–507 (2003b).

Meisel M., Päch M., Wilde L., Wulff-Molder D. Synthese und Kristallstruktur von Vanadium(III)-borophosphat, $V_2[B(PO_4)_3]$, *Z. anorg. allg. Chem.*, 630(7), 983–985 (2004).

Metselaar R. Terminology for compounds in the Si–Al–O–N system, *J. Eur. Ceram. Soc.*, 18(3), 183–184 (1998).

Michor H., Krendelsberger R., Hilscher G., Bauer E., Dusek C., Hauser R., Naber L., Werner D., Rogl P., Zandbergen H.W. Superconducting properties of $La_3Ni_2B_2N_{3-\delta}$, *Phys. Rev. B*, 54(13), 9408–9420 (1996).

Mi J.-X., Zhao J.-T., Mao S.-Y., Huang Y.-X., Engelhardt H., Kniep R. Crystal structure of dichromium monoborotriphosphate, $Cr_2[BP_3O_{12}]$, *Z. Kristallogr., New Cryst. Str.*, 215(2), 201–202 (2000).

Mikami M., Uheda K., Kijima N. First-principles study of nitridoaluminosilicate $CaAlSiN_3$, *Phys. Stat. Sol (a)*, 203(11), 2705–2711 (2006).

Mikami M., Watanabe H., Uheda K., Kijima N. Nitridoaluminosilicate $CaAlSiN_3$ and its derivatives—Theory and experiment, *Mater. Res. Soc. Symp. Proc.*, 1040, 1040-Q10–09 (2008).

Miller L., Kaplan W.D. Solubility limits of La and Y in aluminum oxynitride at 1870°C, *J. Amer. Ceram. Soc.*, 91(5), 1693–1696 (2008).

Mills S.J., Birch W.D., Kampf A.R., Wambeke van L. Kobokoboite, $Al_6(PO_4)_4(OH)_6\cdot11H_2O$, a new mineral from the Kobokobo pegmatite, Democratic Republic of Congo, *Eur. J. Mineral.*, 22(2), 305–308 (2010).

Mitomo M., Kuramoto N. The formation of β-sialon in the system $S_3N_4–SiO_2–AlN$ (in Japanese), *J. Ceram. Soc. Jpn.*, 87(3), 141–146 (1979).

Mitomo M., Kuramoto N., Tsutsumi M., Suzuki H. The formation of single phase Si–Al–O–N ceramics, *J. Ceram. Soc. Jpn.*, 86(11), 526–531 (1978).

Miyake M., Nakamura K., Akiyama T., Ito T. Structural stability of Mn-doped GaInAs and GaInN alloys, *J. Cryst. Growth*, 362, 324–326 (2013).

Mocellin A., Bayer G. Chemical and microstructural investigations of high-temperature interactions between AlN and TiO_2, *J. Mater. Sci.*, 20(10), 3697–3704 (1985).

Molinie P., Brec R., Rouxel J., Herpin P. Structures des amidoaluminates alcalins $MAl(NH_2)_4$ (M = Na, K, Cs). Structure de l'amidogallate de sodium $NaGa(NH_2)$, *Acta Crystallogr.*, B29(5), 925–934 (1973).

Mooney-Slater R.C.L. The crystal structure of hydrated gallium phosphate of composition $GaPO_4\cdot2H_2O$, *Acta Crystallogr.*, 20(4), 526–534 (1966).

Mooney-Slater R.C.L. X-ray diffraction study of indium phosphate dihydrate and isostructural thallic compounds, *Acta Crystallogr.*, 14(11), 1140–1146 (1961).

Moore P.B., Araki T. Trolleite, $Al_4(OH)_3[PO_4]_3$: A very dense structure with octahedral face-sharing dimers, *Amer. Mineralog.*, 59(9–10), 974–984 (1974).

Moore P.B., do Prado Barbosa C., Gaines R.V. Bahianite, $Sb_3Al_5O_{14}(OH)_2$, a new species, *Mineralog. Mag.*, 42(322), 179–182 (1978).

Morgan M.G., Wang M., Mills A.M., Mar A. Lanthanum gallium tin antimonides $LaGa_xSn_ySb_2$, *J. Solid State Chem.*, 167(1), 41–47 (2002).

Mori T., Nishimura T. Thermoelectric properties of homologous p- and n-type boron-rich borides, *J. Solid State Chem.*, 179(9), 2908–2915 (2006).

Morozov V.A., Belik A.A., Stefanovich S.Yu., Grebenev V.V., Lebedev O.I., Van Tendeloo G., Lazoryak B.I. High-temperature phase transition in the whitlockite-type phosphate $Ca_9In(PO_4)_7$, *J. Solid State Chem.*, 165(2), 278–288 (2002).

Morozov V.N., Chernov V.G. Phase equilibria in the InAs–CdTe system [in Russian], *Izv. AN SSSR. Neorgan. Mater.*, 15(8), 1324–1329 (1979).

Morozov V.N., Karnauhova E.N., Skorobogatova L.A., Riazantsev A.A. Apparatus for differential thermal analysis and investigation of InSb–CdTe phase diagram [in Russian], *Izv. Sib. otd. AN SSSR. Ser. Khim. Nauk*, 4(9), 52–56 (1974).

Motrya S.F., Pritz I.P., Potoriy M.V., Tovt V.V., Milyan P.M. Physical-chemical interaction in the $CuInP_2Se_6$–$In_4(P_2Se_6)_3$ system and construction of the isothermal section of the Cu_2Se–In_2Se_3–P_2Se_4 quasiternary system [in Ukrainian], *Visnyk Uzhgorod. nats. un-tu. Ser. Khimiya*, (22), 183–187 (2009).

Mukerji J., Biswas S.K. Synthesis, properties, and oxidation of alumina–titanium nitride composites, *J. Amer. Ceram. Soc.*, 73(1), 142–145 (1990).

Mukerji J. Stabilization of cubic zirconia by aluminum nitride, *J. Amer. Ceram. Soc.*, 72(8), 1567–1568 (1989).

Müller A., Apelt R., Iacobs B., Butter E. Melting diagrams of the quaternary systems Ga–In–As–Ge and Ga–In–As–Sn, *Cryst. Res. Technol.*, 17(10), 1227–1232 (1982).

Müller E.K., Richards J.L. Miscibility of III-V semiconductors studied by flash evaporation, *J. Appl. Phys.*, 35(4), 1233–1241 (1964).

Murashova E.V., Chudinova N.N. Condensed cesium indium phosphates, *Inorg. Mater.*, 37(12), 1521–1524 (2001).

Nahory R.E., Pollack M.A., Beebe E.D., DeWinter J.C., Ilegems M. The liquid phase epitaxy of $Al_yGa_{1-y}As_{1-x}Sb_x$ and the importance of strain effects near the miscibility gap, *J. Electrochem. Soc.*, 125(7), 1053–1058 (1978).

Naik I.K., Gauckler L.J., Tien T.Y. Solid-liquid equilibria in the system Si_3N_4–AlN–SiO_2–Al_2O_3, *J. Amer. Ceram. Soc.*, 61(7–8), 332–335 (1978).

Nakajima K., Kusunoki T., Akita K., Kotani T. Phase diagram of the In–Ga–As–P quaternary system and LPE growth conditions for lattice matching of InP substrates, *J. Electrochem. Soc.*, 125(1), 123–127 (1978).

Nakajima K., Osamura K., Murakami Y. Phase diagram of Al–Ga–In–As quaternary system, *J. Electrochem. Soc.*, 122(9), 1245–1248 (1975).

Nakajima K., Osamura K., Yasuda K., Murakami Y. The pseudoquaternary phase diagram of the Ga–In–As–Sb system, *J. Cryst. Growth*, 41(1), 87–92 (1977).

Nakamura K., Miyake M., Akiyama T., Ito T. Phase stability of Mn-doped GaInAs alloys, *J. Cryst. Growth*, 318(1), 360–362 (2011).

Navrotsky A., Risbud S.H., Liang J.-J., Leppert V.J. Thermochemical insights into rapid solid-state reaction synthesis of β-sialon, *J. Phys. Chem. B*, 101(46), 9433–9435 (1997).

Nazarov A., Vaypolin A.A., Prochukhan V.D., Goryunova N.A. Preparation of the solid solutions along the $2GaAs$–$ZnGeAs_2$, $2GaAs$–$ZnSiAs_2$ sections and some their physicochemical properties [in Russian], *Izv. AN SSSR. Neorgan. Mater.*, 3(12), 2269–2270 (1967).

Neher R., Herrmann M., Fabrichnaya O., Pavlyuchkov D., Seifert H.J. Liquid phase formation in the system AlN–Al_2O_3–Y_2O_3. Part I: Experimental investigations, *J. Eur. Ceram. Soc.*, 33(13–14), 2447–2455 (2013).

Neukirch M., Blaschkowski B., Häberlen M., Meyer H.-J. Rare-earth nickel nitridoborates with (BN) anions: Characterized *RE*Ni(BN) and anticipated *RE*M(BN) compounds, *Z. anorg. allg. Chem.*, 632(10–11), 1799–1803 (2006).

Ng H.N., Calvo C. The crystal structure of $KAlP_2O_7$, *Canad. J. Chem.*, 51(16), 2613–2620 (1973).

Nicolich J.P., Hofer F., Brey G., Riedel R. Synthesis and structure of three-dimensionally ordered graphitlike BC_2N ternary crystals, *J. Amer. Ceram. Soc.*, 84(2), 279–282 (2001).

Nilsson J., Landa-Cánovas A.R., Hansen S., Andersson A. The Al–Sb–V–oxide system for propane ammoxidation: A study of regions of phase formation and catalytic role of Al, Sb, and V, *J. Catal.*, 160(2), 244–260 (1996).

Novikova E.M., Ahverdov O.S., Ershova S.A. Investigation of the Ga–GaAs–ZnSe quasiternary system in high temperature region [in Russian], *Deposited in VINITI*, № 4439–77Dep (1977).

Novikova E.M., Vasil'yev M.G., Krapuhin V.V., Evseev V.A., Ershova S.A. Phase diagram of the Ga–GaAs–ZnSe quasiternary system from the Ga-rich side [in Russian], *Deposited in VINITI*, № 3038–74Dep (1974).

Novikov M.V., Bezhenar M.P., Bozhko S.A. Evolution of the crystal structure of sphaleritic boron nitride at the sintering of $BN_{cф}$–AlN and $BN_{cф}$–TiC components and and its influence on the hardness [in Ukrainian], *Dop. Nats. AN Ukrainy*, (6), 118–122 (1997).

Oberländer A., Kinski I., Zhu W., Pezzotti G., Michaelis A. Structure and optical properties of cubic gallium oxynitride synthesized by solvothermal route, *J. Solid State Chem.*, 200, 221–226 (2013).

Oden L.L., McCune R.A. Contribution to the phase diagram Al_4C_3–AlN–SiC, *J. Amer. Ceram. Soc.*, 73(6), 1529–1533 (1990).

Ohse L., Somer M., Blase W., Cordier G. Verbindungen mit SiS_2-isosteren Anionen $^1_\infty[AlX_{4/2}{}^{3-}]$ und $^1_\infty[InP_{4/2}{}^{3-}]$: Synthesen, Kristallstrukturen und Schwingungsspektren von $Na_3[AlX_2]$, $K_2Na[AlX_2]$ und $K_3[InP_2]$ (X = P, As), *Z. Naturforsch.*, 48B(8), 1027–1034 (1993).

Onabe K. Unstable region in quaternary $In_{1-x}Ga_xAs_{1-y}Sb_y$ calculated using strictly regular solution approximation, *Jpn. J. Appl. Phys.*, 21(6), Pt. 1, 964 (1982a).

Onabe K. Unstable regions in III–V quaternary solid solutions composition plane calculated with strictly regular solution approximation, *Jpn. J. Appl. Phys.*, 21(6), Pt. 2, L323–L325 (1982b).

Onabe K. Calculation of miscibility gap in quaternary InGaPAs with strictly regular solution approximation, *Jpn. J. Appl. Phys.*, 21(5), Pt. 1, 797–798 (1982c).

Onac B.P., Kearns J., Breban R., Pânzaru S.C. Variscite ($AlPO_4·2H_2O$) from Cioclovina Cave (Şureanu mountains, Romania): A tale of a missing phosphate, *Studia universitatis Babeş-Bolyai, Geologia*, 99(1), 3–14 (2004).

Ordan'yan S.S., Vikhman S.V., Osmakov A.S. Activation of the interaction processes in the SiC–AlN system [in Russian], *Zhurn. Prikl. Khim.*, 70(5), 717–721 (1997).

Ouili Z., Leblanc A., Colombet P. Crystal structure of a new lamellar compound: $Ag_{1/2}In_{1/2}PS_3$, *J. Solid State Chem.*, 66(1), 86–94 (1987).

Oyama Y. Solid solution in the system Si_3N_4–AlN–Al_2O_3, *J. Ceram. Soc. Jpn.*, 82(947), 351–357 (1974).

Oyama Y. Solid solution in the ternary system, Si_3N_4–AlN–Al_2O_3, *Jpn. J. Appl. Phys.*, 11(5), 760–761 (1972).

Öztürk S.S., Kokal I., Somer M. Crystal structure of strontium octabarium hexakis(dinitridoborate) and europium octabarium hexakis(dinitridoborate), $MBa_8[BN_2]_6$ (M = Sr, Eu), *Z. Kristallogr., New Cryst. Str.*, 220(3), 303–304 (2005).

Palkina K.K., Maksimova S.I., Chibiskova N.T., Dzhurinskiy B.F., Gokhman L.Z. Synthesis and structure of mixed boratophosphates of rare-earth elements $Ln_7O_6(BO_3)(PO_4)_2$ (Ln–La–Dy) [in Russian], *Izv. AN SSSR. Neorgan. Mater.*, 20(6), 1063–1067 (1984).

Palkina K.K., Maksimova S.I., Chibiskova N.T. The structure of $LiGa(PO_3)_4$ crystals [in Russian], *Izv. AN SSSR. Neorgan. Mater.*, 17(1), 95–98 (1981).

Palkina K.K., Maksimova S.I., Kuznetsov V.G., Chudinova N.N. Structure of crystals of double octametaphosphate $K_2Ga_2P_8O_{24}$ [in Russian], *Dokl. AN SSSR*, 245(6), 1386–1389 (1979).

Panish M.B. Phase equilibria in the system Al–Ga–As–Sn and electrical properties of Sn-doped liquid phase epitaxial $Al_xGa_{1-x}As$, *J. Appl. Phys.*, 44(6), 2667–2675 (1973a).

Panish M.B. The Ga–As–Ge–Sn system: 800°C liquidus isotherm and electrical properties of Ge–Sn-doped GaAs, *J. Appl. Phys.*, 44(6), 2676–2680 (1973b).

Pan M.-Y., Ma Z.-J., Liu X.-C., Xia S.-Q., Tao X.-T., Wu K.-C. $Ba_4AgGa_5Pn_8$ (Pn = P, As): New pnictide-based compounds with nonlinear optical potential, *J. Mater. Chem. C*, 3(37), 9695–9700 (2015).

Pan S., Wu Y., Fu P., Zhang G., Li Z., Du C., Chen C. Growth, structure, and properties of single crystals of $SrBPO_5$, *Chem. Mater.*, 15(11), 2218–2221 (2003).

Pan S., Wu Y., Fu P., Zhang G., Wang G., Guan X., Chen C. The growth of $BaBPO_5$ crystals from $Li_4P_2O_7$ flux, *J. Cryst. Growth*, 236(4), 613–616 (2002).

Pâques-Ledent M.Th., Tarte P. Spectre infa-rouge et structure des composés $GaPO_4 \cdot 2H_2O$ et $GaAsO_4 \cdot 2H_2O$, *Spectrochim. Acta, Part A*, 25(6), 1115–1125 (1969).

Park C.-H., Bluhm K. $Pb_6(AsO_4)[B(AsO_4)_4]$—ein neuartiger Kristallstrukturtyp im System PbO/B_2O_3/As_2O_5 mit einem Beitrag über $Pb(BAsO_5)$, *Z. Naturforsch.*, 51B(3), 313–318 (1996).

Park C.H., Bluhm K. Synthese und Kristallstruktur von Triblei-Diphosphato-Borat-Phosphat, eine Verbindung mit einem $^1_\infty[(PO_4)_2BPO_4]^{6-}$ Anion, *Z. Naturforsch.*, 50B(11), 1617–1622 (1995).

Park D.G., Dong Y., DiSalvo F.J. $Sr(Mg_3Ge)N_4$ and $Sr(Mg_2Ga_2)N_4$: New isostructural Mg-containing quaternary nitrides with nitridometallate anions of $^3_\infty[(Mg_3Ge)N_4]^{2-}$ and $^3_\infty[(Mg_2Ga_2)N_4]^{2-}$ in a 3D-network structure, *Solid State Sci.*, 10(12), 1846–1852 (2008).

Park D.G., Gál Z.A., DiSalvo F.J. Synthesis and structure of $LiSrGaN_2$: A new quaternary nitride with interpenetrating two-dimensional networks, *J. Alloys Compd.*, 353(1–2), 107–113 (2003).

Parry R.W., Schultz D.R., Girardot P.R. The preparation and properties of hexamminecobalt(III) borohydride, hexamminechromium(III) borohydride and ammonium borohydride, *J. Amer. Chem. Soc.*, 80(1), 1–3 (1958).

Pashaev A.M., Ismailov A.M., Aliev I.I., Kuliev K.G. Synthesis and physico-chemical investigation of the $(InSb)_{1-x}(HgTe)_x$ solid solutions [in Russian], *Zhurn. Khim. Problem*, (1), 119–121 (2006).

Pavlyuchkov D., Fabrichnaya O., Herrmann M., Seifert H.J. Thermodynamic assessments of the Al_2O_3–Al_4C_3–AlN and Al_4C_3–AlN–SiC systems, *J. Phase Equilibr. Dif.*, 33(5), 357–368 (2012).

Peacock M.A., Moddle D.A. On a crystal of augelite from California, *Mineralog. Mag.*, 26(175), 105–115 (1941).

Pessetto J.R., Stringfellow G.B. $Al_xGa_{1-x}As_ySb_{1-y}$ phase diagram, *J. Cryst. Growth*, 62(1), 1–6 (1983).

Pfeiff R., Kniep R. Quaternary selenodiphosphates (IV): $M^IM^{III}[P_2Se_6]$, (M^I = Cu, Ag, M^{III} = Cr, Al, Ga, In), *J. Alloys Compd.*, 186(1), 111–133 (1992).

Pfitzner A., Andratschke M., Rau F., Brunklaus G., Eckert H. Präparation, Kristallstruktur und NMR-Spektroskopie an AgAlP2Q6 (Q = S, Se), *Z. anorg. allg. Chem.*, 630(11), 1752 (2004).

Piao X., Machida K.-I., Horikawa T., Hanzawa H., Shimomura Y., Kijima N. Preparation of $CaAlSiN_3$:Eu^{2+} phosphors by the self-propagating high-temperature synthesis and their luminescent properties, *Chem. Mater.*, 19(18), 4592–4599 (2007).

Pieczka A., Evans R.J., Grew E.S., Groat L.A., Ma C., Rossman G.R. Szklaryite, IMA 2012–070. CNMNC Newsletter No. 15, February 2013, page 8, *Mineralog. Mag.*, 77(1), 1–12 (2013).

Pietzka M.A., Schuster J.C. Phase equilibria in the quaternary system Ti–Al–C–N, *J. Amer. Ceram. Soc.*, 79(9), 2321–2330 (1996).

Pietzka M.A., Schuster J.C. Phase equilibria of the quaternary system Ti–Al–Sn–N at 900°C, *J. Alloys Compd.*, 247(1–2), 198–201 (1997).

Pilet G., Grins J., Edén M., Esmaeilzadeh S. $La_{17}Si_9Al_4N_{32-x}O_x$ ($x \leq 1$): A nitridoaluminosilicate with isolated Si/Al–N/O clusters, *Eur. J. Inorg. Chem.*, (18), 3627–3633 (2006).

Poisson S., d'Yvoire F., Nguyen-Huy-Dung, Bretey E., Berthet P. Crystal structure and cation transport properties of the layered monodiphosphates: $Li_9M_3(P_2O_7)_3(PO_4)_2$ (M = Al, Ga, Cr, Fe), *J. Solid State Chem.*, 138(1), 32–40 (1998).

Potoriy M.V., Pritz I.P., Motrya S.F., Milyan P.M., Tovt V.V. Interaction of the components in the Cu_2Se–In_2Se_3–P_2Se_4 quasiternary system and construction of the phase equilibria [in Ukrainian], *Nauk. visnyk Volyns'k. nats. un-tu im. Lesi Ukrayinky*, (30), 27–33 (2010).

Puris T.E., Belaya A.D., Zemskov V.S., Shvarts N.N. Phase equilibria in the In–Sb–Zn–Te system [in Russian], *Izv. AN SSSR. Neorgan. Mater.*, 6(10), 1811–1815 (1970).

Qiu C., Metselaar R. Phase relations in the aluminum carbide–aluminum nitride–aluminum oxide system, *J. Amer. Ceram. Soc.*, 80(8), 2013–2020 (1997).

Radovic M., Ganguly A., Barsoum M.W. Elastic properties and phonon conductivities of $Ti_3Al(C_{0.5},N_{0.5})_2$ and $Ti_2Al(C_{0.5},N_{0.5})$ solid solutions, *J. Mater. Res.*, 23(6), 1517–1521 (2008).

Rafaevich N.B., Tsvetkov V.F., Komov A.N., Losevskaya S.G. Formation of the solid solutions in the SiC–AlN system at the sintering of fine powders [in Russian], *Izv. AN SSSR. Neorgan. Mater.*, 26(5), 973–977 (1990).

Rafaniello W., Cho K., Virkar A.V. Fabrication and characterization of SiC–AlN alloys, *J. Mater. Sci.*, 16(12), 3479–3488 (1981).

Rafaniello W., Plichta M.R., Virkar A.V. Investigation of phase stability in the system SiC–AlN, *J. Amer. Ceram. Soc.*, 66(4), 272–276 (1983).

Raghavan V. Al–As–Ni (aluminum–arsenic–nickel), *J. Phase Equilibr. Dif.*, 27(5), 484–485 (2006).

Raghavan V. Al–C–N–Ti (aluminum–carbon–nitrogen–titanium), *J. Phase Equilibr. Dif.*, 27(2), 166–168 (2006°).

Raghavan V. Al–Cu–N–Ti (aluminum–copper–nitrogen–titanium), *J. Phase Equilibr. Dif.*, 27(2), 171–172 (2006b).

Raghavan V. Phase diagram updates and evaluations of the Al–Fe–P–Zn, Al–Fe–V–Zn, C–Fe–Mn–Nb and C–Fe–N–Ti–V systems, *J. Phase Equilibr. Dif.*, 34(3), 244–250 (2013).

Rahab H., Ouadjaout D., Manseri A., Viraphong O., Chaminade J.P. Growth by the heat exchanger method of $NaBiW_2O_8$ and $Na_5B_2P_3O_{13}$ crystals, *J. Cryst. Growth*, 303(2), 629–631 (2007).

Ramachandran P.V., Gagare P.D. Preparation of ammonia borane in high yield and purity, methanolysis, and regeneration, *Inorg. Chem.*, 46(19), 7810–7817 (2007).

Reckeweg O., Meyer H.-J. Unsymmetrische BN_2^{3-}Ionen in den Strukturen von Ca_2ClBN_2 und Sr_2ClBN_2, *Z. Naturforsch.*, 52B(3), 340–344 (1997).

Reckeweg O., Reiherzer J., Schulz A., DiSalvo F.J. The last missing member of the $AE_2[BN_2]Cl$ series—Synthesis, structural and spectroscopic characterization of $Ba_2[BN_2]Cl$, *Z. Naturforsch.*, 63B(5), 525–529 (2008).

Riazantsev A.A., Telegina M.P. Liquidus surfaces and thermodynamic analysis of InSb–CdTe–In, InSb–CdTe–Cd and InAs–CdTe–In phase diagrams [in Russian], *Zhurn. Neorgan. Khim.*, 23 (8), 2211–2216 (1978).

Riazantsev A.A., Karnauhova E.N., Kuz'mina G.A. Phase diagrams and solubility of components in the A^3B^5–CdTe systems [in Russian], *Zhurn. Neorgan. Khim.*, 25(3), 802–805 (1980).

Rizzol C., Salamakha P.S., Sologub O.L., Bocell G. X-ray investigation of the Al–B–N ternary system: Isothermal section at 1500°C: Crystal structure of the $Al_{0.185}B_6CN_{0.256}$ compound, *J. Alloys Compd.*, 343(1–2), 135–141 (2002).

Robbins M., Lambrecht, V.G.Jr. Preparation and structural investigation of compounds of the type $In_{3-a}Ga_aX_3Y$ where X = S, Se, Y = P, As and a = 1, 2, *Mater. Res. Bull.*, 10(5), 331–334 (1975).

Rogl P., Klesnar H., Fischer P. Neutron powder diffraction of Nb_2BN_{1-x}, *J. Amer. Ceram. Soc.*, 71 (10), C450–C452 (1988).

Rohrer F.E., Nesper R. $Ba_8(BN_2)_5F$: A barium fluoride compound with isolated BN_2^{3-} units, *J. Solid State Chem.*, 142(1), 192–198 (1999a).

Rohrer F.E., Nesper R. M_2BN_2X (M = Ca, Sr; X = F, Cl): New halogenide compounds with isolated BN_2^{3-} units, *J. Solid State Chem.*, 135(2), 194–200 (1998).

Rohrer F.E., Nesper R. Sr_2BN_2I: A strontium iodide compound with isolated BN_2^{3-} units, *J. Solid State Chem.*, 142(1), 187–191 (1999b).

Ruh R., Bentsen L.D., Hasselman D.P.H. Thermal diffusivity anisotropy of SiC/BN composites, *J. Amer. Ceram. Soc.*, 67(5), C83–C84 (1984).

Ruh R., Zangvil A. Compositions and properties of hot-pressed SiC–AlN solid solutions, *J. Amer. Ceram. Soc.*, 65(5), 260–265 (1982).

Rustamov P.G., Aliev I.I., Safarov M.G. As_2S_3–InSe system [in Russian], *Zhurn. Neorgan. Khim.*, 25 (4), 1073–1077 (1980).

Safaraliev G.K., Tairov Yu. M., Tsvetkov V.F. Wide-band $(SiC)_{1-x}(AlN)_x$ solid solutions [in Russian], *Fiz. i tekhn. poluprovod.*, 25(8), 1437–1447 (1991).

Safaraliev G.K., Tairov Yu. M., Tsvetkov V.F., Shabanov Sh.Sh., Pashchuk E.G., Ofitserova N.V., Avrov D.D., Sadykov S.A. Preparation and properties of polycrystalline SiC–AlN solid solutions [in Russian], *Fiz. i tekhn. poluprovod.*, 27(3), 402–408 (1993).

Safarov M.G., Gamidov R.S., Poladov P.M., Bagirova E.M. InS–As$_2$S$_3$–As$_2$Te$_3$ ternary system [in Russian], *Zhurn. Neorgan. Khim.*, 37(6), 1384–1391 (1992).

Safarov M.G., Mekhrabov A.O., Orudzhev N.M. Investigation of the Sb$_2$Te$_3$–GaS system [in Russian], *Azerb. Khim. Zhurn.*, (2), 115–119 (1988).

Safarov M.G., Salmanov S.M. Investigation of the Sb$_2$S$_3$–GaSe system [in Russian], *Azerb. Khim. Zhurn.*, (3), 120–125 (1975).

Saidov M.S., Koshchanov E.A., Dadamukhamedov S., Saidov A.S. Solubility in the liquid phase and growth of Al$_x$Ga$_{1-x}$As epitaxial layers on GaP [in Russian], In: *Poluprovodn. arsenidy i fosfidy elementov III gruppy*. Tashkent: Publ. House of AN UzSSR, 23–40 (1981).

Saidov M.S., Koshchanov E.A., Saidov A.S., Rysaieva V.A. The study of the gallium region of the phase diagram of the Ga–GaP–Al–Cr and Ga–GaAs–Al–Cr four-component systems [in Russian], *Dokl. AN UzSSR*, (11), 34–36 (1979).

Salvador P.S., Fayos J. Some aspects of the structural relationship between «Messbach-type» and «Lucin-type» variscites, *Amer. Mineralog.*, 57(1–2), 36–44 (1972).

Schlenker D., Miyamoto T., Chen Z., Kawaguchi M., Kondo T., Gouardes E., Gemmer J., Gemmer C., Koyama F., Iga K. Inclusion of strain effect in miscibility gap calculations for III-V semiconductors, *Jpn. J. Appl. Phys.*, 39(10), Pt. 1, 5751–5757 (2000).

Schlenker D., Miyamoto T., Pan Z., Koyama F., Iga K. Miscibility gap calculation for Ga$_{1-x}$In$_x$N$_y$As$_{1-y}$ including strain effects, *J. Cryst. Growth*, 196(1), 67–70 (1999).

Schlesinger H.I., Ritter D.M., Burg A.B. Hydrides of boron. X. The preparation and preliminary study of the new compound B$_2$H$_7$N, *J. Amer. Chem. Soc.*, 60(10), 2297–2300 (1938).

Schmid-Beurmann P., Brunet F., Kahlenberg V., Dachs E. Polymorphism and thermochemistry of MgAlPO$_4$O, a product of lazulite breakdown at high temperature, *Eur. J. Mineralog.*, 19(2), 159–172 (2007).

Schmid-Beurmann P., Morteani G., Cemič L. Experimental determination of the upper stability of scorzalite, FeAl$_2$[OH/PO$_4$]$_2$, and the occurrence of minerals with a composition intermediate between scorzalite and lazulite(ss) up to the conditions of the rystalline facies, *Mineral. Petrol.*, 61(1–4), 211–222 (1997).

Schneider G., Gauckler L.J., Petzow G. Phase equilibria in the Si,Al,Be/C,N system, *Ceramurgia Int.*, 5(3), 101–104 (1979).

Schneider G., Gauckler L.J., Petzow G. Phase equilibria in the system AlN–Si$_3$N$_4$–Be$_3$N$_2$, *J. Amer. Ceram. Soc.*, 63(1–2), 32–35 (1980).

Schwarz H., Schmidt L. Neue Verbindungen mit Granatstruktur. IV. Arsenate des Typs {Na$_3$}[M$^{III}_2$](As$_3$)O$_{12}$, *Z. anorg. allg. Chem.*, 387(1), 31–42 (1972).

Schwarz M., Zerr A., Kroke E., Miehe G., Chen I.-W., Heck M., Thybusch B., Poe B.T., Riedel R. Spinel sialon, *Angew. Chem. Int. Ed.*, 41(5), 789–793 (2002).

Schwendtner K., Kolitsch U. Gittinsite-type M^{1+}-M^{3+}-diarsenates (M^{1+} = Li, Na; M^{3+} = Al, Sc, Ga): insights into an unexpected isotypy and crystal chemistry of diarsenates, *Mineralog. Mag.*, 71 (3), 249–263 (2007).

Schwendtner K. TlInAs$_2$O$_7$, RbInAs$_2$O$_7$, and (NH$_4$)InAs$_2$O$_7$: Synthesis and crystal structures of three isotypic microporous diarsenates – representatives of a novel structure type, *J. Alloys Compd.*, 421(1–2), 57–63 (2006).

Seifert H.J., Peng J., Golczewski J., Aldinger F. Phase equilibria of precursor-derived Si–(B–)C–N ceramics, *Appl. Organomet. Chem.*, 15(10), 794–808 (2001).

Selin A.A., Antipin I.Ya. Investigation of the melt—Crystal phase equilibria in the GaAs–GaSb–InAs–InSb–As–Sb secondary system [in Russian], *Zhurn. Neorgan. Khim.*, 36(6), 1572–1574 (1991).

Selin A.A., Gul'gazov V.N., Vigdorovich V.N. Investigation of the phase boundary of the «metal-rich melt—Solid solution» in the Ga–In–As–Sb system [in Russian], *Zhurn. Neorgan. Khim.*, 33(3), 724–728 (1988).

Selin A.A., Vigdorovich V.N., Batura V.P. Construction of the liquidus surface in the In–InP–GaAs–InAs system using simplex-centroid planning [in Russian], *Izv. AN SSSR. Neorgan. Mater.*, 18 (10), 1693–1696 (1982).

Semenova G.V., Sushkova T.P., Shumskaya O.N. Stability and thermodynamic mixing functions of the solid solutions in the InP–InAs–InSb system [in Russian], *Kondens. Sredy i Mezhfaz. Granitsy*, 10(2), 149–155 (2008).

Semenova G.V., Sushkova T.P., Shumskaya O.N. Thermodynamic calculation of the isotherm of the liquidus and solidus surface for the InP–InAs–InSb quasiternary system [in Russian], *Vestn. Voronezh. Gos. Un-ta. Ser. Khimiya. Biol. Farmatsiya*, (2), 96–99 (2006).

Semenova G.V., Sushkova T.P., Shumskaya O.N., Tishin A.V. Phase equilibria in the InP–InAs–InSb quasi-ternary system, *Rus. J. Inorg. Chem.*, 54(3) 472–476 (2009).

Sharma R.C. Srivastava M. Thermodynamic modeling and phase equilibria calculations of the quaternary Al–Ga–In–Sb System, *J. Phase Equilibr.*, 15(2), 178–187 (1994).

Shatruk M.M., Kovnir K.A., Lindsjö M., Presniakov I.A., Kloo L.A., Shevelkov A.V. Novel compounds $Sn_{10}In_{14}P_{22}I_8$ and $Sn_{14}In_{10}P_{21.2}I_8$ with clathrate I structure: Synthesis and crystal and electronic structure, *J. Solid State Chem.*, 161(2), 233–242 (2001).

Shen T.D., Schwartz R.B. Bulk ferromagnetic glasses in the Fe–Ni–P–B system, *Acta Mater.*, 49(5), 837–847 (2001).

Shen Z., Grins J., Esmaeilzadeh S., Ehrenberg H. Preparation and crystal structure of a new Sr containing sialon phase $Sr_2Al_xSi_{12-x}N_{16-x}O_{2+x}$ ($x \approx 2$), *J. Mater. Chem.*, 9(4), 1019–1022 (1999).

Shin N., Kim J., Ahn D., Sohn K.-S. A new strontium borophosphate, $Sr_6BP_5O_{20}$, from synchrotron powder data, *Acta Crystallogr.*, C61(5), i54–i56 (2005).

Shi Y., Liang J., Guo Y., Yang J., Zhuang W., Guanghui R. Phase relations in the system La_2O_3–B_2O_3–P_2O_5, *J. Alloys Compd.*, 242(1–2), 118–121 (1996).

Shi Y., Liang J., Zhang H., Liu Q., Chen X., Yang J., Zhuang W., Rao G. Crystal structure and thermal decomposition studies of barium borophosphate, $BaBPO_5$, *J. Solid State Chem.*, 135(1), 43–51 (1998).

Shi Y., Liang J., Zhang H., Yang J., Zhuang W., Rao G. X-ray powder diffraction and vibrational spectra studies of rare earth borophosphates, $Ln_7O_6(BO_3)(PO_4)_2$ (Ln = La, Nd, Gd, and Dy), *J. Solid State Chem.*, 129(1), 45–52 (1997).

Shreeve-Keyer J.L., Haushalter R.C., Lee Y.-S., Li S., O'Connor C., Seo D.-K., Whangbo M.-H. New layered materials in the K–In–Ge–As system: $K_8In_8Ge_5As_{17}$ and $K_5In_5Ge_5As_{14}$, *J. Solid State Chem.*, 130(2), 234–249 (1997).

Shreeve-Keyer J.L., Haushalter R.C., Seo D.K., Whangbo M.-H. Crystal and electronic structure of a quaternary layered compound from the K–In–Ge–Sb system: $K_9In_9GeSb_{22}$, *J. Solid State Chem.*, 122(1), 239–244 (1996).

Shumilin V.P., Cherviakov A.I., Lobanov A.A. Interaction of Zn and Se at the growth of $(GaP)_x(ZnSe)_{1-x}$ solid solutions by the Chokhralski method and method of liquid epitaxy [in Russian], *Izv. AN SSSR. Neorgan. Mater.*, 13(9), 1560–1564 (1977).

Shumilin V.P., Uglichina G.N., Ufimtsev V.B., Gimel'farb F.A. Phase equilibria in the InAs–ZnTe system [in Russian], *Izv. AN SSSR. Neorgan. Mater.*, 10(8), 1414–1417 (1974).

Singh D.J., Pickett W.E. Electronic and structural properties of $La_3Ni_2B_2N_3$, *Phys. Rev. B*, 51(13), 8668–8671 (1995).

Sirota N.N., Bodnar I.V. Phase diagram of the indium arsenide—Gallium phosphide quasibinary system [in Russian], *Dokl. AN BSSR*, 18(3), 213–215 (1974).

Sirota N.N., Makovetskaya L.A. Lattice parameter and microhardness of the InP–GaAs semiconductor solid solutions [in Russian], *Dokl. AN BSSR*, 7(4), 230–232 (1963).

Sirota N.N., Novikov V.V., Antyukhov A.M. Experimental and theoretical construction of the phase diagram of the GaAs–InP–InAs ternary system [in Russian], *Elektronnaya tekhnika. Ser. 6. Materialy*, [10(195)], 9–12 (1984).

Sirota N.N., Novikov V.V. Phase diagram of the system of gallium and indium arsenide and indium phosphide [in Russian], *Zhurn. Fiz. Khim.*, 59(4), 829–833 (1985).

Sleight A.W., Bouchard R.J. A new cubic $KSbO_3$ derivative structure with interpenetrating networks. Crystal structure of $Bi_3GaSb_2O_{11}$, *Inorg. Chem.*, 12(10), 2314–2316 (1973).

Sleight A.W., Ward R. Compounds of post-transition elements with the ordered perovskite structure, *Inorg. Chem.*, 3(2), 292 (1964).

Smetana V., Vajenine G.V., Kienle L., Duppel V., Simon A. Intermetallic and metal-rich phases in the system Li–Ba–In–N, *J. Solid State Chem.*, 183(8), 1767–1775 (2010).

Sologub O., Mori N. Structural and thermoelectric properties of $Y_{1-x}B_{22+y}C_{2-y}N$, *J. Phys. Chem. Solids*, 74(8), 1109–1114 (2013).

Somer M., Carrillo-Cabrera W., Nuß J., Peters K., Schnering von H.G., Cordier G. Crystal structures of trisodium tristrontium tetrapnictidogallates, $Na_3Sr_3GaP_4$, $Na_3Sr_3GaAs_4$ and $Na_3Sr_3GaSb_4$, *Z. Kristallogr.*, 211(7), 479–480 (1996a).

Somer M., Carillo-Cabrera W., Peters E.-M., Peters K., Schnering von H.G. Crystal structure of dipotassium lithium catena-di-μ-phosphido-aluminate, K_2LiAlP_2, *Z. Kristallogr.*, 210(2), 142 (1995a).

Somer M., Carillo-Cabrera W., Peters E.-M., Peters K., Schnering von H.G. Crystal structure of dipotassium lithium diarsenidoindate, $K_2LiInAs_2$, *Z. Kristallogr.*, 210(12), 959 (1995b).

Somer M., Carillo-Cabrera W., Peters E.-M., Peters K., Schnering von H.G. Crystal structure of disodium lithium diarsenidogallate, $Na_2LiGaAs_2$, *Z. Kristallogr.*, 210(11), 877 (1995c).

Somer M., Carillo-Cabrera W., Peters E.-M., Peters K., Schnering von H.G. Crystal structure of disodium lithium diphosphidoaluminate, $NaLiAlP_2$, *Z. Kristallogr.*, 210(10), 778 (1995d).

Somer M., Carillo-Cabrera W., Peters E.-M., Peters K. Schnering von H.G. Crystal structure of tricaesium phosphido-arsenidoborate, Cs_3BPAs, *Z. Kristallogr.*, 211(3), 192 (1996b).

Somer M., Carillo-Cabrera W., Peters E.-M., Schnering von H.G. Crystal structure of dipotassium lithium diarsenido-gallate, $K_2LiGaAs_2$, *Z. Kristallogr.*, 210(7), 528 (1995e).

Somer M., Carillo-Cabrera W., Peters K., Schnering von H.G. Crystal structure of caesium rubidium di-μ-arsenido-bis(arsenidogallate), $Cs_{5.18}Rb_{0.82}Ga_2As_4$, *Z. Kristallogr.*, 201(3–4), 327–328 (1992a).

Somer M., Carillo-Cabrera W., Peters E.-M., Schnering von H.G. Crystal structure of sodium potassium antimonide triantimonidoaluminate, $Na_3K_6Sb(AlSb_3)$, *Z. Kristallogr.*, 210(7), 527 (1995f).

Somer M., Carillo-Cabrera W., Peters E.-M., Schnering von H.G. Crystal structure of sodium potassium antimonide triantimonido-gallate, $K_6Na_3Sb[GaSb_3]$, *Z. Kristallogr.*, 210(2), 143 (1995g).

Somer M., Carillo-Cabrera W., Peters K., Schnering von H.G. Crystal structure of sodium tetrastrontium tris(dinitridoborate), $NaSr_4[BN_2]_3$, *Z. Kristallogr., New. Cryst. Str.*, 215(2), 209 (2000a).

Somer M., Carillo-Cabrera W., Peters K., Schnering von H.G., Cordier G. Crystal structure of trisodium tricalcium tetraarsenidoaluminate, $Na_3Ca_3[AlAs_4]$, *Z. Kristallogr.*, 211(8), 550 (1996c).

Somer M., Carillo-Cabrera W., Peters K., Schnering von H.G. Crystal structure of tribarium phosphide borate, $Ba_3P(BO_3)$, *Z. Kristallogr.*, 210(6), 449 (1995h).

Somer M., Carillo-Cabrera W., Peters K., Schnering von H.G. Crystal structure of tristrontium phosphide borate, $Sr_3P(BO_3)$, *Z. Kristallogr.*, 210(7), 526 (1995i).

Somer M., Carillo-Cabrera W., Peters K., Schnering von H.G. Darstellung, Kristallstrukturen und Svhwingungsspertren von Verbindungen mit den linearen Dipnictidoborat(3–)-Anionen $[P–B–P]^{3-}$, $[As–B–As]^{3-}$ und $[P–B–As]^{3-}$, *Z. anorg. allg. Chem.*, 626(4), 897–904 (2000b).

Somer M., Gül C., Müllmann R., Mosel B.D., Kremer R.K., Pöttgen R. Vibrational spectra and magnetic properties of $Eu_3[BN_2]_2$ and $LiEu_4[BN_2]_3$, *Z. anorg. allg. Chem.*, 630(3), 389–393 (2004a).

Somer M., Herterich U., Curda J., Carrillo-Cabrera W., Peters K., Schnering von H.G. Ternäre Nitridoborate. 1. $LiMg[BN_2]$ und $Ba_4[BN_2]_2O$, Verbindungen mit dem Anion $[N–B–N]^{3-}$: Darstellung, Kristallstrukturen und Schwingungsspektren, *Z. anorg. allg. Chem.*, 623(1–6), 18–24 (1997).

Somer M., Herterich U., Čurda J., Carrillo-Cabrera W., Zürn A., Peters K., Schnering von H.G. Ternäre Nitridoborate: 2. Darstellung, Kristallstrukturen und Schwingungsspektren neuer ternärer Verbindungen mit dem Anion $[N-B-N]^{3-}$, Z. anorg. allg. Chem., 626(3), 625–633 (2000c).

Somer M., Herterich U., Curda J., Peters K., Schnering von H.G. Crystal structure of lithium tetracalcium tris(dinitridoborate), $LiCa_4(BN_2)_3$, Z. Kristallogr., 209(2), 182 (1994).

Somer M., Herterich U., Curda J., Peters K., Schnering von H.G. Crystal structure of lithium tetrastrontium tris(dinitridoborate), $LiSr_4(BN_2)_3$, Z. Kristallogr., 211(1), 54 (1996d).

Somer M., Herterich U., Curda J., Peters K., Schnering von H.G. Crystal structure of sodium tetrabarium dinitroborate, $NaBa_4(BN_2)_3$, Z. Kristallogr., 210(7), 529 (1995k).

Somer M., Kütükcü M.N., Gil R.C., Borrmann H., Carrillo-Cabrera W. $Mg_2[BN_2]Cl$ and $Mg_8[BN_2]_5I$: Novel magnesium nitridoborate halides—Synthesis, crystal structure and vibrational spectra, Z. anorg. allg. Chem., 630(7), 1015–1021 (2004b).

Somer M., Peters E.M., Peters K., Schnering von H.G. Crystal structure of dipotassium sodium catena-di-μ-arsenido-gallate, $K_2Na(GaAs_2)$, Z. Kristallogr., 193(3–4), 285–286 (1990a).

Somer M., Peters K., Peters E.-M., Schnering von H.G. Crystal structure of dipotassium sodium catenadi-μ-arsenido-indate, $K_2NaInAs_2$, Z. Kristallogr., 195(1–2), 97–98 (1991a).

Somer M., Peters K., Popp Th., Schnering von H.G. Die Antimonid-Triantimonidometallate(III) $Cs_6K_3Sb[AlSb]$ und $Cs_6K_3Sb[GaSb]$, Z. anorg. allg. Chem., 597(1), 201–208 (1991b).

Somer M., Peters K., Schnering von H.G. Crystal structure of dipotassium sodium catena-di-μ-phosphidoindate, K_2NaInP_2, Z. Kristallogr., 194(1–2), 131–132 (1991c).

Somer M., Peters K., Schnering von H.G. Crystal structure of dipotassium sodium diphosphidoborate, K_2NaBP_2, Z. Kristallogr., 194(1–2), 133–134 (1991d).

Somer M., Peters K., Schnering von H.G. Crystal structure of potassium caesium antimonide triantimonidoaluminate, $K_3Cs_6Sb(AlSb_3)$, Z. Kristallogr., 195(3–4), 316–317 (1991e).

Somer M., Peters K., Schnering von H.G. Crystal structure of potassium caesium antimonide triantimonidogallate, $K_3Cs_6Sb(GaSb_3)$, Z. Kristallogr., 195(1–2), 101–102 (1991f).

Somer M., Peters K., Schnering von H.G. Crystal structure of dipotassium sodium diphosphidogallate, K_2NaGaP_2, Z. Kristallogr., 192(3–4), 267–268 (1990b).

Somer M., Peters K., Schnering von H.G. K_2NaGaP_2, Cs_2NaGaP_2, $K_2NaGaAs_2$, K_2NaInP_2 und $K_2NaInAs_2$, Verbindungen mit den SiS_2-isosteren Polyanionen $^1_\infty[MX_{4.2}]^{3-}$ (M = Ga, In; X = P, As), Z. anorg. allg. Chem., 613(7), 19–25 (1992b).

Somer M., Peters K., Thiery D., Schnering von H.G. Crystal structure of caesium potassium di-μ-arsenido-bis(arsenidogallate), $Cs_{5.23}K_{0.77}Ga_2As_4$, Z. Kristallogr., 195(3–4), 312–313 (1991g).

Somer M., Peters K., Thiery D., Schnering von H.G. Crystal structure of caesium potassium di-μ-phosphido-bis(phosphidogallate), $Cs_{5.31}K_{0.69}(Ga_2P_4)$, Z. Kristallogr., 195(1–2), 95–96 (1991h).

Somer M., Yaren Ö., Reckeweg O., Prots Yu., Carrillo-Cabrera W. $Ca_2[BN_2]H$: The first nitridoborate hydride—Synthesis, crystal structure and vibrational spectra, Z. anorg. allg. Chem., 630(7), 1068–1073 (2004c).

Sonomura H., Uragaki T., Miyauchi T. Synthesis and some properties of solid solutions in the GaP–ZnS and GaP–ZnSe pseudobinary systems, Jpn. J. Appl. Phys., 12(7), 968–973 (1973).

Sorokin V.S., Sorokin S.V., Semenov A.N., Meltser B.Ya., Ivanov S.V. Novel approach to the calculation of instability regions in GaInAsSb alloys, J. Cryst. Growth, 208(1–4), 97–103 (2000).

Stefanovich S.Yu., Belik A.A., Azuma M., Takano M., Baryshnikova O.V., Morozov V.A., Lazoryak B.I., Lebedev O.I., Van Tendeloo G. Antiferroelectric phase transition in $Sr_9In(PO_4)_7$, Phys. Rev. B, 70(17), 172103_1–172103_4 (2004).

Sterzer E., Ringler B., Nattermann L., Beyer A., von Hänisch C., Stolz W., Volz K. (GaIn)(NAs) growth using di-tertiary-butyl-arsano-amine (DTBAA), J. Cryst. Growth, 467, 132–136 (2017).

Stoto T., Doukhan J.-C., Mocellin A. Analytical electron microscopy investigation of the AlN–ZrO$_2$ system: Investigation of a quaternary Zr–Al–O–N phase, J. Amer. Ceram. Soc., 72(8), 1453–1457 (1989).

Stringfellow G.B. Miscibility gaps and spinodal decomposition in III/V quaternary alloys of the type $A_xB_yC_{1-x-y}D$, *J. Appl. Phys.*, 54(1), 404–409 (1983).

Stringfellow G.B. Miscibility gaps in quaternary Ill/V alloys, *J. Cryst. Growth*, 58(1), 194–202 (1982a).

Stringfellow G.B. Spinodal decomposition and clustering in III/V alloys, *J. Electron. Mater.*, 11(5), 903–918 (1982b).

Ströbele M., Dolabdjian K., Enseling D., Dutczak D., Mihailova B., Jüstel T., Meyer H.-J. Luminescence matching with the sensitivity curve of the human eye: Optical ceramics $Mg_{8-x}M_x(BN_2)_2N_4$ with M = Al ($x = 2$) and M = Si ($x = 1$), *Eur. J. Inorg. Chem.*, (10), 1716–1725 (2015).

Stuckes A.D., Chasmar R.P. Electrical and thermal properties of alloys of InAs and CdTe, *J. Phys. Chem. Solids*, 25(5), 469–476 (1964).

Sugiyama K., Yu J., Hiraga K., Terasaki O. Monoclinic $InPO_4 \cdot 2H_2O$, *Acta Crystallogr.*, C55(3), 279–281 (1999).

Sun W.-Y., Huang Z.K., Tien T.-Y., Yen T.-S. Phase relationships in the system Y–Al–O–N, *Mater. Lett.*, 11(3–4), 67–69 (1991a).

Sun W.-Y., Ma L.-T., Yan D.-S. Phase relationships in the system Mg–Al–O–N, *Chin. Sci. Bull.*, 35 (14), 1189–1192 (1990).

Sun W.Y., Wu F.Y., Yan D.S. (Yen T.S.). Studies of the formation of α' and α'-β' sialon, *Mater. Lett.*, 6(1–2), 11–15 (1987).

Sun W.-Y., Yan D.-S. (Yen T.S.). Phase relationships in quasi-systems R_2O_3–AlN–Al_2O_3, *Sci. in Cnina (Ser. A)*, 34(1), 105–115 (1991b).

Sun W.Y., Yen T.S. Phase relationships in the system Ca–Al–O–N, *Mater. Lett.*, 8(5), 150–152 (1989a).

Sun W.Y., Yen T.S. Phase relationships in the system Nd–Al–O–N, *Mater. Lett.*, 8(5), 145–149 (1989b).

Sun W.Y., Yen T.S., Tien T.Y. Subsolidus phase relationships in the systems *Re*–Al–O–N (where *Re* = rare earth elements), *J. Solid State Chem.*, 95(2), 424–429 (1991b).

Sveshnikova V.N. Solubility in the Al_2O_3–P_2O_5–H_2O ternary system at 90°C [in Russian], *Zhurn. Neorgan. Khim.*, 5(2), 477–480 (1960).

Swenson D.J., Sutopo, Chang Y.A. Phase equilibria in the system In–Ir–As at 600°C, *Z. Metallkde*, 85 (4), 228–231 (1994).

Tabary P., Servant C., Alary J.A. Effects of a low amount of C on the phase transformations in the AlN–Al_2O_3 pseudo-binary system, *J. Eur. Ceram. Soc.*, 20(12), 1915–1921 (2000).

Takahei T., Nagai H. Instability of In–Ga–As–P liquid solution during low temperature LPE of $In_{1-x}Ga_xAs_{1-y}P_y$ on InP, *Jpn. J. Appl. Phys.*, 20(4), L313–L316 (1981).

Takase A., Tani E. Raman spectroscopic study of β-sialons in the system Si_3N_4–Al_2O_3–AlN, *J. Mater. Sci. Lett.*, 3(12), 1058–1060 (1984).

Takase A., Umebayashi S., Kishi K. Infrared spectroscopic study of β-sialons in the system Si_3N_4–SiO_2–AlN, *J. Mater. Sci. Lett.*, 1(12), 529–532 (1982).

Takayama T., Yuri M., Itoh K., Baba T., Harris, J.S.Jr. Analysis of phase-separation region in wurtzite group III nitride quaternary material system using rystall valence force field model, *J. Cryst. Growth*, 222(1–2), 29–37 (2001).

Takayuki H., Yamane H., Becker N., Dronskowski R. Synthesis and crystal structure of $Ba_{26}B_{12}Si_5N_{27}$ containing [Si_2] dumbbells, *J. Solid State. Chem.*, 230, 390–396 (2015).

Tananaev I.V., Chudinova N.N. On the interaction of gallium phosphate and phosphoric acid [in Russian], *Zhurn. Neorgan. Khim.*, 7(10), 2285–2289 (1962).

Tangen I.-L., Yu Y., Grande T., Høier R., Einarsrud M.-A. Phase relations and microstructural development of aluminium nitride–aluminium nitride polytypoid composites in the aluminium nitride–alumina–yttria system, *J. Amer. Ceram. Soc.*, 89(9), 1734–1740 (2004).

Tang X.-J., Gentiletti M.J., Lachgar A. Synthesis and crystal structure of indium arsenate and phosphate rystallin with variscite and metavariscite structure types, *J. Chem. Thermodyn.*, 31 (1), 45–50 (2001).

Tang X., Lachgar A. The missing link: Synthesis, crystal structure, and thermogravimetric studies of $InPO_4 \cdot H_2O$, *Inorg. Chem.*, 37(24), 6181–6185 (1998).

Tien T.Y., Hummel F.A. Studies in lithium oxide systems: XI, $Li_2O-B_2O_3-P_2O_5$, *J. Amer. Ceram. Soc.*, 44(8), 390–394 (1961).

Todorov I., Chung D.Y., Ye L., Freeman A.J., Kanatzidis M.G. Synthesis, structure and charge transport properties of $Yb_5Al_2Sb_6$: A Zintl phase with incomplete electron transfer, *Inorg. Chem.*, 48(11), 4768–4776 (2009).

Todorov I., Sevov S.C. Synthesis and characterization of $Na_2Ba_4Ga_2Sb_6$ and $Li_{13}Ba_8GaSb_{12}$, *Z. Kristallogr.*, 221(5–7), 521–526 (2006).

Tokarzewski A., Podsiadło S. Formation and thermal decomposition of indium oxynitride compounds, *J. Therm. Anal. Calorim.*, 52(2), 481–488 (1998).

Toy C., Savrun E. Novel composites in the aluminum nitride–zirconia and –hafnia systems, *J. Europ. Ceram. Soc.*, 18(1), 23–29 (1998).

Tranqui D., Hamdoune S., Le Page Y. Synthèse et structure rystalline de $LiInP_2O_7$, *Acta Crystallogr.*, C43(2), 201–202 (1987).

Tran Qui D., Hamdoune S. Structure de $Li_3In_2P_3O_{12}$, *Acta Crystallogr.*, C43(3), 397–399 (1987).

Trigg M.B., Jack K.H. Solubility of aluminium in silicon oxynitride, *J. Mater. Sci. Lett.*, 6(4), 407–408 (1987).

Tsukuma K., Shimada M., Koizumi M. A new compound $Si_3Al_4N_4C_3$ with the wurtzite structure in the system $Si_3N_4-Al_4C_3$, *J. Mater. Sci. Lett.*, 1(1), 9 (1982).

Ufimtseva E.V., Vigdorovich V.N., Pelevin O.V. Phase equilibria in the GaAs–ZnTe system [in Russian], *Izv. AN SSSR. Neorgan. Mater.*, 9(4), 587–591 (1973).

Ufimtsev V.B., Cherviakov A.I., Shumilin V.P. Thermal analysis method of dissociated semiconductor systems [in Russian], *Zavodsk. Laboratoria*, 46(6), 525–527 (1980).

Ufimtsev V.B., Lapkina I.A., Zinov'yev V.G., Sorokina O.V. Influence of zinc on the solubility and segregation coefficients of bismuth in indium antimonide [in Russian], *Izv. AN SSSR. Neorgan. Mater.*, 23(11), 1784–1787 (1987).

Uheda K., Hirosaki N., Yamamoto H. Host lattice materials in the system $Ca_3N_2-AlN-Si_3N_4$ for white light emitting diode, *Phys. Stat. Sol (a)*, 203(11), 2712–2717 (2006a).

Uheda K., Hirosaki N., Yamamoto Y., Naito A., Nakajima T., Yamamoto H. Luminescence properties of a red phosphor, $CaAlSiN_3:Eu^{2+}$, for white light-emitting diodes, *Electrochem. Solid-State Lett.*, 9(4), H22–H25 (2006b).

Urata A., Matsumoto H., Yoshida S., Makino A. Fe–B–P–Cu nanocrystalline soft magnetic alloys with high B_s, *J. Alloys Compd.*, 509, Suppl. 1, S431–S433 (2011).

Ust'yantsev V.M., Zholobova L.S. Subsolidus relations in the $Na_2O-Al_2O_3-P_2O_5$ system [in Russian], *Izv. AN SSSR. Neorgan. Mater.*, 13(8), 1527–1528 (1977).

Van Wambeke L. The uranium-bearing mineral bolivarite: New data and a second occurrence, *Mineralog. Mag.*, 38(296), 418–423 (1971).

Vasil'yev M.G., Novikova E.M. Investigation of equilibria in the Me–ZnSe, Sn–GaAs, ZnSe–GaAs vertical sections of Me–GaAs–ZnSe ternary systems, where Me–Ga, Sn [in Russian], *Deposited in VINITI*, № 971-77Dep (1977).

Vasil'yev M.G., Vigdorovich V.N., Selin A.A., Khanin V.A. Phase equilibria in the Sn–In–P, Sn–Ga–In–As and Sn–Ga–In–As–P systems [in Russian], In: *Legir. poluprovodn. materialy*, Moscow: Nauka, 61–65 (1985).

Vassilev V., Atanassova D., Mihaylova I., Parvanova V., Aljihmani L. Phase equilibria in the $PbSb_2Te_4-InSb$ system, *Thermochim. Acta*, 520(1–2), 80–83 (2011).

Vergasova L.P., Filatov S.K., Gorskaya M.G., Molchanov A.V., Krivovichev S.V., Ananiev V.V. Urusovite, $Cu[AlAsO_5]$, a new mineral from the Tolbachik volcano, Kamchatka, Russia, *Eur. J. Mineral.*, 12(5), 1041–1044 (2000).

Vigdorovich V.N., Selin A.A., Shutov S.G. Investigation of the phase equilibria in the In–P–As–Sb system [in Russian], *Dokl. AN SSSR*, 252(6), 1423–1426 (1980).

Vigdorovich E.N., Sveshnikov Yu.N. Thermodynamic stability of the GaN–InN–AlN system [in Russian], *Neorgan. Mater.*, 36(5), 568–570 (2000).

Voitsehivskiy A.V., Drobyazko V.P. About solid solutions in the InAs–CdS system [in Ukrainian], *Ukr. Fiz. Zhurn.*, 12(3), 460–461 (1967).

Voitsehovskiy A.V., Drobyazko V.P., Mitiurev V.K., Vasilenko V.P. Solid solutions in the InAs–CdS and InAs–CdSe systems [in Russian], *Izv. AN SSSR. Neorgan. Mater.*, 4(10), 1681–1684 (1968).

Voitsehovskiy A.V., Goryunova N.A. Solid solutions in some quaternary semiconductor systems [in Russian], In: *Fizika*, Leningrad: Publish. House of Lening. Inst. Civil Eng., 12–14 (1962).

Voitsehovskiy A.V., Panchenko L.B. About obtaining of $(GaP)_x(ZnS)_{1-x}$ solid solution single crystals [in Russian], In: *Fiz. tv. tela*, Kiev: Kiev. Ped. Inst. Publish., 24–26 (1975).

Voitsehovskiy A.V., Panchenko L.B. Microstructural investigation of crystals of the GaP–ZnS system [in Russian], *Izv. AN SSSR. Neorgan. Mater.*, 13(1), 160–161 (1977).

Voitsehovskiy A.V., Pashun A.D., Mityurev V.K. About interaction of gallium arsenide with II-VI compounds [in Russian], *Izv. AN SSSR. Neorgan. Mater.*, 6(2), 379–380 (1970).

Voitsehovskiy A.V., Stetsenko T.P. About obtaining of single crystals of $(GaP)_x(ZnSe)_{1-x}$ solid solutions by the chemical transport reactions [in Russian], In: *Issled. po molekuliar. fizike i fizike tverdogo tela*, Kiev: Kiev. Ped. Inst. 38–40 (1976).

Vorobyev Yu.P., Shveikin G.P. Composition of two β-$Si_{6-z}Al_zO_zN_{8-z}$. morphological phases [in Russian], *Neorgan. Mater.*, 36(12), 1472–1475 (2000).

Walenta K. Bulachit, ein neues Aluminiumarsenatmineral von Neubulach im noördlichen Schwarzwald, *Der Aufschluss*, 34, 445–451 (1983).

Wang G., Wu Y., Fu P., Liang X., Xu Z., Chen C. Crystal growth and properties of β-Zn_3BPO_7, *Chem Mater.*, 14(5), 2044–2047 (2002).

Wang J., Zhu Y., Su X., Tu H., Liu Y., Wu C. The zinc-rich corner of the Zn–Fe–Al–P quaternary system at 450°C, *CALPHAD: Comput. Coupling Phase Diagrams Thermochem.*, 38, 122–126 (2012).

Wang Q., Yan Y., Ren X., Shu W., Jia Z., Zhang X., Huang Y. The electronic and optical properties of quaternary $B_xGa_{1-x}As_{1-y}Sb_y$ alloys with low boron concentration: A first-principle study, *J. Alloys Compd.*, 563, 18–21 (2013).

Wang X.H., Lejus A.M., Vivien D., Collongues R. Synthesis and characterization of lanthanide aluminum oxynitrides with magnetoplumbite like structure, *Mater. Res. Bull.*, 23(1), 43–49 (1988).

Wang Y., Stoyko S., Bobev S. Quaternary pnictides with complex, noncentrosymmetric structures. Synthesis and structural characterization of the new Zintl phases $Na_{11}Ca_2Al_3Sb_8$, $Na_4CaGaSb_3$, and $Na_{15}Ca_3In_5Sb_{12}$, *Inorg. Chem.*, 54(4), 1931–1939 (2015).

Wang Y., Suen N.-T., Kunene T., Stoyko S., Bobev S. Synthesis and structural characterization of the Zintl phases $Na_3Ca_3TrPn_4$, $Na_3Sr_3TrPn_4$, and $Na_3Eu_3TrPn_4$ (Tr = Al, Ga, In; Pn = P, As, Sb), *J. Solid State Chem.*, 249, 160–168 (2017a).

Wang Y., Zou X., Feng X., Shi Y., Wu L. Quaternary non-centrosymmetric sulfide Y_4GaSbS_9: Syntheses, structures, optical properties and theoretical studies, *J. Solid State Chem.*, 245, 110–114 (2017b).

Watanabe H, Kijima N. Crystal structure and luminescence properties of $Sr_xCa_{1-x}AlSiN_3$:Eu^{2+} mixed nitride phosphors, *J. Alloys Compd.*, 475(1–2), 434–439 (2009).

Watras A., Carrasco I., Pazik R., Wiglusz R.J., Piccinelli F., Bettinelli M., Deren P.J. Structural and spectroscopic features of $Ca_9M(PO_4)_7$ (M = Al^{3+}, Lu^{3+}) whitlockites doped with Pr^{3+} ions, *J. Alloys Compd.*, 672, 45–51 (2016).

Wei C.H., Edgar J.H. Unstable composition region in the wurtzite $B_{1-x-y}Ga_xAl_yN$ system, *J. Cryst. Growth*, 208(1–4), 179–182 (2000).

Weiss J., Greil P., Gauckler L.J. The system Al–Mg–O–N, *J. Amer. Ceram. Soc.*, 65(5), C68–C69 (1982).

Wideman T., Cava R.J., Sneddon L.G. Polymeric precursor synthesis of the superconducting metal boronitride $La_3Ni_2B_2N_3$, *Chem. Mater.*, 8(9), 2215–2217 (1996).

Wild S., Elliot H., Thompson D.P. Combined infra-red and X-ray studies of β-silicon nitride and β'-sialons, *J. Mater. Sci.*, 13(8), 1769–1775 (1978).

Williams D., Kouvetakis J., O'Keeffe M. Synthesis of nanoporous cubic $In(CN)_3$ and $In_{1-x}Ga_x(CN)_3$ and corresponding inclusion compounds, *Inorg. Chem.*, 37(18), 4617–4620 (1998).

Williams D., Pleune B., Kouvetakis J., Williams M.D., Andersen R.A. Synthesis of $LiBC_4N_4$, BC_3N_3 and related C–N compounds of boron: New precursors to light element ceramics, *J. Amer. Chem. Soc.*, 122(32), 7735–7741 (2000).

Wise W.S., Loh S.E. Equilibria and origin of minerals in the system Al_2O_3–$AlPO_4$–H_2O, *Amer. Mineralog.*, 61(5–6), 409–413 (1976).

Wittmann U., Rauser G., Kemmler-Sack S. Über die Ordnung von B^{III} und M^V Perowskiten vom Typ $A^{II}_2B^{III}M^VO_6$ (A^{II} = Ba, Sr; M^V =Sb, Nb, Ta), *Z. anorg. allg. Chem.*, 482(11), 143–153 (1981).

Womelsdorf H., Meyer H.-J. Zur Kenntnis der Struktur von $Sr_3(BN_2)_2$, *Z. anorg. allg. Chem.*, 620(2), 262–265 (1994b).

Woolley J.C., Williams E.W. Cross-substitutional alloys of InSb, *J. Electrochem. Soc.*, 111(2), 210–215 (1964).

Wörle M., Meyer zu Altenschildesche H., Nesper R. Synthesis, properties and crystal structures of α-$Ca(BN_2)_2$ and $Ca_{9+x}(BN_2, CBN)_6$—Two compounds with BN_2^{3-} and CBN^{4-} anions, *J. Alloys Compd.*, 264(1–2), 107–114 (1998).

Xie R.-J., Mitomo M., Xu F.-F., Uheda K., Bando Y. Preparation of Ca-α-sialon ceramics with compositions along the Si_3N_4–$1/2Ca_3N_2$:3AlN line, *Z. Metallkde*, 92(8), 931–936 (2001).

Xiong D.-B., Chen H.-H., Yang X.-X., Zhao J.-T. Low-temperature flux syntheses and characterizations of two 1-D anhydrous borophosphates: $Na_3B_6PO_{13}$ and $Na_3BP_2O_8$, *J. Solid State Chem.*, 180(1), 233–239 (2007).

Xiong Z., Wu G., Chua Y.S., Hu J., He T., Xu W., Chen P. Synthesis of sodium amidoborane ($NaNH_2BH_3$) for hydrogen production, *Energy Environ. Sci.*, 1(3), 360–363 (2008a).

Xiong Z., Yong C.K., Wu G., Chen P., Shaw W., Karkamkar A., Autrey T., Jones M.O., Johnson S.R., Edwards P.P., David W.I.F. High-capacity hydrogen storage in lithium and sodium amidoboranes, *Nat. Mater.*, 7(2), 138–141 (2008b).

Yamane H., Sasaki S., Kubota S., Inoue R., Shimada M., Kajiwara T. Synthesis and structure of $Ba_8Cu_3In_4N_5$ with nitridocuprate groups and one-dimensional infinite indium clusters, *J. Solid State Chem.*, 163(2), 449–454 (2002a).

Yamane H., Sasaki S., Kubota S., Kajiwara T., Shimada M. $Ba_{14}Cu_2In_4N_7$, a new subnitride with isolated nitridocuprate groups and indium clusters, *Acta Crystallogr.*, C58(4), i50–i52 (2002b).

Yamane H., Sasaki S., Kubota S., Shimada M., Kajiwara T. Synthesis and crystal structure analysis of $Sr_8Cu_3In_4N_5$ and $Sr_{0.53}Ba_{0.47}CuN$, *J. Solid State Chem.*, 170(2), 265–272 (2003).

Yeh C.L., Kuo C.W., Wu F.S. Formation of $Ti_2AlC_{0.5}N_{0.5}$ solid solutions by combustion synthesis of Al_4C_3-containing samples in nitrogen, *J. Alloys Compd.*, 508(2), 324–328 (2010).

Yevgen'ev S.B., Kononkova N.N., Lapkina I.A., Sorokina O.V., Ufimtsev V.B. Joint doping of indium antimonide with bismuth and zinc [in Russian], *Izv. AN SSSR. Neorgan. Mater.*, 21(3), 490–492 (1985).

Yilmaz A., Bu X., Kizilyalli M., Kniep R., Stucky G.D. Cobalt borate phosphate, $Co_3[BPO_7]$, synthesis and characterization, *J. Solid State Chem.*, 156(2), 281–285 (2001).

Yim M.F. Solid solutions in the pseudobinary (III-V)–(II-VI) systems and theire optical energy gap, *J. Appl. Phys.*, 40(6), 2617–2623 (1969).

Yin W., Iyer A.K., Li C., Yao J., Mar A. Noncentrosymmetric rare-earth selenides $RE_4InSbSe_9$ (RE = La–Nd), *J. Alloys Compd.*, 710, 424–430 (2017a).

Yin W., Zhou M., Iyer A.K., Yao J., Mar A. Noncentrosymmetric quaternary selenide $Ba_{23}Ga_8Sb_2Se_{38}$: Synthesis, structure, and optical properties, *J. Alloys Compd.*, 729, 150–155 (2017b).

Yi X., Niu J., Nakamaru T., Akiyama T. Reaction mechanism for combustion synthesis of β-SiAlON by using Si, Al and SiO_2 as raw materials, *J. Alloys Compd.*, 561, 1–4 (2013).

Yi X., Watanabe K., Akiyama T. Fabrication of dense β-SiAlON by a combination of combustion synthesis (CS) and spark plasma sintering (SPS), *Intermetallics*, 18(4), 536–541 (2010).

Yoshimoto M., Huang W., Takehara Y., Saraie J., Chayahara A., Horino Y., Oe K. New semiconductor GaNAsBi alloy grown by molecular beam epitaxy, *Jpn. J. Appl. Phys.*, 43(7A), L845–L847 (2004).

Yu Y., Tangen I.-L., Grande T., Høier R., Einarsrud M.-A. HRTEM investigations of new AlN polytypoids in the high-AlN region of the $AlN–Al_2O_3–Y_2O_3$ system, *J. Amer. Ceram. Soc.*, 87 (2), 275–278 (2004).

Zandbergen H.W., Jansen J., Cava R.J., Krajewskii J.J., Peck, Jr W.F. Structure of the 13-K superconductor $La_3Ni_2B_2N_3$ and the related phase LaNiBN, *Nature*, 372(6508), 759–761 (1994).

Zangvil A., Gauchkler L.J., Rühle M. Indexed X-ray diffraction data for the sialon X-phase, *J. Mater. Sci.*, 15(3), 788–790 (1980).

Zangvil A. The structure of the X phase in the Si–Al–O–N alloys, *J. Mater. Sci.*, 13(6), 1370–1374 (1978).

Zangvil A., Ruh R. Phase relationships in the silicon carbide–aluminum nitride system, *J. Amer. Ceram. Soc.*, 71(10), 884–890 (1988).

Zangvil A., Ruh R. Solid solutions and composites in the SiC–AlN and SiC–BN systems, *Mater. Sci. Eng.*, 71, 159–164 (1985).

Zangvil A., Ruh R. The $Si_3Al_4N_4C_3$ and $Si_3Al_5N_5C_3$ compounds as SiC–AlN solid solutions, *J. Mater. Sci. Lett.*, 3(3), 249–250 (1984).

Zbitnew K., Woolley J.C. The quaternary alloy system $Al_xGa_yIn_{1-x-y}Sb$, *J. Appl. Phys.*, 52(11), 6611–6616 (1981).

Zhang E., Zhao S. Znang J., Fu P., Yao J. The β-modification of trizinc borate phosphate, $Zn_3(BO_3)$ (PO_4), *Acta Crystallogr.*, E67(1), i3 (2011).

Zhang F., Leithe-Jasper A., Xu J., Mori T., Matsui Y., Tanaka T., Okada S. Novel rare earth boron-rich solids, *J. Solid State Chem.*, 159(1), 174–180 (2001a).

Zhang F.X., Tanaka T. Crystal structure of tetraaluminium trinitride carbide oxide, Al_4N_3CO, *Z. Kristallogr., New Cryst. Str.*, 218(1), 27–28 (2003).

Zhang F., Xu F., Leithe-Jasper A., Mori T., Tanaka T., Xu J., Sato A., Bando Y. Matsui Y. Homologous phases built by boron clusters and their vibrational properties, *Inorg. Chem.*, 40(27), 6948–6951 (2001b).

Zhang W.-L., Lin C.-S., Geng L., Li Y.-Y., Zhang H., He Z.-Z., Cheng W.-D. Synthesis and characterizations of two anhydrous metal borophosphates: $M^{III}_2BP_3O_{12}$ (M = Fe, In), *J. Solid State Chem.*, 183(5), 1108–1113 (2010).

Zhang S., Langelaan G., Brouwer J.C., Sloof W.G., Brück E., Zwaag van der S., Dijk van N.H. Preferential Au precipitation at deformation-induced defects in Fe–Au and Fe–Au–B–N alloys, *J. Alloys Compd.*, 584, 425–429 (2014).

Zhao D., Geng H., Teng X. Synthesis and thermoelectric properties of $Ga_xCo_4Sb_{11.7}Te_{0.3}$ skutterudites, *Intermetallics*, 26, 31–35 (2012a).

Zhao H.-J. Centrosymmetry vs noncentrosymmetry in $La_2Ga_{0.33}SbS_5$ and Ce_4GaSbS_9 based on the interesting size effects of lanthanides: Syntheses, crystal structures, and optical properties, *J. Solid State Chem.*, 237, 99–104 (2016a).

Zhao H.-J. Synthesis, crystal structure, and NLO property of the chiral sulfide Sm_4InSbS_9, *Z. anorg. allg. Chem.*, 642(1), 56–59 (2016b).

Zhao H.-J., Zhang Y.-F., Chen L. Strong Kleinman-forbidden second harmonic generation in chiral sulfide: La_4InSbS_9, *J. Amer. Chem. Soc.*, 134(4), 1993–1995 (2012b).

Zhizhin M.G., Filaretov A.A., Olenev A.V., Chernyshev V.V., Spiridonov F.M., Komissarova L.N. Synthesis and crystal structure of new double indium phosphates $M^I_3In(PO_4)_2$ (M^I = K and Rb), *Crystallogr. Rep.*, 47(5), 773–782 (2002).

Zhizhin M.G., Morozov V.A., Bobylev A.P., Popov A.M., Spiridonov F.M., Komissarova L.N., Lazoryak B.I. Structure of polymorphous modifications of double sodium and indium phosphate, *J. Solid State Chem.*, 149(1), 99–106 (2000).

Zhukov A.N., Burdina K.P., Semenenko K.N. Investigation of the interactions in the BN–AlN–Mg_3N_2 system at atmospheric and high pressures [in Russian], *Zhurn. Obshch. Khim.*, 66(7), 1070–1072 (1996).

Zhukov A.N., Burdina K.P., Semenenko K.N. Investigation of the phase equilibria in the BN–Si_3N_4–Mg_3N_2 system at atmospheric and high pressures [in Russian], *Zhurn. Obshch. Khim.*, 64(8), 1242–1245 (1994).

Zhu Z., Zhao Y., Zhao M., Yin F., Li Z., Liu Y. The Zn-rich corner of the Zn–Fe–Al–Sb quaternary system at 450°C, *J. Phase Equilibr. Dif.*, 34(6), 474–483 (2013).

Zouihri H., Saadi M., Jaber B., El Ammari L. Silver indium diphosphate, $AgInP_2O_7$, *Acta Crystallogr.*, E67(1), i2 (2011).

Zou J.-P., Guo S.-P., Jiang X.-M., Liu G.-N., Guo G.-C., Huang J.-S. Synthesis, crystal and band structures, and optical properties of a new supramolecular complex: $[Hg_6Sb_4](InBr_6)Br$, *Solid State Sci.*, 11(9), 1717–1721 (2009).

Zou J.-P., Luo S.-L., Tang X.-H., Li M.-J., Zhang A.-Q., Peng Q., Guo G.-C. Synthesis, crystal and band structures, and optical properties of a new framework mercury pnictides: $[Hg_4As_2]$ ($InBr_{3.5}As_{0.5}$) with tridymite topology, *J. Alloys Compd.*, 509(2), 221–225 (2011).

Zykov A.M., Gaydo G.K. Interaction of GaP with ammonia [in Russian], *Izv. AN SSSR. Neorgan. Mater.*, 9(5), 850 (1973).

Index

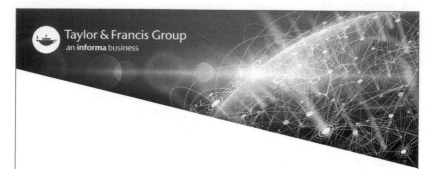